Evolution on Islands

Edited by

PETER R. GRANT
Department of Ecology and Evolutionary Biology
Princeton University
Princeton NJ

Originating from contributions
to a Discussion Meeting
of the Royal Society of London.

Oxford · New York · Tokyo
OXFORD UNIVERSITY PRESS
1998

Oxford University Press, Great Clarendon Street, Oxford OX2 6DP
Oxford New York
Athens Auckland Bangkok Bogota Bombay
Buenos Aires Calcutta Cape Town Dar es Salaam
Delhi Florence Hong Kong Istanbul Karachi
Kuala Lumpur Madras Madrid Melbourne
Mexico City Nairobi Paris Singapore
Taipei Tokyo Toronto Warsaw
and associated companies in
Berlin Ibadan

Oxford is a trade mark of Oxford University Press

Published in the United States
by Oxford University Press Inc., New York

A catalogue record for this book is available from the British Library

Library of Congress Cataloging in Publication Data
Evolution on islands / edited by Peter R. Grant.
Based on contributions to a Royal Society discussion meeting held Dec. 1995.
Includes index.
1. Evolution (Biology) 2. Island ecology. 3. Species.
I. Grant, Peter R., 1936– . II. Royal Society (Great Britain)
QH366.2.E8615 1997 576.8′0914′2–dc21 97-26389

ISBN 0 19 850171 4 (Pbk)
ISBN 0 19 850172 2 (Hbk)

Typeset by Advance Typesetting Ltd, Oxfordshire

Printed in Great Britain by Biddles Ltd, Guildford & King's Lynn

Preface

A major problem in evolutionary biology is to understand why the organic world is as diverse as it is, both in terms of its extraordinary variety in structures, functions, and life histories, and its sheer numbers. One of many approaches to a problem as broad as this is to a seek an understanding of a small and simplified part of it. For example, inspired by this philososphy many investigators re-create a simple version of part of a complex world in the laboratory or in a computer, the better to gain a precise understanding of the workings of its parts. This is the essence of experimentation. Islands provide that simplicity, naturally, because they are discrete pieces of the environment and often very small.

As well as being convenient models of larger realms and larger-scale processes, islands and their inhabitants have special features that command attention from evolutionary biologists. An outstanding feature is their strangeness; many of them are downright weird. Naturalists of the last three centuries, Darwin and Wallace among them, brought back to centres of civilization accounts of strange and unimagined creatures found only on remote islands. Dodos (Fig. P1). *Sphenodon*. The Komodo dragon. Daisies as tall as trees. What is it about islands that promotes such strangeness? In the case of the dodo the strangeness is the evolution of a bird from a presumably standard type of pigeon to something resembling a fat and flightless waddling goose, surely an extraordinary transformation!

A second feature is replicated evolution; numerous organisms of very different origins have gone down the same or very similar evolutionary paths under similar insular conditions. For the dodo the path has led to flightlessness, as it did for many other insular species of birds, most notably kiwis and rails, and many beetles, crickets, and other insects. Large body sizes in some groups of mammals, arborescence of herbs and shrubs, large sizes of seeds, and a predominance of white or dull-coloured flowers are other examples of repeated and independent evolution. Common denominators of repeated trends allow us to sort through and narrow the list of potential causes of evolution. It is easy to see why island populations have been referred to as natural experiments. The experiments have been replicated, though not controlled.

A third feature is the apparent rapidity of diversifying evolution on islands. Outstanding examples are the nearly 1000 species of *Drosophila* that were formed in the Hawaiian archipelago apparently in only a few million years, and the hundreds of cichlid fish that were formed by rapid speciation in the African Great Lakes in less, perhaps much less, than a million years. We still do not fully understand how a combination of ecological circumstances (opportunities) and population genetic structures and processes has fostered such rapid evolution in islands and island-like settings. Nevertheless, even though they may have no strict counterpart in continental regions, these radiations are more suitable than most for examining general questions of evolutionary rates and their variation.

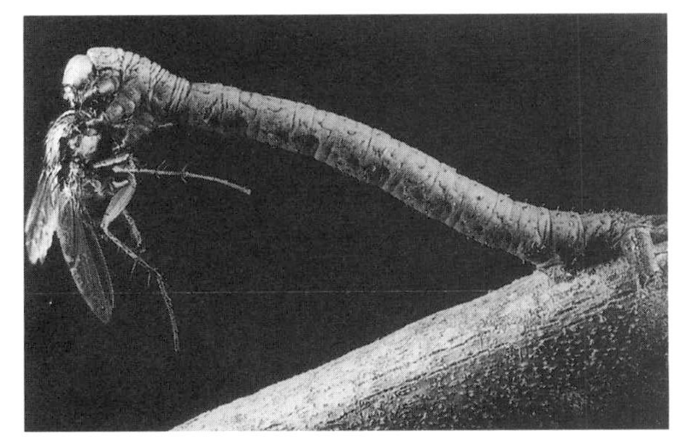

Fig. P1 The dodo; symbol of insular specialization, evolutionary simplification, and vulnerability to extinction (top left; K. T. Grant); tuatara (*Sphenodon*), sole representative of the reptilian family Rhynchocephalia and restricted to offshore New Zealand islands (top right; K. T. Grant); and *Eupithicia*, a unique genus of carnivorous lepidopteran larvae in Hawaii; lower left, *E. staurophragma* eating a *Drosophila heteroneura*, and lower right *E. niphoreas* eating a dolichopodid fly (reprinted with permission of S. L. Montgomery, and F. G. Howarth and W. P. Mull from *Hawaiian insects and their kin*, University of Hawaii Press, Honolulu, 1992).

If islands are to tell us much more than we know already there is some urgency in studying their inhabitants: they are being lost at a higher rate than their continental relatives, and at an unprecedentally high rate too, owing to human misuse of their environments (Fig. P2). More than 100 species of birds restricted to islands have become extinct in the last four centuries, and humans have had a hand in almost all, if not all, of these extinctions. Being restricted, these populations were especially vulnerable. The dodo, for example, was observed by few westerners before it went extinct. One head and a couple of feet, preserved in museums, are all that remain of the once living birds.

And extinction continues today. For example one of the contributors to this book, Bryan Clarke, has lived through a period during which the endemic *Partula* snail fauna of seven species on the island of Moorea has been completely eliminated by an introduced predator, another snail species. Variations on this theme of extinction can be told for one group after another from around the world. Extinction is regrettably important because it is depriving us of much of the evidence we need to both reconstruct and interpret evolutionary patterns on islands. We have already lost much valuable material for the experimental testing of evolutionary hypotheses we might wish to design now or undertake in the future.

These are the main reasons why it is appropriate at this time to publish a book on island evolution. One further reason deserves to be mentioned. Our knowledge of evolution in the past is being transformed by one of the quieter manifestations of the molecular revolution. Molecular genetic data are enabling us to reconstruct phylogenies in a more objective way than was possible in the pre-electrophoresis era. Nuclear, mitochondrial, and chloroplast genes constitute signals of phylogenetic or taxonomic affinities of populations, to a large extent independent of the morphological traits that were previously used to classify most of them. When the signals are clearly identified and the phylogeny reconstructed the direction of trait evolution can be determined, and the evolution itself interpreted. This may be easier on islands than on continents, because islands are relatively free from the ebb and flow of genes, a form of noise that tends to obscure the phylogenetic signal of divergence among demes. Phylogenetic analyses of evolutionary radiations on islands are one of the exciting fields of modern evolutionary biology.

The chapters fall into three groups. The first group is concerned with microevolution, the second with the question of how species are formed, and the third with adaptive radiations and patterns of variation displayed by clusters of species. Chapters are grouped by scale for convenience, and not because the groups are discrete: all chapters are interrelated. A recurring theme, for example, is how natural selection and genetic drift interact and jointly cause the patterns of variation we observe. Each group of chapters is introduced by a summary chapter which gives the background to the particular topic of that group, explains the main evolutionary issues, and makes explicit the connections among the component chapters. The book concludes with an epilogue which discusses some outstanding questions, and gives some pointers for future research.

Beyond the groupings is an inherent design. First, there is an examination of general principles applying to the evolution of all organisms on islands (Barton, Chapter 7). Second, other chapters complement the search for generality by examining evolutionary

Fig. P2 *Metrosiderus-Freycinetia* forest at about 800 m on Hawaii, at the turn of the century when some species of endemic honeycreeper finches went extinct as a result of various human activities. From D. Sharp (ed.), *Fauna Hawaiiensis*, vol. 1, 1899–1913. Reprinted with permission of Cambridge University Press.

features specific to particular organisms and specific to particular environments. Third, the book is designed to achieve a balance between generality and specificity. The balance is not perfect. Some organisms and environments are examined more thoroughly than others, and some (parasites and marine organisms for example) are missing altogether. Nevertheless there is coverage of plant taxa and animal taxa; of vertebrates and invertebrates; of morphological traits (Berry, Chapter 3; Grant and Grant, Chapter 9) and reproductive systems (Barrett, Chapter 2); of temperate islands and tropical islands; of true islands and island-like habitats such as lakes (Schluter, Chapter 10; Rüber and colleagues, Chapter 14) and insular patches of terrestrial habitat (Prance, Chapter 15); of small-scale processes of genetics (Clarke and colleagues, Chapter 11; Hollocher, Chapter 8; Pemberton and colleagues, Chapter 4) and ecology (Thorpe and Malhotra, Chapter 5), and their translation to large-scale patterns and processes of biogeography (Losos, Chapter 13; Mallet and Turner, Chapter 16; Givnish, Chapter 17).

The book is based on contributions to a Royal Society Discussion Meeting in December 1995, organized by Bryan Clarke and myself and published as papers in the *Philosophical Transactions of the Royal Society London B*, **351**, 723–854 (1996). Most of the chapters have been modified and parts rewritten in the light of new developments since then, and Chapters 1, 6, 12, 17, and 18 were written specifically for this book. I am grateful to the authors for making the changes so promptly and for reviewing these extra chapters, to the Royal Society for providing us with the forum to meet, discuss and synthesize our work, to the Ciba Foundation for providing facilities for a further discussion on speciation, to Jerry Coyne for participating in that meeting, and to Bryan Clarke for his help in organizing both of these meetings.

Princeton P.R.G.
January 1997

Contents

List of contributors xiii

1. Patterns on islands and microevolution 1
   Peter R. Grant
2. The reproductive biology and genetics of island plants 18
   Spencer C. H. Barrett
3. Evolution of small mammals 35
   R. J. Berry
4. The maintenance of genetic polymorphism in small island populations: large mammals in the Hebrides 51
   Josephine M. Pemberton, Judith A. Smith, Tim N. Coulson, Tristan C. Marshall, Jon Slate, Steve Paterson, Steve Albon, and Tim Clutton-Brock.
5. Molecular and morphological evolution within small islands 67
   Roger S. Thorpe and Anita Malhotra

6. Speciation 83
   Peter R. Grant
7. Natural selection and random genetic drift as causes of evolution on islands 102
   N. H. Barton
8. Island hopping in *Drosophila*: genetic patterns and speciation mechanisms 124
   Hope Hollocher
9. Speciation and hybridization of birds on islands 142
   Peter R. Grant and B. Rosemary Grant
10. Ecological speciation in postglacial fishes 163
    Dolph Schluter
11. How 'molecular leakage' can mislead us about island speciation 181
    Bryan Clarke, M. S. Johnson, and James Murray

12. Radiations, communities, and biogeography 196
    Peter R. Grant
13. Ecological and evolutionary determinants of the species–area relationship in Caribbean anoline lizards 210
    Jonathan B. Losos
14. Lake level fluctuations and speciation in rock-dwelling cichlid fish in Lake Tanganyika, East Africa 225
    Lukas Rüber, Erik Verheyen, Christian Sturmbauer, and Axel Meyer

15. Islands in Amazonia 241
Ghillean T. Prance
16. Biotic drift or the shifting balance—did forest islands drive the diversity of warningly coloured butterflies? 262
James L. B. Mallet and John R. G. Turner
17. Adaptive plant evolution on islands: classical patterns, molecular data, new insights 281
Thomas J. Givnish

18. Epilogue and questions 305
Peter R. Grant

Index 321

Contributors

S. D. Albon Institute of Zoology, Zoological Society of London, Regent's Park, London NW1 4RY, UK

S. C. H. Barrett Department of Botany, University of Toronto, Toronto, Ontario, Canada M5S 3B2

N. H. Barton Institute of Cell, Animal and Population Biology, University of Edinburgh, West Mains Road, Edinburgh EH9 3JT, UK

R. J. Berry Department of Biology, University College London, Gower Street, London WC1E 6BT, UK

B. Clarke Department of Genetics, University of Nottingham, Queens Medical Centre, Clifton Boulevard, Nottingham NG7 2UH, UK

T. H. Clutton-Brock Department of Zoology, University of Cambridge, Downing Street, Cambridge CB2 3EJ, UK

T. N. Coulson Institute of Zoology, Zoological Society of London, Regent's Park, London NW1 4RY, UK

T. J. Givnish Department of Botany, University of Wisconsin, Madison, WI 53706, USA

P. R. Grant Department of Ecology and Evolutionary Biology, Princeton University, Princeton, NJ 08544–1003, USA

B. R. Grant Department of Ecology and Evolutionary Biology, Princeton University, Princeton, NJ 08544–1003, USA

H. Hollocher Department of Ecology and Evolutionary Biology, Princeton University, Princeton, NJ 08544–1003, USA

M. S. Johnson Department of Zoology, University of Western Australia, Nedlands 6907, Western Australia

J. B. Losos Department of Biology, Box 1137, Washington University, St Louis, MO 63130–4899, USA

A. Malhotra School of Biological Sciences, University of Wales, Bangor, Gwynedd LL57 2UW, UK

J. L. B. Mallet Galton Laboratory, Department of Biology, University College London, 4 Stephenson Way, London NW1 2HE, UK

T. C. Marshall Institute of Cell, Animal and Population Biology, University of Edinburgh, West Mains Road, Edinburgh EH9 3JT, UK

A. Meyer Department of Ecology and Evolution, State University of New York, Stony Brook, NY 11794–5245, USA

J. Murray Department of Biology, University of Virginia, Charlottesville, VA 22901, USA

S. Paterson Department of Genetics, University of Cambridge, Downing Street, Cambridge CB2 3EH, UK

J. M. Pemberton Institute of Cell, Animal and Population Biology, University of Edinburgh, West Mains Road, Edinburgh EH9 3JT, UK

G. T. Prance Royal Botanic Gardens, Kew, Richmond, Surrey TW9 3AB, UK

L. Rüber Section of Taxonomy and Biochemical Systematics, Royal Belgian Institute of Natural Sciences, Vautierstraat 29, 1000 Brussels, Belgium

D. Schluter Department of Zoology and Centre for Biodiversity Research, University of British Columbia, Vancouver, British Columbia, Canada V6T 1Z4

J. Slate Institute of Cell, Animal and Population Biology, University of Edinburgh, West Mains Road, Edinburgh EH9 3JT, UK

J. A. Smith Department of Genetics, University of Cambridge, Downing Street, Cambridge CB2 3EH, UK

C. Sturmbauer Department of Zoology, University of Innsbruck, Technikerstrasse 25, 6020 Innsbruck, Austria

R. S. Thorpe School of Biological Sciences, University of Wales, Bangor, Gwynedd LL57 2UW, UK

J. R. G. Turner Department of Genetics, University of Leeds, Leeds LS2 9JT, UK

E. Verheyen Section of Taxonomy and Biochemical Systematics, Royal Belgian Institute of Natural Sciences, Vautierstraat 29, 1000 Brussels, Belgium

1

Patterns on islands and microevolution

Peter R. Grant

1.1 Introduction

The two basic questions in evolutionary biology are (1) how does evolution occur and (2) why does evolution occur? The first is a question of mechanisms and the second is a question of influences on those mechanisms. By studying evolution in the relative simplicity of island environments we hope to be able to answer both questions, using the full range of techniques of modern biology: description, experimentation, modelling, statistical analysis, and the ordering of data to make inferences about evolution in the past.

This chapter provides a brief overview of some of the main evolutionary trends on islands, then summarizes ideas on how those trends are thought to begin. The trends may be considered as the major patterns to be explained, and microevolution as laying the foundation for their explanation in terms of mechanisms. Thus the chapter provides the broad context for the four chapters on microevolution that follow.

1.2 Island features and evolutionary trends

We contrast islands and mainlands as opposites, yet they differ only arbitrarily in scale. The biological interest of islands lies in the fact that many are relatively small, often well isolated, and owe their faunas and floras to long-distance dispersal from mainlands or other large islands. Oceanic islands are difficult to reach, and since organisms of different taxa have different dispersal abilities it is inevitable that these islands will possess a non-representative sample of species from the mainland. Moreover, chance must play a strong role in governing which species arrive on such islands, when, and in what numbers. Difficulties of establishment will further influence the composition of the island community by favouring some types of colonists over others. The ecological stage is thus set for divergent evolution to take place in a physically and biotically novel environment. What are those changes and how are they explained?

Changes are both idiosyncratic, that is specific to particular organisms and particular locations, and general and widespread. Wallace (1880, pp. 433–4) provides an outstanding example of a pronounced local trend (see also Wallace 1865):

> Even more remarkable are the peculiarities of shape and colour in a number of Celebesian butterflies of different genera. These are found to vary all in the same manner, indicating some

general cause of variation able to act upon totally distinct groups, and produce upon them all a common result. Nearly thirty species of butterflies, belonging to three different families, have a common modification in the shape of their wings, by which they can be distinguished at a glance from their allies in any other island or country whatever; and all these are larger than the representative forms inhabiting most of the adjacent islands. No such remarkable local modification as this is known to occur in any other part of the globe; and whatever may have been its cause, that cause must certainly have been long in action, and have been confined to a limited area. ...There is no other example on the globe of an island so closely surrounded by other islands on every side, yet preserving such a marked individuality in its forms of life; while, as regards the special features which characterise its insects, it is, so far as yet known, absolutely unique.

Common features in a group of organisms in the same environment may be due to shared inheritance and/or shared environmental pressures. In contrast, similar trends among different organisms occupying disparate islands are more revealing about general evolutionary forces. If animals and plants are thought of as adaptive compromises moulded by conflicting selection pressures, then the reason for there being repeated evolutionary patterns on islands is that the same shifts in the compromise occur, as a result of repeated and similar alterations of the factors that give rise to the selection pressures. What are these factors? The primary ones are mild equable climate of many islands, year-round availabilty of food, availability of ecological niches, and relative scarcity of predators and ecological competitors as a simple consequence of the reduction in total number of species (for biogeographic reasons: MacArthur and Wilson 1967; Williamson 1981). Secondary factors, dependent upon the primary ones, include high population densities and intraspecific competitive pressures.

These factors, or a subset of them, occur in other environments that are not traditionally thought of as islands; caves, bogs, lakes, and mountain tops, for example. Like islands, such environments are isolated and generally small. Not surprisingly, therefore, evolutionary trends that are seen on islands are also seen in island-like settings. Beetles, for example, are apt to be wingless on the tops of mountains as well as on islands, especially at high altitudes, most likely for the same reasons (Darlington 1943; Roff 1990; see below). With a scarcity of animal pollinators at the highest altitudes plants tend to be wind pollinated and self-compatible (Carlquist 1974; Berry and Calvo 1989), trends that are seen also on several islands.

The most common and repeated evolutionary responses of both plants and animals to the altered pressures from insular environments are first, a reduction in traits that help them to disperse over long distances, and second, the development of greater stature.

1.3 Reduced dispersal

Organisms must have had exceptional powers of dispersal to get to the most distant islands, unless they accompanied the islands when they moved imperceptibly to their remote position on a drifting tectonic plate. Long-distance dispersal of plants by wind, water, or birds is aided by small size of spores and seeds, flotation devices, hooks, adhesive surfaces, and colourful and succulent material surrounding the seed that is attractive to birds. A repeated evolutionary trend is the reduction of these morphological traits. Carlquist (1974) gives examples of the structural alterations of seeds resulting in a lack of buoyancy, and large size, in some species of *Canavalia* (Fabaceae) and *Zanthoxylum* (Rutaceae) in Hawaii. The main explanation for these trends invokes natural selection and adaptation. Selection for dropping seeds in locally favourable areas for germination, mainly in forests, would result in the evolution of large and sedentary seeds from small and wind-blown ones, or from hooked, barbed, and floating ones brought to the island by sea or marine birds. A recent study on small offshore islands of western Canada has shown that a reduction in dispersal potential of island colonists may arise in just a few generations (Cody and Overton 1996).

A parallel trend is produced in animals by a similar set of selective factors. Long-distance dispersal in animals is aided by small size or powerful wings, the latter being characteristic of birds, bats, and insects. Flightlessness is a repeatedly evolved trait in island populations of birds and insects. As in plants, the advantages of sedentariness may increase as the advantages of dispersal decrease. An evolutionary trend resulting from this shift in advantages is from a single, broadly adapted, generalist species that colonizes an archipelago to several, locally adapted, specialist and restricted species as a result of speciation and adaptation: a taxon cycle of evolution, eventually leading to extinction and replacement (Wilson 1961; Ricklefs and Cox 1972; Pregill and Olson 1981).

There is an energy saving to be made by birds with reduced wings (McNab 1994*a, b*), both in the manufacture of the equipment and in its use, when conditions permit. Those conditions include a permanent habitat with a local year-round food supply, obviating migration by flight from one suitable habitat to another, and the absence (or scarcity) of predation, especially mammalian predation. Some islands provide those conditions for some species. Thus rails have evolved flightlessness many times (McNab 1994*a*; Roff 1994; Steadman 1995). Predisposing factors for this group were long-distance dispersal to find suitable habitat (finding the islands), and walking without much use of flight while exploiting a habitat. Exploitation of habitats in the manner of large herbivores (dodos, moas, ratites) provides similar opportunities (e.g. absence of herbivorous and predatory mammals) and similar selective pressures.

Comparable factors are responsible for the evolution of flightlessness in a large variety of insects (e.g. see Darlington 1943; Roff 1990; Peck 1996; Shaw 1996). The evolution of flightlessness takes such bizarre forms as the fusing of the elytra in beetles (Fig. 1.1). Flightlessness has evolved many times on islands and mountain tops, on high-elevation islands especially, like Madeira (Wollaston 1854), Cuba, and Jamaica (Darlington 1943), and more in ground-living than arboreal or water-associated Carabid beetles (Darlington 1943): in other words, in stable habitats of limited heterogeneity, especially woodland (Roff 1990, 1994), where energy devoted to wing production can be reallocated to reproduction (Roff 1990).

There are two more factors that may be important; wind and cold. An idea attributed to Darwin is that winglessness is selectively favoured because winged insects are more likely than wingless ones to be blown off islands (the wind hypothesis). This is an extreme variation on the general theme of the adaptiveness of staying put in a restricted and stable habitat, but it has not been supported by careful comparisons of beetles in contrasting habitats (Darlington 1943; Roff 1990). A second idea is that flying is difficult

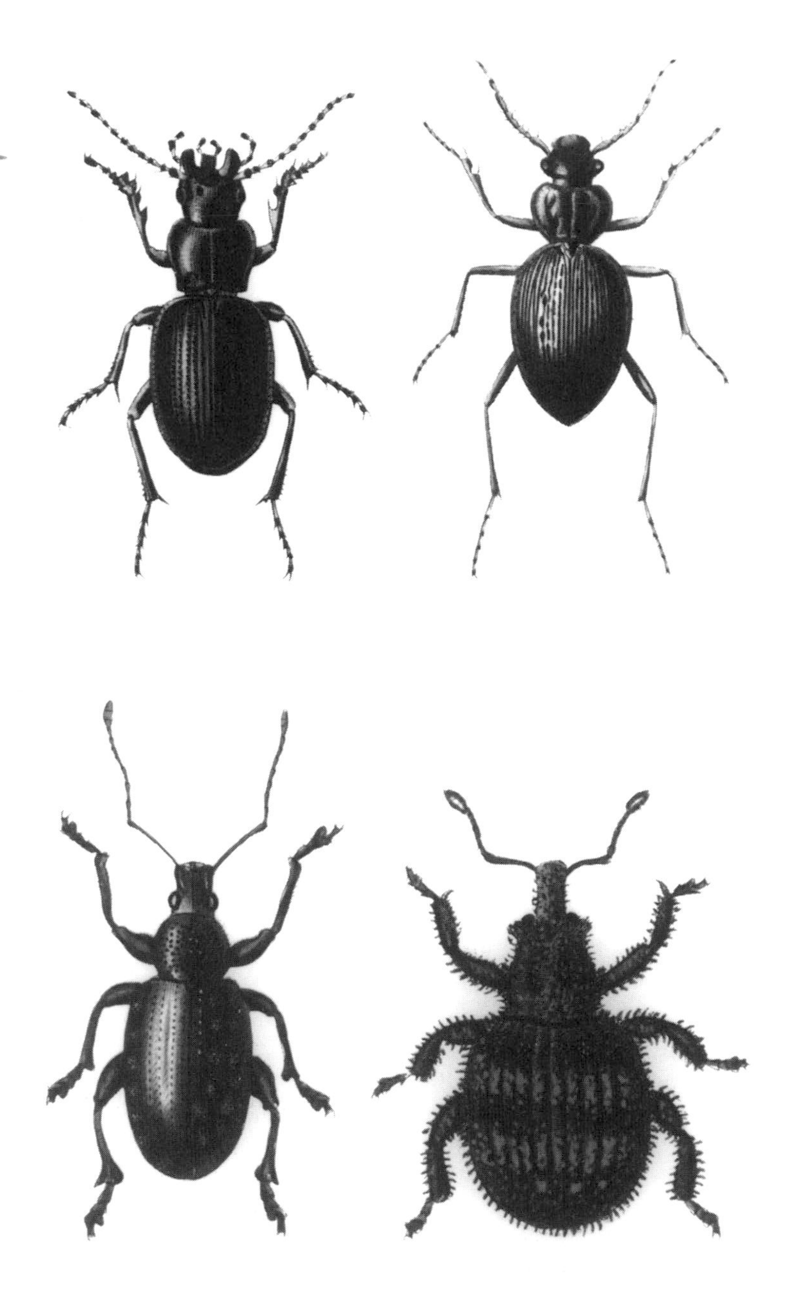

Fig. 1.1 Flightlessness: four species of flightless beetles from Madeira (from Wollaston 1854). *Eurygnathus latreillei* (Carabidae) upper left, *Loricera wollastoni* (Carabidae) upper right, *Atlantis vespertinus* (Curculionidae) lower left, *Echinosoma porcellus* (Curculionidae) lower right.

at low temperatures (the cold hypothesis). It has been used to explain the high proportion of flightless insects in Antarctica (Diptera, Lepidoptera, and Hymenoptera), especially on Campbell Island (Gressitt 1970). Tristan da Cunha has a cold climate and is also notable for the high proportion of wingless insects (Kuschel 1962). In contrast to Darwin's hypothesis this has received support from modern analyses; statistical analyses show an association between the incidence of flightlessness and low temperatures (Roff 1990). Roff's (1990) analyses also show that flightlessness should not be thought of as a peculiarity of insular environments, even though it has evolved numerous times in such environments. The incidence of flightlessness on islands and mainlands is no different at comparable latitudes and altitudes. Nevertheless, if the initial colonization of remote islands and archipelagos is biased towards winged insects the frequent occurrence of wingless forms on islands represents a directional evolutionary trend. Phylogenies of individual groups are needed to determine evolutionary trends.

1.4 Size changes

Plants evolve new growth forms on islands, and there is a trend towards greater stature, from herbs and shrubs to trees, especially to rosette trees (Mabberley 1979). Insular 'woodiness' has been known for more than a century, but its evolutionary significance is still not fully understood. Woodiness and tree growth-form have repeatedly evolved in several angiosperm families, including the Asteraceae, Boraginaceae, Campanulaceae, Lobeliaceae, and Euphorbiaceae (Böhle *et al.* 1996), and Apiaceae and Brassicaceae (see Section 17.3), though by different anatomical means. These features are particularly prevalent in the Asteraceae. Within this family (but in different tribes) woodiness is found on the Canary Islands, Juan Fernandez Islands, Hawaiian Islands, and St Helena, as well as others. Arborescence is also a feature of some plants in the insular-like setting of the high Andes (Carlquist 1974) and the East African highlands (giant senecios and lobelias; Knox *et al.* 1993; Knox and Palmer 1995, 1996).

Carlquist (1974) identified moderate temperatures, rainfall (>1000 mm/year), and humidity as factors fostering the evolution of woodiness, without explaining why woodiness would evolve. Other facilitating factors are a shift from weedy open habitat to forest, absence of large herbivores, and release from seasonality.

Darwin believed woodiness evolved under intraspecific competitive pressure (crowding), and Wallace thought it was favoured because it increased longevity, and longevity was favoured through the advantages of repeated breeding. Both may have been right. Nevertheless Carlquist argued against Darwin on the grounds that Hawaiian woody herbs are dispersed in forests and not clumped; and against Wallace largely on the grounds that his assumption of a paucity of pollinating insects (as a driving force in the evolution of longevity) is dubious. These ideas are still debated (Section 17.3). Böhle *et al.* (1996) favour a modified version of Wallace's hypothesis to explain the evolution of woodiness in the genus *Echium* (Boraginaceae) on the north atlantic islands of Macaronesia (Fig. 1.2). The modification is in laying stress on a connection between the selective advantage of outbreeding and the persistence of conspicuous flowers on a perennial (and hence woody) plant. Racine and Downhower (1974) argued more along ı's lines (see also Section 17.3) in explaining the tree growth form of *Opuntia* , on those Galápagos islands where competition with other woody vegetation for lignt ıs likely to occur (Fig. 1.2).

Since remote islands are colonized to a disproportionate extent by small animals, especially small insects (Zimmerman 1972) and small snails (Vagvolgyi 1975), we should not be surprised that there is a tendency for such island animals to evolve larger size. Analogously Cope's rule, that is the tendency for independent mammalian lineages to evolve large size, has been explained as stemming from the invasion of a new adaptive zone by small generalist species (Stanley 1973). The easiest direction to take is upwards. That direction has been taken by the vast majority of island populations of rodents surveyed by Foster (1964) and Lomolino (1985). Herbivorous mammals, on the other hand, are larger and show an opposite trend, so too do carnivorous mammals, as well as snakes (Case 1978).

A general explanation for the size trends is not simple, and involves an interaction of food resources, competition, and predation (Case 1978; Heaney 1978; Lomolino 1985; see also Chapter 3). In the absence of their predators (and competitors) all three groups of mammals, like lizards (Kramer 1946; Case 1983), tend to reach high densities. They are then in danger of reducing their food supply, thereby setting up selective pressures for large individual size through dominance in the case of rodents (Foster 1964; Williamson 1981) and several lizards (Kramer and Mertens 1938; Case 1978; Petren and Case 1997), and small individual body size for energetic reasons under food limitation in the case of the other groups of mammals (e.g. Klein 1968; Heaney 1978; Lomolino 1985; McNab 1994*b*) and snakes (Case 1978).

1.5 Other trends

Size changes and reduced dispersal do not exhaust the trends that have been reported to occur on islands. Briefly, here are seven others that have been established in various degrees of thoroughness. Modern comparative methods are needed to establish (1) whether these truly are general trends and not taxonomically or geographically restricted, (2) whether they have arisen through evolution on islands or through biased colonization, and (3) whether two or more trends covary through correlated evolution or have arisen independently (e.g. see Losos 1990, 1996; Harvey and Pagel 1991; Roff 1994; Wagner and Funk 1995). These methods should be extended to determine if there are parallel trends in island-like habitats (e.g. mountain tops, lakes, and caves).

Floral colours and pollinators

Comparisons of mainland and island floras have shown repeated tendencies for island plants to have small and often inconspicuous flowers, with white, green, and yellow colours (in that order) predominating on New Zealand, Juan Fernandez Islands, Hawaiian Islands and St Helena (Carlquist 1974). Similarly the principal colours of Galápagos flowers are white and yellow. There is little doubt that the trend is general.

Fig. 1.2 Evolution of woodiness on islands. In the left panel the herbaceous continental *Echium creticum* is contrasted with the cylindrical and woody *E. simplex* from Tenerife, Canary Islands (Original photos: *H. H. Hilger* and *H. Kuerschner*, Berlin; graphics *S. B. Tautz*, Braunschweig). (Reprinted from Böhle *et al.* (1996). Copyright 1996 Proceedings of the National Academy of Sciences, USA.) On the right the ~10 m tall *Opuntia echios* from Isla Santa Cruz, Galápagos, contrasts with the ~2 m shrub of the same species on Isla Daphne Major, below (*P. R. Grant*).

The paucity of conspicuous flowers in the flowers of the Galápagos Islands and New Zealand is correlated with poverty of butterflies and bees (Wallace 1891). The correspondence arises partly from selective colonization, and partly from evolution on the islands. Carlquist (1974) stresses that small size classes of insects have been most successful at colonizing islands, and this has favoured the establishment of angiosperms with small entomophilous flowers that are pollinated by generalist pollinators. The evolutionary component has not been well documented quantitatively. It is discussed in Chapter 2.

The breeding systems of plants

Baker (1955, 1967) hypothesized that colonizing plant species on islands would be self-compatible as a rule in view of the obvious advantages to self-compatibility in the early stages of establishment on islands by, presumably, a small number of immigrants. Nevertheless, various means of outcrossing tend to be highly developed in island plants, and selection for outcrossing in small populations seems to be strong (e.g. see Böhle *et al.* 1996). Although self-incompatibility and heterostyly are rare on islands, dioecism (separate male and female plants) is exceptionally common in certain insular floras, and reaches its highest level in the Hawaiian archipelago. The high incidence of dioecism here can be accounted for by (1) colonization of initially dioecious species and subsequent speciation, and (2) an evolutionary shift to dioecy in lineages derived from monoecious colonists, which accounts for one-third of dioecious species in the archipelago (Sakai *et al.* 1995*a*). Thus Baker's rule gains some, but not universal, support. In this archipelago dioecy is associated with woodiness and the possession of small and pale-coloured flowers (Sakai *et al.* 1995*b*).

Other floral conditions that tend to promote outcrossing on islands include gynodioecism, protandry, protogyny, and wind pollination. The main floral trends on islands are discussed in Chapter 2.

Hybridization

Hybridization occurs on islands, and is especially common in some island groups of plants. For example it is relatively common in plant taxa in New Zealand (Rattenbury 1962; Raven 1972). Carlquist (1974, p.534) wrote 'Reports of hybridization in insular floras depend on the tendency of systematists to recognize the presence of hybrids or to neglect them. There seems no question that ... we may expect the roster of known and probable hybrids on islands to lengthen greatly.' It has.

Introgressive hybridization has also been implicated in the evolution of the silversword alliance, *Bidens* (Asteraceae), *Scaevola* (Goodeniaceae), and *Pipturus* (Urticaceae) in the Hawaiian Islands (Gillet 1972; Baldwin *et al.* 1990; Carr 1995), and in the endemic plant genus *Argyranthemum* (Asteraceae) in Macaronesia (Azores, Canaries, Cape Verde, Madeira, and Selvagens; Francisco-Ortega *et al.* 1996). It is not known if a high incidence of hybridization can be considered an island trend because the appropriate quantitative and comparative analysis has not been performed. Since archipelagos are forums for recent adaptive radiations, and hybridization occurs in the early stages of the differentiation of a taxon, an unusually high incidence of hybridization on islan[illegible] in fact occur (Sections 9.6 and 11.5), especially on geologically unstable islands [illegible] to frequent ecological disturbance (Carr 1995). Introgressive hybridization [illegible] contribute to morphological trends on the same island, such as the trends in the wings of butterflies on Celebes discovered by Wallace (1880) and quoted on p.1.

Enhanced population variation

Some island populations are exceptionally variable in morphological features. Examples include moths in the Hawaiian archipelago (Perkins 1913; Howarth and Mull 1992), Darwin's finches on the Galápagos (Grant 1986; Grant and Grant 1989) and several groups of plants in Hawaii and New Zealand (Carlquist 1974; Wagner and Funk 1995). Whether this feature deserves to be considered an island trend is an open question. Is enhanced variation indicative of higher rates of mutation, and if so why is this manifested in island settings? Is the cause to be sought in geophysical factors such as mutagenic radiation, biological factors such as hybridization and a release of variation that this produces, or some combination of the two?

Clutch sizes

Clutch sizes of island wall lizards (*Lacerta sicula*) off the coast of Italy are small compared with the clutch sizes of their mainland relatives. Kramer (1946) was able to show that the difference was at least partly due to an evolutionary divergence by performing a cross-breeding experiment in captivity and finding that the hybrids had intermediate clutch sizes. Small clutches have been reported for island lizards elsewhere (e.g. Case 1983), as well as for birds (Grant and Grant 1989; Martin 1992). A trend towards reduced 'birth' rate on islands appears to be general among terrestrial vertebrates, but a systematic and broad-scale analysis that controls for latitudinal effects on clutch or litter size has not yet been performed. If the trend is confirmed it may reflect one aspect of an evolutionary shift away from high reproductive rate and short life span towards the opposite combination of life history traits. An important factor in such a shift is intraspecific competition under conditions of sustained high densities in the absence (or scarcity) of predators and heterospecific competitors (Kramer 1946; MacArthur and Wilson 1967; Boyce 1984).

Broad ecological niches

A large literature suggests that many island vertebrates (lizards, mammals, and birds), freed from the constraints of mainland competitors, expand the range of habitats occupied or dimensions of their feeding niches (Williams 1972; Case 1978; Roughgarden 1995). Ecological release from competitive constraints is inferred mainly from island–mainland (or island–island) comparisons of observed behavior and ecology (e.g. Feinsinger and Swarm 1982), as well as from the measurement of ecologically significant morphological traits. As one example of the latter, island birds from around the world and from a variety of taxonomic groups tend to have large beaks (Murphy 1938; Grant

1965). The trend is at least partly independent of body size: body size of birds does not show the same trend. Since birds with large beaks have a wider range of food sizes available to them than do birds with small beaks, the trend has been interpreted as reflecting a wider ecological niche facilitated by the absence of ecological competitors and favoured under occasional conditions of food shortage (Grant 1965).

The same problem of interpretation arises with morphological features as with clutch sizes. Could the differences between island and mainland features be simply different phenotypic expressions of the same genotypes raised under different environmental conditions? Only one published study has investigated whether morphological differences between island and mainland species or subspecies of birds have a genetic basis (Alatalo and Gustafsson 1988). In the absence of larger congeneric species from the Swedish mainland, coal tits (*Parus ater*) on Gotland are larger than their mainland relatives, and forage like the missing species. A cross-fostering experiment showed that the size difference is due to genetic, probably polygenic, factors.

On large islands and in archipelagos, ecological generalization is not necessarily the endpoint of an evolutionary trend. Rather, the opposite is likely to be true in those situations permitting speciation, with efficient specialists being produced from an original generalist (Lack 1947). Their ultimate fate is believed to be extinction caused by counter-adaptations from the biota and competition from a newly arriving generalist. Thus ends a taxon cycle and the beginning of a new cycle of diversification and specialization (Wilson 1961; Ricklefs and Cox 1972). There are ecological limits to which such specialization can occur within a group of similar ecological niches. When those limits are reached, if not beforehand, a shift to another niche type may open up yet further opportunities and trigger another bout of speciation, initially perhaps producing a generalist and subsequently specialists. In this way radiations are built up sequentially, by the accumulation of different mini-radiations produced in bursts within somewhat restricted and different ecological zones; a taxon spiral more than a cycle. Some of these ideas on speciation, specialization and adaptive radiation are discussed further in Chapters 6, 9, 10, 12, 13, 17, and 18.

The reason why a broadening of niches appears to be an island trend is that there are many more small islands than large ones that are aggregated into well-isolated archipelagos, and the populations of relatively small islands are kept in the initial stages of evolutionary diversification by limited ecological opportunities, and possibly by frequent extinctions (Mayr 1965).

Tameness of vertebrates

Many descriptions have been given of the tameness and apparent fearlessness of island animals, particularly of birds. Darwin, for example, described how he was able to approach a Galápagos hawk close enough to displace it from a branch with a stick. An uncontroversial view of tameness is that fear of humans has been lost in the absence of large, human-like, predators on islands. The balance between exploratory behaviour in the search for information about the environment and for food, on the one hand, and the avoidance of dangerous organisms, on the other hand, is shifted on islands towards the former (Fig. 1.3). This does not mean that all fear has been lost. Darwin's finches are

Fig. 1.3 Island tameness. A hawk, *Buteo galapagoensis*, on the head of the author; Isla Marchena, Galápagos (*G. Seutin*).

also approachable (by humans) but are fearful of Galápagos hawks and short-eared owls (Curio 1965), which prey upon them, and will fly up into trees or even on to the head and shoulders of a human observer when one of these predators flies over.

1.6 Microevolutionary beginnings of evolutionary trends

I have described the major evolutionary trends in the phenotypic traits of organisms on islands, and the main explanations offered for them. The standard explanations are framed in terms of selective (environmental) factors that produce adaptations. Adaptations involve a reallocation of the resources devoted to growth and reproduction under the altered opportunities and hazards of island life. Adaptive interpretations of island evolution are strongest when a trend is statistically demonstrable, the features of interest are known through phylogenetic analysis to have arisen independently and repeatedly in unrelated taxa, and moreover are repeated in geographically separated areas.

Explanations for insular phenomena are incomplete without considering the genetic structure and dynamics of populations. Stochastic processes become important when the attempt is made to understand the distribution of genetic variation among individuals in a population. A population may be stranded on a piece of land that becomes insular, or colonize an existing island. In the former case no immediate change in the now insular population is expected, unless the island is very small, whereas immediate change is expected when an island is newly colonized from elsewhere because the founding of a new population is likely to take place by just a few individuals, and these will not carry with them alleles at all loci in exactly the frequency at which they occur in the source population. Unequal breeding by the founders may further distort allele frequencies, either by chance or for reasons of unequal phenotypic properties, and thereafter the frequencies may change yet further through drift and selection. Population fluctuations will occasionally result in numbers being reduced to something approaching the starting population size, or lower, and the process may begin all over again, and be repeated.

It is in this context that the next four chapters focus on different genetic aspects of populations and microevolutionary change. All are concerned with the interplay of selection and drift in governing changes in the phenotypic attributes of small populations on islands.

The fourth member of this group adopts a different method, correlating morphology with ecology, to expose the role of natural selection.

The first of the chapters provides a link with the major evolutionary trends. Barrett (Chapter 2) describes enquiries into the reproductive biology of plants on islands, paying particular attention to genetics where possible. Stochastic processes are important, both in the founding of insular plant populations and subsequently. This is inferred from evidence of developmental instability, a loss of genetic (allozymic) variation within island populations coupled with differentiation among them, and the argument that genetically effective sizes of selfing populations are often very small.

Systematic processes are revealed in several ways. In accordance with Baker's rule, the establishment of plant species on islands is biased in the direction of self-compatible forms. Self-compatibility is maintained where pollinators are sparse, and selfing is enhanced by alterations in floral traits that promote autonomous self-pollination. Nevertheless many plant species, especially in the Hawaiian archipelago, have evolved sexual dimorphisms that promote outcrossing. What determines this dichotomy of trends? Recent studies (Sakai *et al.* 1995*a*, *b*) have identified a combination of instrumental factors in the evolution of outcrossing; inbreeding depression resulting from selfing, and inadequate pollinator service. Barrett offers an explanation for the dichotomy in terms of two forms of pollination inadequacy; insufficient and inferior pollination service. The first leads to the evolution of selfing, whereas the second is a factor in the evolution of sexual dimorphism. The suggested distinction opens up new possibilities for fieldwork on pollinator-limited plant populations on islands.

The next chapter (Chapter 3) by Berry addresses the question of how island populations become differentiated, using extensive data from studies of small mammals on islands in the British Isles. Island populations in three groups of rodents display marked morphological, skeletal, and genetic variation. The groups are *Microtus* (voles), *Apodemus* (typically woodland mice), and *Mus* (typically commensal mice). Berry argues that the most important factor causing the differentiation is the biased composition of the colonists, in other words founder effects in the strict sense (e.g. see Mayr 1942, 1954). The next most important factor is post-colonization selection, which may favour the directional change in body size discussed in the section on trends. Surprisingly, the least important factor is genetic drift, though it must be acknowledged that this is not easily determined directly in field studies. Furthermore a strong trend towards loss of allozymic diversity and reduced heterozygosity displayed by island populations of small mammals lends itself to an interpretation of loss of alleles through drift (for a comparable situation in plants see Section 2.4).

Against the expectation of loss of genetic variation through drift, Pemberton and colleagues show in Chapter 4 how genetic variation may be maintained by selection, or at least the rate of loss slowed; conceivably the rate might be so slow as to be balanced by recurrent mutation. Two fairly small populations of individually large mammals have been studied intensively for many years on two Scottish islands; Red deer (*Cervus elaphus*) on the island of Rhum and Soay sheep (*Ovis aries*) on Hirta (St Kilda group). Intensive monitoring has allowed a detailed investigation of genetic and phenotypic changes in fluctuating populations, well after the initial founding event occurred. Like Berry and Murphy (1970) in a study of feral mice (*Mus musculus*) on the island of Skokholm, Pemberton and colleagues find evidence of oscillating directional selection, as well as heterozygote advantage, at a protein-coding locus (or at tightly linked loci). The oscillating selection arises from the fact that a particular genotype that has an advantage at high density, when juvenile survival is an important component of fitness for example, has a disadvantage at low density when juvenile reproduction becomes a more important determinant of relative success. An important feature of this study is an exceptional attention to ecological detail that makes it possible to provide a mechanism for the maintenance of genetic variation.

In the final chapter (Chapter 5) in this group on microevolution Thorpe and Malhotra assess the role of natural selection in a modern multivariate version of an ancient method — correlation analysis. They take advantage of the fact that populations of lizards can vary quite substantially in morphology across the terrain of a single island, even on small islands. Understanding such variation can provide insights into factors responsible for the evolution of differences between populations on different islands, or between island and mainland populations. This chapter provides an example of the importance of scale; ecotypic variation is conspicuous in these relatively sedentary lizards, as in plants (Section 2.3), and absent or minor in the more mobile small and large mammals discussed in Chapters 3 and 4.

Three lines of evidence support the hypothesis of adaptation to local conditions. First, their analyses of variation in Canary Island and Caribbean lizards repeatedly find strong associations between morphological variables such as colour, body size, and scalation patterns and environmental variables such as habitat structure and climate.

They find little residual influence of phylogenetic history on morphology, as indicated by a lack of association with DNA-based phylogenetic reconstructions. Second, in both archipelagos related species on different islands show similar, parallel, geographical variation in morphology in relation to the same environmental gradients. As they point out, ecology is shared but history is not; in contrast ecology and history may be shared and difficult to disentangle when different taxa display parallel patterns on the same island, as in the case of butterflies on Celebes (Sulawesi) described by Wallace (Section 1.2). Third, field experiments on the Caribbean island of Dominica with *Anolis oculatus* to test their adaptive hypothesis provided generally confirming results. The main conclusion is that natural selection arising from current ecological conditions is a primary force influencing morphological population differentiation, irrespective of phylogenetic history. Climatic and biotic factors (predation) are implicated as the agents of natural selection. It would be interesting to know to what extent dietary factors, sexual selection and random genetic drift have contributed to the patterns of ecotypic variation that have been so well documented. Some of the colour variation may be influenced by sexual selection (Section 5.2).

The mechanisms of microevolutionary change discussed in the next four chapters form the basis of speciation, which is the topic that follows in Chapters 6–11.

References

Alatalo, R. V. and Gustafsson, L. (1988). Genetic component of morphological differentiation in coal tits under competitive release. *Evolution*, **42**, 200–3.

Baker, H. G. (1955). Self-compatibility and establishment after 'long-distance' dispersal. *Evolution*, **9**, 347–9.

Baker, H. G. (1967). Support for Baker's law — as a rule. *Evolution*, **21**, 853–6.

Baldwin, B. G., Kyhos, D. W., and Dvorák, J. (1990). Chloroplast DNA evolution and adaptive radiation in the Hawaiian silversword alliance (Asteraceae–Madiinae). *Annals of the Missouri Botanical Gardens*, **77**, 96–109.

Berry, P. E. and Calvo, R. N. (1989). Wind pollination, self-incompatibility, and altitudinal shifts in pollination systems in the high andean genus *Espeletia* (Asteraceae). *American Journal of Botany*, **76**, 1602–14.

Berry, R. J. and Murphy, H. M. (1970). The biochemical genetics of an island population of the house mouse. *Proceedings of the Royal Society of London* B, **176**, 87–103.

Böhle, U.-R., Hilger, H. H., and Martin, W. F. (1996). Island colonization and evolution of the insular woody habit in *Echium* L. (Boraginaceae). *Proceedings of the National Academy of Sciences USA*, **93**, 11740–5.

Boyce, M. S. (1984). Restitution of *r*- and *K*-selection as a model of density dependent natural selection. *Annual Review of Ecology and Systematics*, **15**, 427–47.

Carlquist, S. (1974). *Island Biology*. Columbia University Press, New York.

Carr, G. D. (1995). A fully fertile intergeneric hybrid derivative from *Argyroxiphium sandwicense* ssp *macrocephalum* × *Dubautia menziesii* (Asteraceae) and its relevance to plant evolution in the Hawaiian islands. *American Journal of Botany*, **79**, 1574–81.

Case, T. (1978). A general explanation for insular body size trends in terrestrial vertebrates. *Ecology*, **59**, 1–18.

Case, T. (1983), The reptiles: ecology. In *Island biogeography in the Sea of Cortez*, (ed. T. Case and M. L. Cody), pp. 159–209. University of California Press, Los Angeles.

Cody, M. L. and Overton, J. McC. (1996). Short-term evolution of reduced dispersal in island plant populations. *Journal of Ecology*, **84**, 53–62.

Curio, E. (1965). Funktionsweise und stammesgeschichte des flugfeinderkennens einiger Darwinfinken (Geospizidae). *Zeitschrift für Tierpsychologie*, **26**, 394–487.

Darlington, P. J., Jr. (1943). Carabidae of mountains and islands: data on the evolution of isolated faunas and on atrophy of wings. *Ecological Monographs*, **13**, 37–61.

Feinsinger, P. and Swarm, L. A. (1982). 'Ecological release', seasonal variation in food supply, and the hummingbird *Amazilia tobaci* on Trinidad and Tobago. *Ecology*, **63**, 1574–87.

Foster, J. B. (1964). Evolution of mammals on islands. *Nature*, **202**, 234–5.

Francisco-Ortega, J., Jansen, R. K., and Santos-Guerra, A. (1996). Chloroplast DNA evidence of colonization, adaptive radiation, and hybridization in the evolution of the Macaronesian flora. *Proceedings of the National Academy of Sciences USA*, **93**, 4085–90.

Gillet, G. W. (1972). The role of hybridization in the evolution of the Hawaiian flora. In *Taxonomy, Phytogeography and Evolution*, (ed. D. H. Valentine), pp. 205–19. Academic Press, London.

Grant, B. R. and Grant, P. R. (1989). *Evolutionary dynamics of a natural population: the large cactus finch of the Galápagos*. University of Chicago Press.

Grant, P. R. (1965). The adaptive significance of some size trends in island birds. *Evolution*, **19**, 355–67.

Grant, P. R. (1986). *Ecology and evolution of Darwin's finches*. Princeton University Press.

Gressitt, J. L. (1970). Subantarctic entomology and biogeography. *Pacific Insects Monograph*, **23**, 295–374.

Harvey, P. H. and Pagel, M. D. (1991). *The comparative method in evolutionary biology*. Oxford University Press.

Heaney, L. R. (1978). Island area and body size of insular mammals: evidence from the tri-colored squirrel (*Callosciurus prevosti*) of southwest Asia. *Evolution*, **32**, 29–44.

Howarth, F. G. and Mull, W. P. (1992) *Hawaiian insects and their kin*. University of Hawaii Press, Honolulu.

Klein, D. R. (1968). The introduction, increase and crash of reindeer on St. Matthew Island. *Journal of Wildlife Management*, **32**, 350–67.

Knox, E. B. and Palmer, J. D. (1995). Chloroplast DNA variation and the recent radiation of the giant senecios (Asteraceae) in the tall mountains of Eastern Africa. *Proceedings of the National Academy of Sciences USA*, **92**, 10349–53.

Knox, E. B. and Palmer, J. D. (1996). The origin of *Dendrosenecio* within the Senecioneae (Asteraceae) based on chloroplast evidence. *American Journal of Botany*, **82**, 1567–73.

Knox, E., Downie, S. R., and Palmer, J. D. (1993). Chloroplast genome rearrangements and the evolution of giant lobelias from herbaceous ancestors. *Molecular Biology and Evolution*, **10**, 414–30.

Kramer, G. (1946). Veranderungen von nachkommenziffer und nachtkommengrosse sowie der altersverteilung von inseleidechsen. *Zeitschrift für Naturforschungsgeschichte*, **1**, 700–10.

Kramer, G. and Mertens, R. (1938). Rassenbildung bei westistrianischen Inseleidechsen in abhangegheit von Isolierungsalter und Arealgrosse. *Archiv für Naturgeschichte*, **7**, 189–234.

Kuschel, G. (1962). The Curculionidae of Gough Island and the relationships of the weevil fauna of the Tristan da Cunha group. *Proceedings of the Linnean Society of London*, **173**, 69–78.

Lack, D. (1947). *Darwin's finches*. Cambridge University Press.

Lomolino, M. V. (1985). Body size of mammals on islands: the island rule reexamined. *American Naturalist*, **125**, 310–16.

Losos, J. B. (1990). A phylogenetic analysis of character displacement in Caribbean *Anolis* lizards. *Evolution*, **44**, 558–69.

Losos, J. B. (1996). Phylogenetic perspectives on community ecology. *Ecology*, **77**, 1344–54.

Mabberley, D. J. (1967). Pachycaul plants and islands. In *Plants and islands*, (ed. D. Bramwell), pp. 259–77. Academic Press, London.
Mabberley, D. J. (1979). Pachycaul plants and islands. In *Plants and islands*, (ed. D. Bramwell), pp. 259–77. Academic Press, New York.
MacArthur, R. H. and Wilson, E. O. (1967). *The theory of island biogeography*. Princeton University Press.
Martin, J.-L. (1992) Niche expansion in an insular bird community: an autecological perspective. *Journal of Biogeography*, **19**, 375–81.
Mayr, E. (1942). *Systematics and the origin of species*. Columbia University Press, New York.
Mayr, E. (1954). Change of genetic environment and evolution. In *Evolution as a process*, (ed. J. Huxley, A. C. Hardy, and E. B. Ford), pp. 157–80. Allen and Unwin, London.
Mayr, E. (1965). Avifauna: turnover on islands. *Science*, **150**, 1587–8.
McNab, B. H. (1994*a*). Energy conservation and the evolution of flightlessness in birds. *American Naturalist*, **144**, 628–42.
McNab, B. H. (1994*b*). Resource use and the survival of land and freshwater vertebrates on oceanic islands. *American Naturalist*, **144**, 643–60.
Murphy, R. C. (1938). The need for insular exploration as illustrated by birds. *Science*, **88**, 533–9.
Peck, S. B. (1996). Diversity and distribution of the orthopteroid insects of the Galápagos islands, Ecuador. *Canadian Journal of Zoology*, **74**, 1497–1510.
Perkins, R. C. L. (1913). Introduction. In *Fauna Hawaiiensis*, (ed. D. Sharp), pp. xv–ccxxviii. Cambridge University Press.
Petren, K. and Case, T. J. (1997). A phylogenetic analysis of body size evolution and biogeography in chuckwalla (*Sauromalus*) and other iguanines. *Evolution*, **51**, 206–19.
Pregill, G. K. and Olson, S. L. (1981). Zoogeography of West Indian vertebrates in relation to Pleistocene climatic cycles. *Annual Review of Ecology and Systematics*, **12**, 75–98.
Racine, C. H. and Downhower, C. H. (1974). Vegetative and reproductive strategies of Opuntia (Cactaceae) in the Galapagos islands. *Biotropica*, **6**, 175–86.
Rattenbury, J. (1962). Cyclic hybridization as a survival mechanism in the New Zealand forest flora. *Evolution*, **16**, 348–63.
Raven, P. H. (1972). Evolution and endemism in the New Zealand species of *Epilobium*. In *Taxonomy, phytogeography and evolution*, (ed. D. H. Valentine), pp. 259–74. Academic Press, London.
Ricklefs, R. E. and Cox, G. W. (1972). Taxon cycles in the West Indian avifauna. *American Naturalist*, **106**, 195–219.
Roff, D. (1990). The evolution of flightlessness in insects. *Ecological Monographs*, **60**, 389–421.
Roff, D. (1994). The evolution of flightlessness: is history important? *Evolutionary Ecology*, **8**, 639–57.
Roughgarden, J. (1995). *Anolis lizards of the Caribbean: ecology, evolution and plate tectonics*. Oxford University Press.
Sakai, A., Wagner, W. L., Ferguson, D. M., and Herbst, D. R. (1995*a*). Origins of dioecy in the Hawaiian flora. *Ecology*, **76**, 2517–29.
Sakai, A., Wagner, W. L., Ferguson, D. M., and Herbst, D. R. (1995*b*). Biogeographical and ecological correlates of dioecy in the Hawaiian flora. *Ecology*, **76**, 2530–43.
Shaw, K. L. (1996). Sequential radiations and patterns of speciation in the Hawaiian cricket genus *Laupala* inferred from DNA sequences. *Evolution*, **50**, 237–55.
Stanley, S. M. (1973). An explanation of Cope's rule. *Evolution*, **27**, 1–26.
Steadman, D. W. (1995). Prehistoric extinctions of Pacific island birds: biodiversity meets zooarchaeology. *Science*, **267**, 1123–31.
Vagvolgyi, J. (1975). Body size, aerial dispersal, and origin of the Pacific land snail fauna. *Systematic Zoology*, **24**, 465–88.

Wagner, W. L. and Funk, V. A. (1995). *Hawaiian biogeography: evolution on a hot-spot archipelago.* Smithsonian Institution Press, Washington, D. C.

Wallace, A. R. (1865). I. On the phenomena of variation and geographical distribution as illustrated by the Papilionidæ of the Malayan region. *Transactions of the Linnean Society*, **25**, 1–71 and eight plates.

Wallace, A. R. (1880). *Island life: or, the phenomenon and causes of insular faunas and floras, including a revision and attempted solution of the problem of geological climates.* Macmillan, London.

Wallace, A. R. (1891). *Natural selection and tropical nature.* Macmillan, London.

Williams, E. E. (1972). The origin of faunas. Evolution of lizard congeners in a complex island fauna: a trial analysis. *Evolutionary Biology*, **6**, 47–89.

Williamson, M. (1981). *Island populations.* Oxford University Press.

Wilson, E. O. (1961). The nature of the taxon cycle in the melanesian ant fauna. *American Naturalist*, **95**, 169–93.

Wollaston, T. V. (1854). *Insecta Maderensia.* van Voorst, London.

Zimmerman, E. C. (1972). Adaptive radiation in Hawaii with special reference to insects. *Biotropica*, **2**, 32–8.

2

The reproductive biology and genetics of island plants

Spencer C. H. Barrett

2.1 Introduction

Islands have long held a fascination for biologists and those interested in natural history. The distinctive biotas of oceanic islands were of considerable significance to Darwin and Wallace in developing their ideas on evolution and the recognition that islands act as 'evolutionary laboratories' has stimulated considerable modern work on the systematics, genetics, and ecology of island groups. This research has provided some of the clearest evidence for the mechanisms responsible for evolutionary diversification and has contributed significantly to the development of ecological and evolutionary theory. Why do islands provide such a rich source of biological novelty for evolutionary enquiry? The answer to this question is not always obvious, although their unsaturated habitats, non-equilibrial communities, unique biotas, and isolation from recurrent gene flow have all been implicated as playing a role in directing evolutionary responses distinct from those observed in related continental taxa. In addition, stochastic forces involving founder events and genetic drift have also been frequently invoked to account for the unusual patterns of variation that occur in many island groups.

Most work on island plants this century has been systematic or biogeographic in focus, addressing issues concerned with endemism, adaptive radiation, and the phylogenetic history of island taxa (Carlquist 1974; Bramwell 1979; Wagner and Funk 1995). Despite their capacity for long-distance dispersal, comparative investigations of the ecology and genetics of continental and island plant populations are few compared with the sizable literature for many animal groups (reviewed in Frankham 1997). Studies of this type are important for assessing whether the genetic systems of island populations are distinct from those on the mainland and for testing theories concerned with the population genetic consequences of long-distance migration and the role of genetic bottlenecks in evolution (Carson and Templeton 1984; Barton 1989). The paucity of such comparisons may be because following establishment many plant groups, particularly those on oceanic islands, diversify rapidly in morphology and ecology as a result of adaptive radiation (e.g. Carr *et al.* 1989; Ganders 1989). It is therefore often difficult to determine ancestral relationships and explicit continent–island comparisons are usually not possible. Archipelagos that are less isolated and closer to the mainland can sometimes offer more rewarding model systems for intraspecific studies on the microevolutionary processes associated with island colonization.

Successful colonization and establishment on islands for any group of organisms will be influenced by their life histories and reproductive systems. The predominantly hermaphrodite condition of most plant species combined with self-compatibility has been proposed as an advantage in enabling establishment following long-distance dispersal (Baker 1955). The requirement for specialized animal pollinators by some species may, in contrast, be an impediment to island establishment if particular pollinators are absent. Species with floral syndromes capable of being serviced by a range of pollinators may be more likely to succeed on some islands. These issues raise the question of how much the distinctive nature of some island floras reflects selection of particular traits during establishment or whether characteristics of island plants have evolved autochthonously due to special ecological circumstances on islands.

In this chapter I review the growing literature concerned with the reproductive biology and genetics of island plants. Several questions often raised when considering the evolutionary biology of island groups are addressed. Are the pollination biology and mating systems of island plants different from the mainland, and if so why? What role have stochastic forces played in the reproductive biology and genetics of island plants? Do island populations of plants contain less genetic diversity than those from the mainland? Islands vary enormously in size, geographical isolation, topography, climate, and ecology and are therefore by no means a homogeneous group. As a result, answers to the above questions may sometimes be equivocal. Despite this caveat I hope this review, by pointing out gaps in our knowledge, will draw attention to future work that needs to be undertaken on the reproduction and genetics of island plants.

2.2 Pollination biology of island plants

Islands usually support fewer animal species than occur on comparable mainland areas (MacArthur and Wilson 1967). Pollinator faunas on islands are often depauperate with important groups of pollinators either absent or poorly represented. In the Hawaiian Islands, for example, only 15% of known families of insects are represented, with only six native species of hawkmoths, two species of butterflies, and no bumble bees (Howarth and Mull 1992). Small moths and flies are disproportionately represented as pollinators. Among the Galápagos Islands few insects are reported as pollinators and only one species of pollinating bee occurs (*Xylocopa darwinii*; McMullen 1987). Long-tongued bees and hawkmoths are entirely absent from New Zealand and no genuine case of a plant adapted to butterfly pollination has been reported (Lloyd 1985). Instead, flies are the predominant pollinators with beetles, wasps, and small moths also playing some role. Such groups usually forage promiscuously visiting a broad range of taxa, thus reducing opportunities for specialized plant–pollinator interactions to evolve. The disproportionate representation of small insects on oceanic islands probably reflects their superior migratory abilities.

The pollination biology of mainland and island populations of individual species has seldom been compared. The limited data that are available indicate that patterns of pollinator visitation usually differ. Spears (1987) found significant differences in pollinator diversity and visitation rates between island and mainland populations of *Centrosema virginianum* and *Opuntia stricta* on the west coast of Florida. Differences in the quantity and quality of pollinators visiting flowers of the two species were shown to influence components of female and male reproductive success. Similarly, Inoue *et al.* (1996) found reduced pollinator diversity and visitation rates between mainland and island populations of *Campanula* taxa, with declines in both with distance of the island from the Japanese mainland. Differences in the diversity and rates of floral visitation by hummingbird species between Trinidad and Tobago, two islands which differ in size and distance from the mainland, have also been shown to have important reproductive consequences to conspecific taxa occurring on them (Feinsinger *et al.* 1982).

Floral biology

What evidence is there that the floral biology of island plants differs significantly from mainland sources as a result of altered pollination conditions? Several authors have drawn attention to the low representation of brightly-coloured, tubular, zygomorphic flowers on oceanic islands, and instead a high frequency of small, white or green non-showy flowers with simple bowl-shaped corollas (Carlquist 1974). Nowhere is this contrast more evident than in the floras of Australia and New Zealand (Webb and Kelly 1993). Many Australian species possess showy complex flowers while those in New Zealand are commonly dull, white, and unspecialized. These differences are often maintained within genera that occur in both areas and are probably associated with the relative importance of promiscuous flies and other generalist flower visitors for pollination in the two regions. The contributions of allochthonous (elsewhere) versus autochthonous (*in situ*) origins in accounting for the small, unspecialized flowers of island floras such as New Zealand is not known. Lloyd (1985) developed a simple method to estimate the relative importance of these two processes based on the character states present in related taxa in the presumed source region. Modern methods of phylogenetic reconstruction offer an alternative and more sophisticated approach for inferring the origins of reproductive traits.

Microevolutionary investigations of taxa with island and mainland populations provide opportunities to determine whether the reproductive consequences of altered pollination conditions have resulted in evolutionary responses. Inoue *et al.* (1996) found that variation in flower size among mainland and island populations of *Campanula* taxa was associated with different pollinators visiting the populations. A significant positive relation was observed between the mean flower width of populations and the mean body size of pollinators. In mainland and offshore island populations, which have larger flowers, the predominant pollinators were bumblebees and megachilid bees. In contrast, on islands further from the mainland where populations possess smaller flowers, halictid bees were the main pollinators. Experimental studies on the preferences of these bees demonstrated that pollen-collecting halictids were indifferent to flower size whereas nectar-feeding megachilids and bumblebees preferred flowers of larger size. These results are consistent with the hypothesis that pollinator-mediated selection maintains variation in flowers size among mainland and island populations of *Campanula*.

Wind pollination

The apparently high incidence of wind pollination on some oceanic islands has prompted several authors to propose adaptive hypotheses for the benefits of wind pollination on islands (Carlquist 1974; Ehrendorfer 1979). These involve the following ideas: (1) wind-pollinated taxa are favoured over animal-pollinated taxa during island establishment because of their independence from pollinators; (2) strong winds and inclement weather conditions on some islands favour wind pollination over animal pollination; (3) wind pollination is more effective than animal pollination in promoting pollen dispersal and the associated benefits of outcrossing. Before adaptive hypotheses are tested it is important to determine whether wind pollination is indeed more frequent on islands compared with mainland source areas. This is by no means clear. On the Galápagos Islands, for example, several authors have commented on the paucity of wind-pollinated taxa (McMullen 1987). Moreover, because wind pollination on islands is often associated with life-history features such as woodiness and dioecy (Lloyd 1985; Sakai *et al.* 1995*a*), it is important that these potentially confounding factors are assessed using comparative approaches. However, where the autochthonous evolution of wind pollination from animal pollination has occurred (e.g. *Schiedea*, Weller *et al.* 1995), excellent opportunities for investigating the selective forces responsible for this shift in pollination system are provided. Studies on this topic are of general significance as the change from animal to wind pollination is one of the most important evolutionary trends in flowering plants, yet little is known about the microevolutionary processes that are responsible.

2.3 Evolution of mating systems

Studies of reproduction in island plants have been dominated by two central questions in mating-system biology: the evolutionary forces responsible for different amounts of self- versus cross-fertilization in populations; and the selection of combined versus separate sexes. Several island groups have been used as experimental systems for addressing these problems.

Evolution of self-fertilization

Baker's rule (Baker 1955) proposes that self-compatible rather than self-incompatible plants will be favoured in establishment following long-distance dispersal to islands. This is because a single self-compatible immigrant is sufficient to initiate a sexually reproducing colony. Surveys of the compatibility status of island plants are rudimentary, but what data do exist strongly support the prediction that self-incompatible species are disadvantaged during island colonization and establishment. The floras of New Zealand (Webb and Kelly 1993), Hawaii (Carr *et al.* 1986), and the Galápagos (McMullen 1987) are deficient in taxa possessing homomorphic or heteromorphic incompatibility compared with continental areas. The predominance of self-compatible plants on the Galápagos Islands probably results from the scant insect fauna and availability of extensive pioneer habitats. These ecological factors are known to favour the evolution of self-compatibility.

Insufficient pollination caused by a paucity of pollinators favours floral traits that promote increased levels of selfing in self-compatible island colonists. Several comparative studies of mainland and island taxa have documented reductions in flower size and the loss of floral adaptations that promote cross-pollination associated with island colonization (e.g. Barrett 1985; Barrett and Shore 1987; Inoue *et al.* 1996). In the Mediterranean the 14 *Nigella* species in Greece and Turkey are largely outcrossing but on Aegean Islands the two endemic species have small flowers and are predominantly selfing (Strid 1969). A revealing contrast involves the distribution of the outcrossing *N. degenii* and selfing *N. doerfleri* among the Kikladhian Islands. While the outcrossing species is confined to large and more topographically complex mesic islands, the selfer occurs on these islands and a series of much smaller, low, and more arid islands. This pattern would be expected if repeated cycles of colonization and extinction on smaller islands favoured selfing immigrants because of their ability to establish colonies without the requirements of insect pollinators.

In the Caribbean similar biogeographical patterns involving island size and the distribution of selfing and outcrossing forms are evident in *Turnera ulmifolia,* a complex of perennial weeds composed of self-incompatible distylous and self-compatible homostylous forms (Barrett and Shore 1987). In this group, island colonization favours selfing homostylous forms, but on larger islands such as Jamaica selection for increased outcrossing through herkogamy in homostyled colonists may explain the complex patterns of floral variation that occur. The open bowl-shaped flowers of *Turnera* are relatively unspecialized and levels of pollinator visitation on Jamaica are sufficient to enable moderate levels of outcrossing to occur (Belaoussoff and Shore 1995).

In the self-compatible, tristylous, annual aquatic *Eichhornia paniculata*, a particularly striking biogeographical pattern of mating-system variation is evident between continental and island populations (Barrett *et al.* 1989; Barrett and Husband 1990*a*). In north-east Brazil, the centre of the range for the species, most populations are tristylous (Fig. 2.1), serviced by long-tongued anthophorid bees, and outcrossing. In contrast, in Jamaica pollinator visitation to *E. paniculata* is extremely low and selfing is high. Unlike *Turnera*, the flowers of *E. paniculata* are tubular, zygomorphic and for effective cross-pollination require long-tongued pollinators. Two floral morphs of *E. paniculata* occur in Jamaica (Fig. 2.1) with very different frequencies. The dominant morph is a self-pollinating, mid-styled variant, while the other morph occurs sporadically in approximately one-third of the island's populations and is an unmodified long-styled morph. The morphs differ dramatically in mating patterns and fertility owing to their differences in floral morphology. The close proximity of sexual organs in the mid-styled variant promotes high selfing and ensures abundant seed set. In the long-styled morph, anthers and stigmas are widely separated promoting high outcrossing but preventing autonomous self-pollination. Because in Jamaica levels of pollinator visitation to *E. paniculata* are so low this morph is unable to found colonies, as indicated by the absence of populations fixed for this form, and in dimorphic populations it suffers severe reductions in fertility compared with the mid-styled

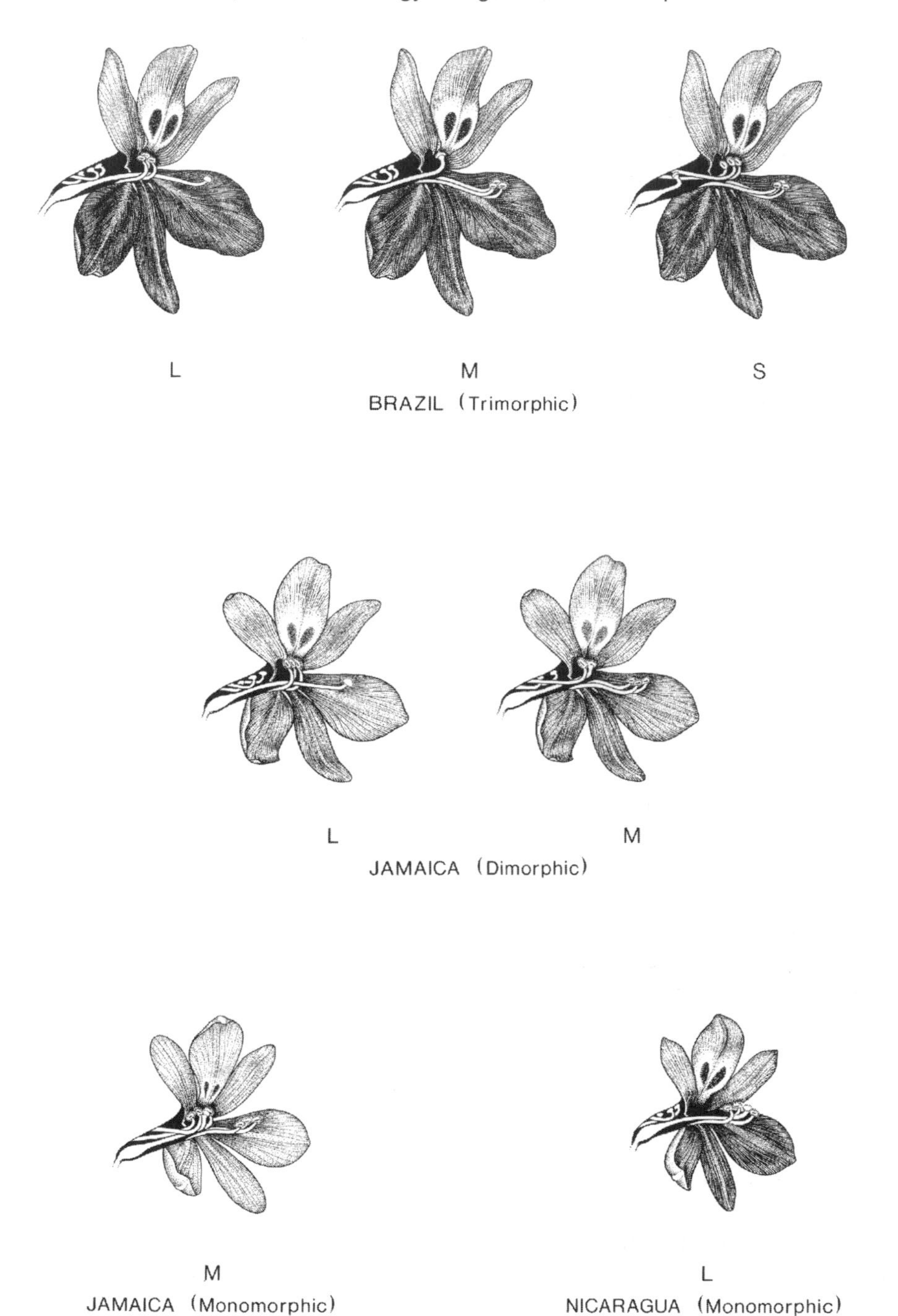

Fig. 2.1 Evolutionary breakdown of tristyly in *Eichhornia paniculata* populations on Jamaica. L, M, and S refer to the long-, mid-, and short-styled morph, respectively. See text for further details.

variant. If these fitness asymmetries persist, the long-styled morph should become extinct on Jamaica. However, because this phenotype is governed by recessive alleles (see below) selective elimination could be quite protracted.

If selfing is adaptive under Jamaican conditions, why have mating-system modifier genes not spread in the long-styled morph? Studies on the inheritance of the stamen modifications that cause autonomous selfing in mid-styled variants indicate that a small number of recessive genes are responsible (Fenster and Barrett 1994). These modifiers have no phenotypic effects when transferred into long-styled plants, presumably because they are morph-limited in expression. Does this imply that genetic constraints prevent the long-styled morph from evolving a selfing habit ? The recent discovery of small disjunct populations of *E. paniculata* in Nicaragua and Mexico composed exclusively of an autogamous variant of the long-styled morph (Fig. 2.1) suggest otherwise. Preliminary studies on the inheritance of this phenotype indicate that the floral modifications responsible for selfing are polygenically controlled (Barrett, unpublished data).

The contrasting patterns of floral variation in *E. paniculata* may be explained by founder events and the alternative genetic pathways to selfing that occur in the long- and mid-styled morphs. Allozyme studies (Husband and Barrett 1991) indicate that populations on Jamaica are descended from two separate colonization events, with at least one involving a mid-styled plant heterozygous at the *M* locus governing mid (*Mm* or *MM*) versus long styles (*mm*). In contrast, in Central America the mid-styled morph is missing, implying that migration only involved the long-styled morph. Because of homozygosity at the *M* locus, the mid-styled phenotype cannot arise through segregation from long-styled plants. The contrasting morph structure of populations in the two regions is likely to mean that the selection dynamics of selfing genes will be quite different. Selection for selfing in Jamaica by a small number of favourable recessive modifiers has probably occurred much more rapidly than in Central America where polygenic variation is responsible. It is possible that because of these different modes of inheritance, the likelihood of selfing evolving in the long-styled morph is conditional on whether the mid-styled morph is present or absent from populations.

An important issue for understanding the evolution of mating systems in Jamaican populations of *E. paniculata* concerns their genetic load. Theoretical models predict that inbreeding depression should evolve with the mating system, with selfing populations maintaining significantly less inbreeding depression than outcrossing populations due to selective purging of deleterious recessives in inbred populations (Charlesworth and Charlesworth 1987). Experimental work on populations of *E. paniculata* with known rates of selfing support these predictions (Barrett and Charlesworth 1991). Studies on six predominantly inbreeding populations sampled in Jamaica have failed to detect inbreeding depression for both vegetative and reproductive traits when grown in the glasshouse and compared under both competitive and non-competitive conditions (Fig. 2.2; Toppings and Barrett unpublished data). These results are in contrast to several recent studies that have found moderate to high levels of inbreeding depression in selfing species (see Lande *et al.* 1994).

The failure to detect inbreeding depression in Jamaican populations of *E. paniculata* may be because comparisons were made under glasshouse rather than field conditions.

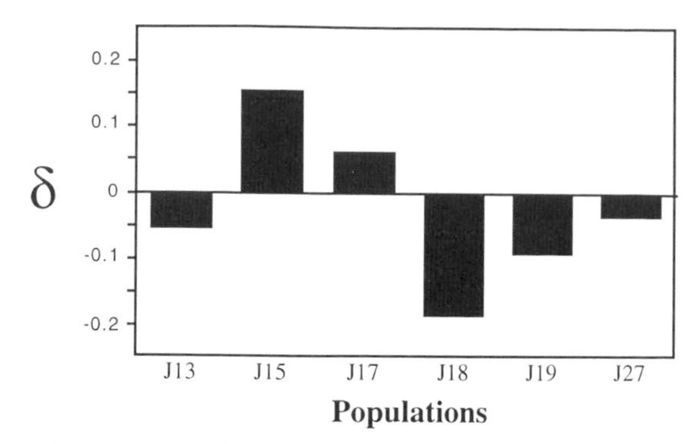

Fig. 2.2 Estimates of inbreeding depression (δ) in six predominantly selfing populations of *Eichhornia paniculata* from Jamaica. Estimates are derived from a multiplicative measure: the product of (a) parental seed set following selfing or outcrossing, (b) per cent seed germination, and (c) flower production of selfed or outcrossed offspring. All comparisons were made in the glasshouse and involved 6–10 families per population and 6–10 offspring per pollination treatment grown under non-competitive conditions.

Although an attempt at ecological realism was made by using competitive treatments, more intense inbreeding depression is usually found when fitness comparisons are made under natural conditions (see Barrett and Kohn 1991). Genetic bottlenecks during island colonization resulting in severe reductions in genetic load might also account for the observed results. However, several recent studies of partially selfing taxa occurring on islands have demonstrated significant inbreeding depression (Sakai *et al.* 1989; Belaoussoff and Shore 1995; Schultz and Ganders 1996). Presumably, bottlenecks have also occurred during their evolutionary history. Another possibility is that the genome-wide mutation rate for deleterious genes is lower in *E. paniculata* than in other species examined to date, perhaps due to reduction in the mutation rate associated with evolution of predominant selfing (Kondrashov 1995 and personal communication). Information on the genetic architecture of inbreeding depression and the factors controlling genomic mutation rates for lethal and deleterious genes is required to interpret patterns of inbreeding depression in plant species (Lande *et al.* 1994). For island taxa knowledge of their colonization history, residency period on islands, and the types of selection they are exposed to will also be important.

Evolution of outcrossing

While the advantages of self-fertilization for island colonization are apparent, a dominant theme in island plant reproductive biology concerns the selection of 'out-crossing mechanisms' following establishment on islands (Carlquist 1974; Ehrendorfer 1979; Thomson and Barrett 1981). 'Escape from homozygosity' is often seen as an essential prerequisite for subsequent radiation and diversification in island groups. Wind pollination, various diclinous sexual systems (e.g. dioecy, gynodioecy), and floral traits such as herkogamy (spatial separation of anthers and stigmas within a flower) and dichogamy (different timing of anther dehiscence and stigma receptivity within a flower) have all been interpreted as different means of achieving outcrossing in island plants. It has even been proposed that the occurrence of weak isolating mechanisms in many island groups promotes evolutionary flexibility by providing high levels of heterozygosity through hybridization (Rattenbury 1962; Carlquist 1974).

There are several issues that need to be addressed in interpreting the adaptive significance of outcrossing, not the least of which is that group rather than individual selection is implied in the case of the supposed benefits of hybridization in island plants. First, some controversy surrounds the particular advantages of the genetic consequences of this pattern of mating. Biotic selection imposed by pest, parasite, and disease pressures may play an important role in maintaining outcrossing (Levin 1975). Paradoxically, however, many islands appear to have reduced diversity and biotic interactions may be more relaxed in comparison with the mainland. The ecological benefits of genetic variation therefore need to be critically assessed for island taxa. Second, alternative perspectives on the adaptive significance of reproductive traits that emphasize selective factors such as fitness gain through male function and pollen–stigma interference need to be considered before assuming that their primary function is to promote outcrossing (Barrett and Harder 1996). Finally, before assuming that the special ecological conditions on islands have resulted in the selection of 'outcrossing mechanisms' it is of importance, as discussed earlier, to determine the place of origin of reproductive traits and whether island immigrants were likely to have possessed them on arrival. This particular issue has been of considerable importance in interpreting the factors responsible for the high incidence of sexual dimorphism in the floras of the Hawaiian Islands and New Zealand.

Recent biogeographical analyses of the Hawaiian flora by Sakai *et al.* (1995*a, b*) have provided new insights into the origins of sexual dimorphism and the conditions favouring the evolution of dioecy. By using a lineage-by-lineage analysis of the entire flora these authors obtained the following results:

1. Of the 971 native species, 14.7% are dioecious and 20.7% are sexually dimorphic, proportions that are the highest of any flora studied.
2. Of the 291 inferred colonists, 10% were dimorphic with 55.2% of all current dimorphic species arising from these lineages.
3. Autochthonous evolution of dimorphism from monomorphism occurred in at least 12 lineages giving rise to approximately one-third of the current dimorphic species.
4. Sexual dimorphism is significantly associated with woodiness and small green flowers and among dioecious woody species wind pollination is disproportionately represented.
5. Among colonists sexual dimorphism is associated with fleshy fruits at the generic level.

These results clearly show that the high incidence of dioecy in the Hawaiian Islands results from a variety of factors. While broad generalities await explicit phylogenetic analyses of individual groups, it is clear that dioecy has not acted as a constraint to island colonization as Baker's rule originally implied. In some lineages bird dispersal via multiseeded fleshy fruits and unspecialized requirements for pollination probably favoured establishment of sexually dimorphic taxa on islands (see Lloyd (1985) for similar arguments for New Zealand). In other groups, sexual dimorphism has arisen autochthonously through the spread of unisexual individuals in hermaphrodite populations. These latter cases have provided model systems for investigations of the outcrossing-advantage hypothesis.

Of the 27 taxa of *Bidens* endemic to the Hawaiian Islands, 13 are gynodioecious (separate hermaphrodite and female plants), a condition unknown in the 200 species occurring elsewhere. Schultz and Ganders (1996) studied the evolution and maintenance of gynodioecy in *Bidens sandvicensis* by attempting to find the source(s) of female advantage. With nuclear control of male sterility and where outcrossing advantage is the only source of female superiority then the product of the selfing rate and inbreeding depression must exceed 0.5 (Charlesworth and Charlesworth 1978). However, the occurrence of intense inbreeding depression and high selfing is controversial, in part, because high selfing should purge deleterious recessive alleles thus eliminating inbreeding depression. In gynodioecious *B. sandvicensis,* selfing rates of hermaphrodites exceeded 0.57 and inbreeding depression measured in the field averaged 0.94, among the highest documented values for an angiosperm species. Since the product of the selfing rate and inbreeding depression is greater than 0.5 the conditions required for the maintenance of a stable polymorphism are met. Similar results involving extremely high inbreeding depression and partial selfing have also been found in *Schiedea* (Caryophyllaceae), an endemic Hawaiian genus with both sexually monomorphic and dimorphic taxa (Sakai *et al.* 1989). In both genera there is therefore evidence supporting the outcrossing-advantage hypothesis for the evolution of sexual dimorphisms in island plants. Further work is required, however, to determine the ecological and genetic factors that enable stable coexistence of severe inbreeding depression and high selfing. Inferior pollinator service has been implicated as a factor causing increased selfing in both groups.

2.4 Genetics of island populations

Since migration to islands is inevitably associated with periods of small population size, especially during the establishment phase, stochastic forces involving founder events and genetic drift have been recognized as playing a dominant role in determining patterns of genetic variation in animal groups colonizing islands (e.g. Giddings *et al.* 1989; Brakefield 1990). What evidence exists that stochasticity has influenced the patterns and amount of genetic diversity in island versus mainland populations of plants?

Stochastic processes

Island isolation and opportunities for gene flow from source populations are of prime importance in assessing the likelihood of stochastic influences. For most plant groups

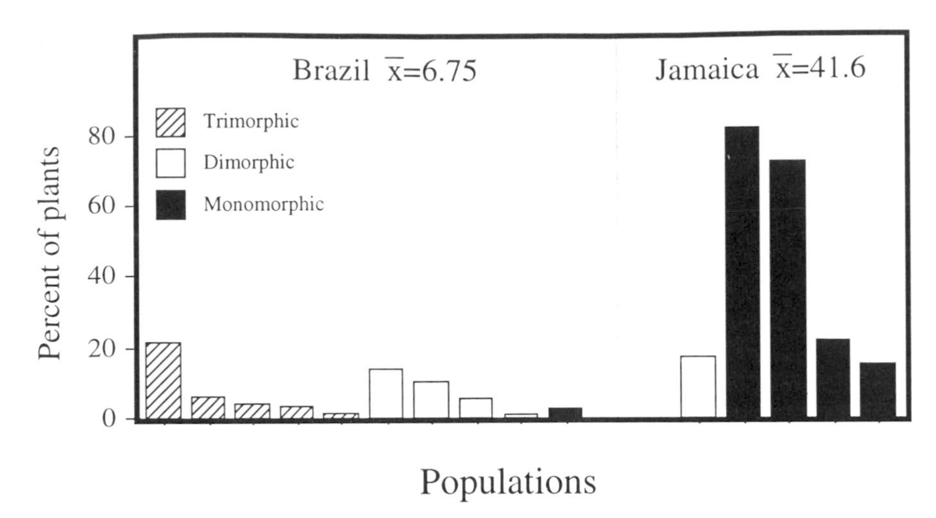

Fig. 2.3 Differences in developmental instability of flowers between ten continental (north-east Brazil) and five island (Jamaica) populations of *Eichhornia paniculata*. All flowers open on a single inflorescence in 25 plants per population were scored for abnormal tepal development. Populations from Jamaica are highly selfing, those from Brazil vary from outcrossing (trimorphic) to those with mixed mating systems (dimorphic and monomorphic).

on remote oceanic islands extreme genetic bottlenecks must have been involved in their evolutionary history. However, the occurrence of such events and their evolutionary significance is poorly understood. Molecular studies provide opportunities for detecting historical bottlenecks, but as yet these approaches have not been applied to questions concerned with island colonization.

What empirical evidence is there that present-day patterns of variation in island plant populations have been influenced by genetic drift? Apparently non-adaptive differentiation of several morphological traits in Aegean island populations of taxa in *Erysimum* section Cheiranthus was interpreted by Snogerup (1967) as resulting from drift operating in populations of these rare cliff plants. Only 107 populations were found and estimates of population size indicated that about 10% of populations contained only one to two plants and >50% had 50 or fewer individuals. The occurrence of non-adaptive and in some cases even maladaptive traits in island plants is most easily explained by the random fixation of recessive genes exposed through inbreeding in small populations. High levels of developmental instability in *E. paniculata* flowers from island compared with mainland populations also appear to have arisen in this manner (Fig. 2.3; Richards and Barrett 1992). Some of the striking morphological differentiation that characterizes island taxa displaying 'adaptive' radiation may result from stochastic processes operating during founder events and periods of small population size.

Founder events and genetic drift have been invoked to explain the absence of the short-styled morph from Jamaican populations of *E. paniculata* (Barrett *et al.* 1989). Stochastic theory indicates that this morph is more vulnerable to loss following bottlenecks due to a constraint imposed by the genetic system governing the inheritance of the tristylous polymorphism (Barrett 1993). Stochastic forces have also played an important role in reducing genetic variation at allozyme loci compared with those on the mainland (Glover and Barrett 1987) and in structuring the patterns of genetic diversity within and among island populations (Husband and Barrett 1991). Allele frequencies at polymorphic loci are highly asymmetrical in most Jamaican populations with a relatively small number of different multilocus genotypes dominating. This pattern involving high levels of gametic disequilibrium is expected for neutral loci if populations are founded by a small number of genotypes and restricted gene flow and inbreeding act to preserve particular allelic combinations.

Genetic variation in island populations

Theoretical work on population structure has used the continent-island and island models as paradigms for understanding how interactions between gene flow, effective population size, and genetic drift influence patterns of genetic variation. Under most scenarios theory predicts that island populations should be less variable and more genetically differentiated relative to source populations. Few studies of plant populations have, however, explicitly tested models of population structure and only a small number of comparisons of genetic diversity in island and mainland plant populations of the same species are available. What data do exist generally indicate that island populations contain less variation than those from mainland areas (Fig. 2.4; reviewed in

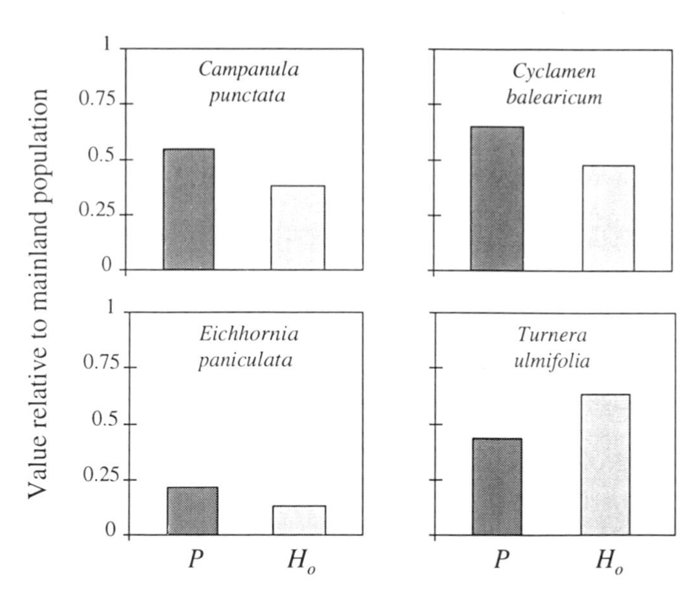

Fig. 2.4 Reduced genetic variation at allozyme loci in island versus mainland populations of four plant species. Values for the percentage of loci polymorphic (P) and observed heterozygosity (H_o) for each species are calculated as the percentage reduction from mainland populations. Sources: *Eichhornia* and *Turnera*, see Barrett and Husband (1991); *Campanula* and *Cyclamen*, see Inoue and Kawahara (1990) and Affre *et al.* (1996), respectively.

Barrett and Husband 1990*b*; Frankham 1997). Where exceptions do occur these arise either because the species involved has little or no allozyme diversity in either area (Ledig and Conkle 1983; Mosseler *et al.* 1991), the islands sampled were not isolated from the mainland (Gauthier *et al.* 1992), or known historical factors are involved (Tsumura and Ohba 1993).

Interpretation of the factors causing particular patterns of allozyme variation on islands can be complicated because colonization is often confounded with evolutionary modifications to the genetic system (e.g. diploidy to polyploidy in *Turnera*, outcrossing to selfing in *Campanula* and *Eichhornia*). These changes directly affect patterns of genetic variation and can also influence interactions with stochastic factors. For example, a loss of neutral genetic variation is expected in selfing organisms due to reductions in effective population size, hitch-hiking, and background selection (Charlesworth *et al.* 1993). While bottlenecks also reduce variation it is unclear whether these various processes can be distinguished by their genetic consequences. Geographical surveys of allozyme variation are insufficient for disentangling the individual and combined effects of these factors on genetic diversity. These problems are exacerbated because no detailed theoretical treatment is available for the influence of bottlenecks on variability in partially selfing populations, although Jarne (1995) has recently made a start on this problem.

Restricted gene flow, inbreeding, and genetic drift will strongly influence the extent of diffentiation among populations. Not unexpectedly, island plant populations often exhibit a higher degree of genetic differentiation both among populations on a single island (*Eichhornia*) and among populations on different islands (*Campanula*) in comparison with corresponding mainland samples. However, exceptions to this pattern occur, as in the Mediterranean endemic *Cyclamen balearicum*. Affre *et al.* (1996) found population differentiation at allozyme loci to be significantly higher among mainland populations in southern France than among those occurring on the Balearic Islands off the north coast of Spain. They suggested that glaciation and habitat fragmentation due to human land use patterns resulted in greater reductions in effective population size and more isolation in southern France than on the Balearic Islands, where populations are large and found over a wider range of ecological conditions. This interpretation should remind us of the increasing number of 'terrestrial habitat islands' that are being formed because of ecosystem modifications through agriculture, forestry, and urbanization. Studies of true island populations may assist in predicting the genetic consequences of such anthropogenic changes.

2.5 Conclusions

A recurring theme throughout this chapter is that many general problems in evolutionary biology can be addressed through investigations of the reproductive biology and genetics of island plants. The adaptive benefits of selfing and outcrossing, selection of combined versus separate sexes, the role of stochastic processes in evolution, and models of population structure can each be investigated by exploiting island plant taxa as model systems. Also, genera displaying adaptive radiation frequently exhibit weak isolating mechanisms and can provide unsurpassed opportunities to investigate the genetic basis of adaptation and of the speciation process itself. Unfortunately, island populations are more vulnerable to extinction than those in continental regions and many endemic plant taxa, including several discussed above, are rare or threatened. It would be regrettable if this rich biological heritage was lost and with it opportunities for increased understanding of evolutionary processes.

2.6 Summary

Island and mainland populations of plants often differ in their reproductive biology and genetics. The differences become more pronounced the further islands are from mainland sources. Altered pollination conditions have influenced the floral biology and mating systems of island plants in distinct ways. Insufficient pollination has favoured selection of floral traits promoting selfing. In contrast, inferior pollinator service resulting in selfing and inbreeding depression appears to be a factor involved in the evolution of sexual dimorphism. Stochastic forces play a major role in governing patterns of genetic variation. Island populations are usually more differentiated and contain less diversity than comparable mainland samples. Many general issues in evolutionary biology can be addressed by studies of reproduction and genetics in island plants.

Acknowledgements

I thank Stefan Andersson, Richard Frankham, Ann Sakai, Stewart Schultz, John Thompson, and Stephen Weller for help with references and for making available unpublished manuscripts, Nick Barton for comments on the manuscript, William Cole for assistance in the preparation of the figures and the Natural Sciences and Engineering Research Council of Canada for research funding.

References

Affre, L., Thompson, J. D., and Debussche, M. (1996). Genetic structure of continental and island populations of the Mediterranean endemic *Cyclamen balearicum* Willk. (Primulaceae). *American Journal of Botany*, **84**, 437–51.

Baker, H. G. (1955). Self-compatibilty and establishment after 'long-distance' dispersal. *Evolution*, **9**, 347–9.

Barrett, S. C. H. (1985). Floral trimorphism and monomorphism in continental and island populations of *Eichhornia paniculata* (Spreng.) Solms. (Pontederiaceae). *Biological Journal of the Linnean Society*, **25**, 41–60.

Barrett, S. C. H. (1993). The evolutionary biology of tristyly. *Oxford Surveys in Evolutionary Biology*, **9**, 283–326.

Barrett, S. C. H. and Charlesworth, D. (1991). Effects of a change in the level of inbreeding on the genetic load. *Nature*, **352**, 522–4.

Barrett, S. C. H. and Harder, L. D. (1996). Ecology and evolution of plant mating. *Trends in Ecology and Evolution*, **11**, 73–9.

Barrett, S. C. H. and Husband, B. C. (1990*a*). Variation in outcrossing rate in *Eichhornia paniculata*: the role of demographic and reproductive factors. *Plant Species Biology*, **5**, 41–56.

Barrett, S. C. H. and Husband, B. C. (1990*b*). The genetics of plant migration and colonization. In *Plant population genetics, breeding, and genetic resources*, (ed. A. H. D. Brown, M. T. Clegg, A. L. Kahler, and B. S. Weir), pp. 254–77. Sinauer, Sunderland, MA.

Barrett, S. C. H. and Kohn, J. R. (1991). The genetic and evolutionary consequences of small population size in plants: Implications for conservation. In *Genetics and conservation of rare plants*, (ed. D. Falk and K. E. Holsinger), pp. 3–30. Oxford University Press.

Barrett, S. C. H. and Shore, J. S. (1987). Variation and evolution of breeding systems in the *Turnera ulmifolia* complex (Turneraceae). *Evolution*, **41**, 340–54.

Barrett, S. C. H., Morgan, M. T., and Husband, B. C. (1989). The dissolution of a complex polymorphism: The evolution of self-fertilization in tristylous *Eichhornia paniculata* (Pontederiaceae). *Evolution*, **43**, 1398–416.

Barton, N. H. (1989). Founder effect speciation. In *Speciation and its consequences*, (ed. D. Otte and J. A. Endler), pp. 229–56. Sinauer, Sunderland, MA.

Belaoussoff, S. and Shore, J. S. (1995). Floral correlates and fitness consequences of mating-system variation in *Turnera ulmifolia*. *Evolution*, **49**, 545–56.

Brakefield, P. M. (1990). Genetic drift and patterns of diversity among colour-polymorphic populations of the homopteran *Philaenus spumarius* in an island archipelgo. *Biological Journal of the Linnean Society*, **39**, 219–37.

Bramwell, D. (ed.) (1979). *Plants and islands*. Academic Press, London.

Carlquist, S. (1974). *Island biology*. Columbia University Press, New York.

Carr, G. D., Powell, E. A., and Kyhos, D. W. (1986). Self-Incompatibility in the Hawaiian Madiinae (Compositae): An exception to Baker's Rule. *Evolution*, **40**, 430–4.

Carr, G. D., Robichaux, R. H., Witter, M. S., and Kyhos, D. W. (1989). Adaptive radiation of the Hawaiian Silversword alliance (Compositae-Madiinae): A comparison with Hawaiian Picture-Winged *Drosophila*. In *Genetics, speciation and the founder principle*, (ed. L. V. Giddings, K. Y. Kaneshiro, and W. W. Anderson), pp. 79–97. Oxford University Press.

Carson, H. L. and Templeton, A. R. (1984). Genetic revolutions in relation to speciation phenomena: the founding of new populations. *Annual Review of Ecology and Systematics*, **15**, 97–131.

Charlesworth, B. and Charlesworth, D. (1978). A model for the evolution of dioecy and gynodioecy. *American Naturalist*, **112**, 975–97.

Charlesworth, B., Morgan, M. T., and Charlesworth, D. (1993). The effect of deleterious mutations on neutral molecular variation. *Genetics*, **134**, 1289–1303.

Charlesworth, D. and Charlesworth, B. (1987). Inbreeding depression and its evolutionary consequences. *Annual Review of Ecology and Systematics*, **18**, 273–98.

Ehrendorfer, F. (1979). Reproductive biology in island plants. In *Plants and islands*, (ed. D. Bramwell), pp. 293–306. Academic Press, London.

Feinsinger, P., Wolfe, J., and Swarm, L. A. (1982). Island ecology: reduced hummingbird diversity and the pollination biology of plants, Trinidad and Tobago, West Indies. *Ecology*, **63**, 494–506.

Fenster, C. B. and Barrett, S. C. H. (1994). Inheritance of mating-system modifier genes in *Eichhornia paniculata* (Pontederiaceae). *Heredity*, **72**, 433–45.

Frankham, R. (1997). Do island populations have less genetic variation than mainland populations? *Heredity*, **78**, 291–307.

Ganders, F. R. (1989). Adaptive radiation in Hawaiian *Bidens*. In *Genetics, speciation and the founder principle*, (ed. L. V. Giddings, K .Y. Kaneshiro, and W. W. Anderson), pp. 99–112. Oxford University Press.

Gauthier, S., Simon, J.-P., and Bergeron, Y. (1992). Genetic structure and variability in Jack Pine populations: effects of insularity. *Canadian Journal of Forestry Research*, **22**, 1958–65.

Giddings, L. V., Kaneshiro, K.Y., and Anderson, W. W. (eds.) (1989). *Genetics, speciation and the founder principle*. Oxford University Press.

Glover, D. E. and Barrett, S. C. H. (1987). Genetic variation in continental and island populations of *Eichhornia paniculata* (Pontederiacae). *Heredity*, **59**, 7–17.

Howarth, F. G. and Mull, W. P. (1992). *Hawaiian insects and their kin*. University of Hawaii Press, Honolulu.

Husband, B. C. and Barrett, S. C. H. (1991). Colonization history and population genetic structure of *Eichhornia paniculata* in Jamaica. *Heredity*, **66**, 287–96.

Inoue, K. and Kawahara, T. (1990). Allozyme differentiation and genetic structure in island and mainland populations of *Campanula punctata* (Campanulaceae). *American Journal of Botany*, **77**, 1440–8.

Inoue, K., Maki, M., and Masuda, M. (1996). Evolution of *Campanula* flowers in relation to insect pollinators on islands. In *Floral biology: studies on floral evolution in animal-pollinated plants*, (ed. D. G. Lloyd and S. C. H. Barrett), pp. 377–400. Chapman and Hall, New York.

Jarne, P. (1995). Mating system, bottlenecks and genetic polymorphism in hermaphrodite animals. *Genetic Research*, **65**, 193–207.

Kondrashov, A. S. (1995). Modifiers of mutation-selection balance: general approach and the evolution of mutation rates. *Genetic Research*, **66**, 53–69.

Lande, R., Schemske, D. W., and Schultz, S. T. (1994). High inbreeding depression, selective interference among loci, and the threshold selfing rate for purging deleterious recessive lethal mutations. *Evolution*, **48**, 965–78.

Ledig, F. T. and Conkle, M. P. (1983). Genetic diversity and genetic structure in a narrow endemic, Torrey pine (*Pinus torreyana* Parry ex Carr). *Evolution*, **37**, 79–85.

Levin, D. A. (1975). Pest pressure and recombination systems in plants. *American Naturalist*, **109**, 437–51.

Lloyd, D. G. (1985). Progress in understanding the natural history of New Zealand plants. *New Zealand Journal of Botany*, **23**, 707–22.

MacArthur R. H. and Wilson, E. O. (1967). *The theory of island biogeography*. Princeton University Press.

McMullen, C. K. (1987). Breeding systems of selected Galápagos Islands angiosperms. *American Journal of Botany*, **74**, 1694–705.

Mosseler, A., Innes, D. J., and Roberts, B. A. (1991). Lack of allozyme variation in disjunct Newfoundland populations of red pine (*Pinus resinosa*). *Canadian Journal of Forestry Research*, **21**, 525–8.

Rattenbury, J. A. (1962). Cyclic hybridization as a survival mechanism. *Evolution*, **16**, 348–63.

Richards, J. H. and Barrett, S. C. H. (1992). The development of heterostyly. In *Evolution and function of heterostyly*, (ed. S. C. H. Barrett), pp. 85–127. Springer, Berlin.

Sakai, A. K., Karoly, K., and Weller, S. G. (1989). Inbreeding depression in *Schiedea globosa* and *S. salicaria* (Caryophyllaceae), subdioecious and gynodioecious Hawaiian species. *American Journal of Botany*, **76**, 437–44.

Sakai, A., Wagner, W. L., Ferguson, D. M., and Herbst, D. R. (1995*a*). Biogeographical and ecological correlates of dioecy in the Hawaiian Flora. *Ecology*, **76**, 2530–43.

Sakai, A., Wagner, W. L., Ferguson, D. M., and Herbst, D. R. (1995*b*). Origins of dioecy in the Hawaiian flora. *Ecology*, **76**, 2517–29.

Schultz, S. T. and Ganders, F. R. (1996). Evolution of unisexuality in the Hawaiian Islands: A test of microevolutionary theory. *Evolution*, **50**, 842–55.

Snogerup, S. (1967). Studies in the Aegean Flora IX. *Erysimum* sect. Cheiranthus. B. Variation and evolution in the small-population system. *Opera Botanica*, **14**, 5–86.

Spears Jr., E. E. (1987). Island and mainland pollination ecology of *Centrosema virginianum* and *Opuntia stricta*. *Journal of Ecology*, **75**, 351–62.

Strid, A. (1969). Evolutionary trends in the breeding system of *Nigella* (Ranunculaceae). *Botaniska Notiser*, **122**, 380–97.

Thomson, J. and Barrett, S. C. H. (1981). Selection for outcrossing, sexual selection and the evolution of dioecy in plants. *American Naturalist*, **118**, 443–9.

Tsumura, Y. and Ohba, K. (1993). Genetic structure of geographical marginal populations of *Cryptomeria japonica*. *Canadian Journal of Forestry Research*, **23**, 859–63.

Wagner, W. L. and Funk, V. (eds.) (1995). *Hawaiian biogeography: evolution on a hot-spot archipelago*. Smithsonian Institution, Washington, D. C.

Webb, C. J. and Kelly, D. (1993). The reproductive biology of the New Zealand flora. *Trends in Ecology and Evolution*, **8**, 442–7.

Weller, S. G., Wagner, W. L., and Sakai, A. L. (1995). A phylogenetic analysis of *Schiedea* and *Alsinidendron* (Caryophyllaceae: Alsinoideae): Implications for the evolution of breeding systems. *Systematic Botany*, **20**, 315–37.

3

Evolution of small mammals

R. J. Berry

Oceanic islands are to the naturalist what comets and meteorites are to the astronomer, and even that pregnant doctrine of the origin and succession of life which we owe to Darwin, and which is to us what spectrum analysis is to the physicist, has not proved sufficient to unravel the tangled phenomena they present.

Sir Joseph Hooker, 1866, q.v. Williamson (1984)

3.1 Introduction

Less than six years after the publication of the *Origin of species*, Joseph Hooker lectured on 'Insular floras' at a British Association meeting, describing four common characteristics in them: endemicity, impoverishment, dispersal, and disharmony. In a section added before the text was published the following year, Hooker added the problem of the interactions between life form, taxonomic status, and abundance. It was the first comprehensive account of island biology in an evolutionary context (Williamson 1984). All Hooker's points have remained of interest, and all of them are implicitly discussed in this volume.

In the peroration of his lecture, Hooker points out that island floras can only be understood within an evolutionary framework. In his time, he could discuss only the end products of evolutionary change; it is relatively recently that we have begun to be able to distinguish the effects of evolution *on* islands from the consequences of the genes carried *to* islands by colonizers. This chapter is largely about this distinction, as illustrated by small mammal differentiation and the explanations given for it.

R. C. Lewontin has compared the fanciful fiction of Kipling's 'just-so' stories to evolutionary explanations in the sense that they are 'informed speculative rationalizations of history' (Skelton 1993, p. 745). It was the wide credence of such speculations to explain adaptation that prompted the notorious 'critique of the adaptationist programme' by Gould and Lewontin (1979). Island endemics have provided (and probably will continue to provide) a fertile bank of evolutionary 'just-so' stories. Charles Darwin's vaunted Road to Damascus conversion to evolution when he encountered the remarkable animals of the Galápagos was in fact nothing of the sort: by the time the *Beagle* put into the Galápagos, Darwin was bored and somewhat homesick (Sulloway 1982; Berry 1984). Soon after his return to England, he published a note revealing that he retained only very vague memories of the islands. All he could say about the finches (later to be eponymized as 'Darwin's finches') was that 'their general resemblance in character and the circumstance of their indiscriminately associating in large flocks, rendered it almost impossible to study the habits of particular species.... They appeared to subsist on seeds' (Darwin 1837).

3.2 Small mammal 'just-so' stories

Many small mammals have distinct island forms, sometimes recognized as races or subspecies, sometimes accorded full specific status. The commonest explanation for their differentiation is that they are relicts of previously widespread forms which have been isolated on small islands and subsequently undergone drift (see Chapter 2) or adaptation to the environment thereon. Such stories are often supported by dubious embellishments about land bridges and/or sea-level changes, not infrequently put forward by geologists impressed by biological claims that such features must have occurred (Beirne 1952). This in turn has led to biologists being persuaded that their biological speculations have been endorsed by geologists.

Evolutionary just-so stories as they relate to the Orkney vole (*Microtus arvalis orcadensis*), the long-tailed field mouse (*Apodemus sylvaticus* = *A. hebridensis, hirtensis, fridariensis*) of the islands to the north and west of Scotland, and the house mouse (*Mus musculus faeroensis*) of the Faroes have been reviewed by Berry (1986, 1996*a*) with the likely history of the various island races; details about them must be sought therein.

3.3 The founder effect

The death knell to the just-so stories about island evolution as far as British small mammals are concerned was a short paper by Corbet (1961), in which he pointed out that neither geological nor distributional data support the assumptions of differentiation following isolation, and that most of the races must have originated through human agency. Corbet's argument can be applied generally. For example, there are both 'native' and introduced rats on the Galápagos archipelago. There are three groups of native rats: the extinct genus *Megaoryzomys*, apparently derived from a thomasomyine of mainland South America; *Nesoryzomys*, which has no close relatives among the mainland rats nearest to it (which are the highly diverse rice rats *Oryzomys*); and a single living species of *Orzomys, O. bauri,* which is morphologically and allozymically virtually identical with the mainland *O. xantheolus.* (Another island species *O. galapagoensis*, has become extinct within historical times.) This implies two successful introductions in the far distant past, plus a more recent introduction of *Oryzomys* (Patton 1984).

Allozymic, morphometric, and non-metrical skeletal traits all show the same set of inter-island relationships in the 'introduced rat' *Rattus rattus*, and can be interpreted as indicating another three successful colonizations, corresponding to different periods of human activity in the islands: the first in the late 1600s, the most recent during World War II. After the primary introductions, gene flow between the islands seems to have been slight (Patton *et al.* 1975). Analysis of mitochondrial DNA polymorphisms in nine subspecies of *Peromyscus maniculatus* on the Californian Channel Islands indicates at least four separate colonizations from the mainland, all probably within the last 500 000 years (Ashley and Wills 1987).

'Natural', 'native', or 'endemic' are imprecise terms, describing little more than long-ago time. The Polynesian or Maori rat, *Rattus exulans*, is frequently described as 'native' on Pacific islands, yet it must have been introduced either intentionally (it is eaten by the Maori in New Zealand) or accidentally in pre-European influence days (Lever 1985; King 1990). Although there are many examples of relict populations of small mammals (e.g. Stewart and Baker 1992), most island populations which have been studied in detail seem to have been the result of colonization at some time in the not too far distant past (Crowell 1986; Patterson and Atmar 1986). One assumes that such colonizers will be few in number and, with mammals at least, that multiple colonizations are uncommon (cf. MacArthur and Wilson 1967, p. 155), although it is rarely possible to be definite. The colonizing event means that the population will go through a bottleneck in numbers with potentially major effects on genetic variation. A founding event results in intermittent genetic drift (so named by Waddington (1957, p. 86)), which is likely to be much more drastic than the more normally assumed persistent drift, where alleles change in frequency relatively slowly (at a rate inversely dependent on the effective breeding size of the population). The whole future reaction and adjustment of the colonizing group will depend largely on the alleles and their frequencies in the original members. Even if further immigration brings in fresh variation, by the time it occurs the original founders are likely to have increased in number and range, reducing the genetic impact of fresh individuals.

The importance of the initial founding population has been consistently undervalued by population biologists. The dominating influence on island biology in recent years has been MacArthur and Wilson's (1963, 1967) theory of island biogeography. But they were interested almost entirely in diversity, not differentiation. Their 1967 book contains a chapter on 'Evolutionary changes following colonization' (36 pages out of a total 183), but it is almost entirely concerned with adaptive adjustment; in this they followed Mayr (1954), who saw the founder effect as disturbing the genetic cohesion generally assumed to hold species together. Wilson (1969) continued this bias, concentrating on changes in species composition with time and on interspecific relationships. Without denigrating or challenging these effects, and accepting that natural selection may affect island biotas as much as—if not more than—continental ones, it is nonetheless difficult to avoid the conclusion that the main differentiation of island forms is usually the result of the chance colonizers of each population, and only secondarily due to subsequent adaptation.

This conclusion has been repeatedly challenged (most determinedly by E. B. Ford, arguing particularly from genetical changes in *Maniola jurtina* populations in the Scillies (Ford 1975, pp. 61–77); Ford's ideas have persisted, because, as MacArthur and Wilson (1967, p. 156) note, they were based on 'one of the few adequate analyses of ecogenetic variation in insular populations' when modern views of island evolution were forming). Williamson (1981, pp. 131–8) points out that founder effect differentiation depends on anecdotal suppositions (which form ideal bases for 'just-so' stories); he suggested that the differentiation of island mouse races could be explained by 'adaptation to the colder and damper situation found on islands'. This was tested by Davis (1983) by comparing house mice from the mainland of Great Britain with mice from the Orkney, Shetland, and Faroe archipelagos. His null hypothesis was that a major 'maritime island' effect would result in island populations being more similar to each other than to mainland populations. Using a discriminant factor analysis similar to that employed by Williamson, Davis found that the island groups were very different from each other, but more like Caithness (north-east Scotland) than other British mainland samples. In other words, his data showed a regional geographic influence rather than an island effect, thus effectively disproving his null hypothesis. Another study comparing populations with different chromosomal (Robertsonian) fusions in Caithness and several Orkney islands found that there was an overall genetic similarity between all the populations, despite the fact that some of the populations differed considerably in their chromosomal constitutions (Nash *et al.* 1983). Genetic (chromosomal) differences which in some cases must have arisen since isolation, did not obscure the general relationships between the populations in the area.

Island small mammals are almost always less variable than mainland populations of the same species (Table 3.1); most data apply to allozymic comparisons, but the small amount of information from mtDNA and nuclear DNA studies gives a similar result (Frankham 1996). Leberg (1992) has shown experimentally in mosquito fish (*Gambusia holbrooki)* and argued theoretically that allele frequencies and incidence of polymorphisms are more sensitive indications of a bottleneck in numbers than mean heterozygosity, but that does not alter the argument. This reduced variability would be

Table 3.1 Inherited allozymic variation (heterozygosity) in island and mainland populations of small mammals of the same species (after Kilpatrick 1981; Berry 1986; Frankham 1997). (*N* is number of populations studied.)

	Mainland		Island		Reduction in heterozygosity (%)
	N	Mean heterozygosity per locus (range)	*N*	Mean heterozygosity per locus (range)	
Macrotus waterhousii	3	0.021(0–0.043)	1	0.040	–
Macaca fuscata	13	0.019(0–0.035)	5	0.013(0.03–0.018)	31.6
Spermophilus spilosoma	12	0.090(0.049–0.160)	1	0.009	90.0
Peromyscus eremicus	44	0.040(0.006–0.079)	2	0.009(0–0.022)	77.5
Peromyscus leucopus	3	0.080(0.076–0.084)	3	0.071(0.052–0.078)	11.3
Peromyscus maniculatus	22	0.088(0.054–0.124)	11	0.068(0.010–0.131)	22.7
Peromyscus polionotus	26	0.063(0.050–0.086)	4	0.052(0.018–0.086)	17.5
Sigmodon hispidus	4	0.022(0.017–0.025)	1	0.021	4.6
Microtus pennsylvanicus	4	0.142(0.120–0.171)	9	0.056(0.023–0.114)	60.6
Mus domesticus	16	0.091(0.032–0.114)	20	0.041(0–0.079)	55.0
Rattus fuscipes	3	0.047(0.020–0.100)	9	0.011(0–0.040)	76.6
Rattus rattus	1	0.031	11	0.026(0.008–0.056)	16.1

expected as a direct consequence of a founding event by a small number of indi\
but could also arise as a result of post-colonization drift. Kilpatrick (1981) has an.
the contribution of different loci to the reduction in variation. He found that island populations were often homozygous for the more common mainland allele and that the major differences between mainland and island were usually through an increase in the most common mainland alleles. He also compared the amount of differentiation between so-called fast and slow evolving loci (Sarich 1977): 83% more genetic variation occurred in island populations among 'fast' as opposed to 'slow' evolving loci; in mainland populations this figure was 36%. The difference was largely due to increased levels of genetic differentiation among insular populations at 'slow' loci. This again would seem to be a founder effect, since all loci will be affected equally by a colonization bottleneck, but it also indicates the possible operation of post-colonization adaptive changes. Kilpatrick concluded that 'accumulated evidence suggests founder effect as the major evolutionary force responsible for the reduction of genetic variation and differentiation of insular populations (but) additional data are needed to substantiate the relative importance of founding effects and to determine the relative importance of other evolutionary forces'.

3.4 Colonization and establishment

If one accepts that most island populations arise from colonization rather than through survival as relicts, the hazards of ecological establishment and adjustment have to be seen as important factors, stressing individuals and hence acting as likely selective agents (Berry 1996*b*). Entry into an area free of competitors, predators, and presumably parasites is easier than into a habitat already occupied by species likely to interact with the invaders. There are plenty of examples of small-mammal colonies being established and thriving to make labouring this point unnecessary (e.g. Crowcroft 1966; Lidicker 1976). Of more evolutionary interest are situations where a population has either failed to establish itself or become extinct. Grant (1972, 1978) reviewed a large number of distributional and experimental studies of interactions between rodent species and was wholly convinced that competitive interaction (mainly) for space was a general phenomenon. Even such an efficient colonizer as the house mouse may be affected in this way: Berry *et al.* (1982) describe two attempts to establish populations on Scottish islands which were unsuccessful due to lack of food in one case and competition from *Rattus norvegicus* in the other. Detailed case studies have reported the disappearance of flourishing island house mouse populations as a result of competition from *Microtus californicus* (Lidicker 1966) and *Apodemus sylvaticus* (Berry and Tricker 1969); both resulted from a failure to recruit young individuals into the population, rather than through increased adult mortality. In the latter case, the extinct mice were the only British house mice to have been given taxonomic differentiation, as *Mus (musculus) muralis* (Barrett-Hamilton 1899). Dueser and Porter (1986) examined competition between seven small-mammal species on a large (60 km long) island off the Maryland coast. They found that each species was fairly closely confined to specific habitats. They estimated the intensity of competition between pairs of species on the basis of changes in population density of the two in areas where they met. On this measure, house mice were 'at the bottom of the competitive hierarchy ... *Mus* is an inferior competitor in comparison with native rodents in natural habitats' and was found most commonly in marginal rodent habitat, 'xeric grassland on and immediately behind the seaward foredunes where there is sparse cover, low plant and structural diversity, low biomass and productivity, a shifting substrate and frequent disturbance'. Notwithstanding, the species was found on at least five of the small islands in the vicinity (all of which were occupied by at least one additional species) whereas two strongly competitive species (*Peromyscus leucopus* and *Zapus hudsonius*) were absent from nine of the ten islands for which there were data (Dueser and Brown 1980). Dueser and Brown (1980) concluded that 'extinction rates are relatively high and colonization rates are relatively low for small mammals in this physically rigorous environment'.

Hanski and his colleagues have studied the persistence (including both colonizing ability and extinction rate) of three species of shrews (*Sorex araneus, S. caecutiens*, and *S. minutus*) on 108 islands in three lakes in Finland. They found the species differed in their migration capability, the larger species having an advantage, apparently because of larger body size (higher swimming rate, longer starvation time: Pettonen and Hanski 1991). However, on arrival at an empty island, the smallest species (*S. minutus*) proved to have the largest colonization success, for reasons that are not known but may include their smaller per capita food requirements. Populations of the larger species (*S. araneus* and *S. caecutiens*) had a higher survival rate than populations of the small species (*S. minutus*) on small islands, probably a consequence of the short starvation time of small species and their consequent vulnerability to environmental stochasticity (Hanski 1992). Hanski (1993) has extended his analysis of island occupancy patterns to other sets of data, such as those of Lomolino (1993) on small mammals on islands in Lake Huron, USA, and shown that similar conclusions emerge. He bases his methodology on combining MacArthur and Wilson's (1967) equilibrium model of island species number with models based on metapopulation dynamics (Hanski and Gilpin 1991).

Studies such as these are almost always carried out without any genetic characterization of the individuals involved; Hanski and Kuitunen (1986) provide one of the few exceptions to this, and their data show differentiation between island but not mainland populations, even though their island populations were small and short-lived. Small mammals provide ideal material for combining genetical with ecological studies; perhaps a better integration of disciplines may come from the growing interest in life-history evolution (Berry and Bronson 1992).

3.5 Founder effect: principle and speciation

The introduction of the 'founder effect' into evolutionary literature was due to Ernst Mayr in his *Systematics and the origin of species*, one of the books which forged the neo-Darwinian synthesis. He wrote, 'The reduced variability of small populations is not always due to accidental gene loss, but sometimes to the fact that the entire population was started by a single pair or by a single fertilized female. These "founders" of the population carried with them only a very small proportion of the variability of the parent population. This "founder" principle sometimes explains even the uniformi rather large populations, particularly if they are well isolated and near the borders of the range of the species' (Mayr 1942, p. 237). This founder suggestion was Mayr's addition to the received wisdom of half a century ago that island differentiation was the result of drift in small populations. He saw the most important property of a founder population as its 'sudden conversion from an open to a closed population ... at once completely emancipated from the parental population' and hence subject to radically new selection pressures (Mayr 1963, p. 532). He believed this would lead to a 'genetic revolution', producing a new species. My contention is different: that the key significance of founding populations is that they provide a mechanism for rapid (= sudden) changes in gene frequencies and genome content, and this gives a new platform for subsequent adjustment. Sewall Wright seems to be the only person to have argued similarly. In a letter to Victor McKusick (24 May 1977), he wrote, 'The effects attributed to the "founder" principle by Mayr (gene loss, reduced variability) are the most obvious but the least important of the three I had stressed. I attributed most significance to wide random variability of gene frequencies (*not fixation or loss*) expected to occur simultaneously in tens of thousands of loci...' (Provine 1989, p. 57). This contrasts with the common discussions about founder effects, which tend to focus on allele loss. For example, MacArthur and Wilson (1967, p. 154) wrote, 'The founder principle is actually no more than the observation that a propagule should contain fewer genes than the entire mother population.' In view of these differences of emphasis, it is probably useful to follow Halkka *et al.* (1974) and use the term 'founder principle' to include a *founder event* (which leads to genetic impoverishment, elimination of immigration, and—almost certainly—changed gene frequencies) and subsequent *founder selection*.

It is the effects of founder selection (or genetic revolution *sensu* Mayr) that have attracted most debate. For example Barton and Charlesworth (1984) have sought to define conditions by which a species (defined as an equilibrium gene pool, stable towards the introgression of foreign genes) can move from one 'adaptive peak' to another. They conclude that 'the generally small chance of achieving reproductive isolation or marked phenotypic change in a single founder event means that founder effects themselves probably do not provide the explanation (for speciation). It is impossible to separate the effects of isolation, environmental differences, and continuous change by genetic drift from the impact of population bottlenecks in these cases. Since all of these factors promote divergence by a variety of processes, it is not clear that the additional influence of founder effects need be invoked.' (Barton and Charlesworth 1984, p. 158). In other words, the founder effect might contribute in particular instances, but is not necessary for speciation. Although Barton and Charlesworth ask 'what is the mechanism driving divergence?', they do not consider the allele-changing effect of the founder (colonizing) event itself (see Chapter 7).

Notwithstanding, it seems worth insisting that founder events may often provide the conditions to facilitate speciation (Wright 1942, p. 244). One of the difficulties about the sort of models of speciation considered by Barton and Charlesworth is that 'our imagination is limited by viewing populations as being at equilibrium, and by neglecting the complicated geometry that channels evolution' (Barton 1989). It is this latter neglect that Mayr (1954) and Waddington (1957) in very different (and admittedly

Table 3.2 Change in size of island mammals in comparison with their mainland nearest relative. Data are numbers of species. (After Foster 1964.)

	Smaller	Same	Larger
Insectivores	4	4	1
Lagomorphs	6	1	1
Rodents	6	3	60
Carnivores	13	1	1
Artiodactyls	9	2	0

qualitative) ways were attempting to remedy. Interestingly, Mayr cited Charles Elton (1930) in his original 'founder statement' (which is, of course, based on biology rather than theory). Elton derived many of his ideas from the study of microtine fluctuations (Sheail 1987). He believed that speciation could result from bottlenecks produced by fluctuations in numbers—which is, of course, exactly what a founding propagule is (a possibly comparable situation is described by Rüber and colleagues in Chapter 14). He was also innately distrustful of theoretical models. He wrote about the assumption of ecological equilibrium, 'It has the disadvantage of being untrue. The "balance of nature" does not exist....' (Elton 1930, p. 17).

It seems worthwhile pursuing the idea that the start given to differentiation by a small founding group may affect the genetic possibilities of their descendants, and sometimes lead to a new species. There is no doubt about the first, and plenty of evidence for subsequent founder selection, although not in the detail suggested by Mayr (1954). For example, one of the commonest characteristics of island mammals is a change in size relative to their mainland relatives: in general, large animals get smaller and small animals get bigger (Table 3.2). The usual explanation for this is that large animals are chronically food restricted on islands, whereas small ones can increase to a more physiologically efficient size when not constrained by ground predators or competitors. Angerbjörn (1986) has shown that the body size of *Apodemus sylvaticus* on European islands is not affected by climate, island size, or distance from the mainland but is greater if either the usual competitors (*A. flavicollis* and *Clethrionomys glareolus*) or ground predators (*Mustela erminea, M. nivalis, M. foina* or *Vipera berus*) are absent.

The size difference between island and mainland forms can be quite marked: *A. sylvaticus* from St Kilda are on average about twice the weight of mainland British mice. Moreover the size adjustment may be rapid. House mice on the island of Skokholm were about 25% heavier than their mainland ancestors after about 60 generations in isolation, and this increase could have occurred much earlier (Berry 1964). The island size is inherited (Wallace 1981). House mouse populations contain a great deal of variance for size (Crowcroft and Rowe 1961).

It is impossible without knowing the genetic composition of the colonizers to distinguish between the chance characteristics of the founding propagule and subsequent adaptation. For example, in his original description of two of the Faroese house mouse populations, Clarke (1904) spoke of the 'remarkable coarseness' of their feet, 'a modification which has been probably brought about by the rough nature of their haunts'. This falls into the 'just-so' explanation category. Hebridean field mice vary from 'reddish' on Rhum to 'greyish' on St Kilda. There are no obvious differences in background coloration on different islands, and the most likely explanation is that different alleles were represented in the founders on the different islands. However, an informative parallel emerges from a laboratory study of nest-building in house mice, one of the few unequivocally adaptive traits in small mammals which has been investigated (Lynch 1992). Laffen (cited by Lynch 1994) has shown that the genetic determinants of nest building efficiency are different in different selected lines derived from the same stock; in other words adaptation is serendipitous as well as pragmatic.

The evidence for the persistence of founder heterogeneity is much firmer in human populations. For example, retinitis pigmentosa in Tristan da Cunha, porphyria variegata in South Africa, Huntington's disease and nephrogenic diabetes insipidus in New England, and other inherited conditions, can all be traced back to alleles carried by the colonizing groups (reviewed by Berry 1972). A small-mammal example is the high frequency (over 60%) of midline skeletal defect (= spina bifida occulta) in Skokholm house mice: this is rare in most wild-living mice, but occurs nowadays in about 10% of the mice in the area from which the original founders came (Berry 1964).

As far as founder selection *sensu stricto* is concerned, there is almost no evidence for this in small mammals, although there are plenty of examples of strong selection in island populations (Endler 1986; Berry 1987). An attempt to detect and measure founder selection in a fully genetically characterized foundation population by Berry *et al.* (1982) was a complete failure because of the inability to establish viable populations. However, an experiment in which house mice were introduced into an existing population on a small (60 ha) island in which certain inherited traits were absent showed an increase in these traits (allozyme alleles, Robertsonian translocations, mtDNA, and Y-chromosome variation) to a stable level after about four years (Table 3.3)

Table 3.3 Allele frequencies in a *Mus domesticus* population after an experimental release on the Isle of May; the population on the island contained none of the alleles before the introduction. (From Berry *et al.* 1991.)

	Eday (source of introduced animals)	Frequencies in released animals	Estimated averages post-introduction*	Frequencies six years post-introduction
Hbb[d]	0.214	0.170	0.010	0.071
Car-2[b]	0.423	0.221	0.020	0.171
Ada[b]	0.187	0.160	0.020	0.021
Gda[b]	0.536	0.620	0.040	0.338
Es-3[b]	1.000	1.000	0.070	0.613
Es-10[a]	0.100	0.080	0.005	0.029
N	96	77	—	120

*The resident population at the time of release was estimated to be about 1000 mice.

(Berry *et al.* 1991; Jones *et al.* 1995). It was not possible to identify the factors affecting fitness, but there can be no doubt that the observed genetical changes were not random. The increase and spread of the Robertsonian translocations was unexpected because Robertsonian heterozygotes have a reduced fertility and the rate of non-disjunction was doubled in wild-caught heterozygotes when compared with the parental groups (Scriven 1992; cf. Winking 1986; Searle 1993). There were clearly genetical forces operating, which involve many parts of the genome, and which may approach a genetic revolution *sensu* Mayr. Bonhomme *et al.* (1989) believe that studies on the stable hybrid zone between *Mus* (semi)species in Europe and lack of genome homogeneity between wild-caught mouse samples in Japan indicate the effectiveness of intragenomic cohesion.

Conversely, there is a marked lack of evidence for random genetic changes in island small mammals, apart from the stochastic changes of founding colonization. Indeed it is worth pointing out how rare are well-documented examples of genetic drift in any wild population. Berry *et al.* (1992) describe fluctuations at an allozyme locus (*Hbb*) in a house mouse population on a small (250 ha) Orkney island, but this was a very small population rarely exceeding a hundred individuals and probably on the verge of ecological viability. Petras and Topping (1983) present calculations to show that models incorporating strong selection explain well the distribution of alleles at the same locus in both Skokholm and Canadian populations, contrary to their earlier assumption of randomness from the same data.

Direct evidence of reproductive barriers which might indicate incipient speciation are sparse, but have not often been sought. Godfrey (1958) crossed British *Clethrionomys glareolus* races and discovered hybrid vigour in the F_1 but small F_2 and backcross litters, as would be expected if some sort of coadaptation was important (see also Grant 1974). He found that animals from different races discriminated in favour of mates of their own race on the ground of scent, i.e. racial integrity was maintained (or strengthened) by behaviour. A rather similar result was obtained by Zimmerman (1959) when he hybridized *Microtus arvalis* from Orkney and Germany: the F_2 offspring were smaller and less vigorous than both the F_1 and the parental forms. Inbreeding in small mammals has been reviewed by Smith (1993); it is not uniformly deleterious. Mayr (1963, p. 534) has pointed out that a founder event may remove undesirable recessive alleles and hence reduce the common effects of inbreeding.

3.6 Significance and conservation

Island races are particularly subject to extinction: Ceballos and Brown (1995) record that 81% (65/80) of known mammal extinctions over the past 500 years have been on islands; Reid and Miller (1989) note that island species form 22% (48/216) of endangered or vulnerable mammals. This sensitivity has been analysed for a number of situations (see Heaney and Patterson 1986). However, it is something which ought to be expected if islands are, as it were, testing grounds for new genetic combinations rather than the result of more definitive evolution. Williamson (1981) calculated that only one colonization every 25 000 years leads to speciation in Hawaiian *Drosophila*. Mayr

(1963, p. 513) has taken a similar line, but extrapolated it to a belief in the evolutionary significance of island races:

Do peripheral isolates frequently (or usually) produce new species and evolutionary novelties ? No.

Are new species and evolutionary novelties usually produced by peripheral isolates ? Yes—since peripheral isolates are produced fifty to five hundred times as frequently as new species. Hence most peripheral isolates do *not* evolve into a new species, but *when* a new species evolves, it is almost invariably from a peripheral isolate.

The evidence from island small mammals is that significant differentiation often arises as a result of a founder event. From a conservation point of view, this differentiation is little more than an ecological accident, produced by the vagary of sampling. But it does not mean that island races should be treated merely as curiosities. Mayr (1967) has written : 'Islands are an enormously important source of information and an unparalleled testing ground for various scientific theories.' Kilpatrick (1981) adds an experimental edge, 'Manipulations of insular populations with partially known genomes and repeated sampling of these populations prior and subsequent to manipulation should allow determination of the relative importance of founder effect, genetic drift, gene flow and selection in determining the genetic structure on insular populations.' Almost certainly conservationists have over-reacted to variation loss (Berry 1997). Bryant and his colleagues have shown that manipulating house fly populations so that they go through small bottlenecks in number nevertheless leaves substantial potential variation remaining for evolutionary adaptation (Bryant *et al.* 1990; see also Chapter 7). Brakefield (1991) comments, 'the existence of coadaptation between genes and of forms of non-additive genetic contributions to quantitative variation are likely to make variability within populations more resistant to loss than would be expected on the basis of theory developed largely from the perspectives of genes acting independently from each other and in a purely additive manner.'

It would be premature to generalize too much from such experiments, but it is worth emphasizing that island small mammals provide ideal opportunities for similar manipulation. Sadly most are still virgin territory, observed from afar by geneticists, ecologists, theoreticians secure on their own pedestals; the main linkage to them remains the corrosive and persisting mists of evolutionary story tellers. Wright, Fisher, and Mayr all saw the main barrier to speciation as genetic cohesion in natural populations (Provine 1989, p. 62). Founder differention may be a key factor in disrupting this. But, as MacArthur and Wilson (1967, p. 179) point out, 'the actual contribution of the "founder effect" to evolution can be assessed only by empirical field studies,' a plea repeated by many of the contributors to this volume (most strongly by Schluter in Chapter 10).

Darwin (1871) commented that 'False facts are highly injurious to the progress of science, for they often endure long; false views, if supported by some evidence do little harm, for everyone takes a salutary pleasure in proving their falseness; when this is done, one path towards error is closed and the road to truth is often at the same time opened.' He also wrote (1872, p. 395), 'Great is the power of steady misinterpretation.' Research has helped considerably to clarify our understanding of insular evolution, but there is much work still to be done, particularly on the persistence of 'founding genomes' and on the importance and strength of intragenomic interactions.

3.7 Summary

Differentiation of small mammals on islands has traditionally attracted some vivid story-telling, usually involving isolation (as a relict) followed by adaptation and/or random genetic changes. Studies of voles on Orkney, long-tailed field mice on the Hebrides and Shetland, and house mice on the Faroe archipelago show that the main factor in differentiating island races from their mainland ancestors is the chance genetic composition of the founding animals. Subsequent change has necessarily to be based on the genes and frequencies carried by this colonizing group. Probably most post-colonization change is adaptive, although possibly limited in extent both by the initial paucity of variation and by the conservative effect of intragenomic interactions. It is helpful to recognize that the 'founder' effect or principle commonly invoked in discussions about evolution on islands involves a founder *event*, followed by founder *selection*. Island differentiation is not necessarily a precursor to speciation, although the wide occurrence of island endemics suggests that founder effects should not be rejected as a driving force initiating speciation. Notwithstanding, island forms provide a valuable 'laboratory' for testing new genetic combinations, a small proportion of which may prove evolutionarily exciting. Only more empirical studies will uncover their evolutionary importance.

Acknowledgements

My thanks are due to comments on particular points from Raymond Dueser, Ilka Hanski, and Andrew Pomiankowski.

References

Angerbjörn, A. (1986). Gigantism in island populations of wood mice (*Apodemus*) in Europe. *Oikos*, **47**, 47–56.

Ashley, M. and Wills, C. (1987). Analysis of mitochondrial DNA polymorphisms among Channel Island deer mice. *Evolution*, **41**, 854–63.

Barrett-Hamilton, G. E. H. (1899). On the species of the genus *Mus* inhabiting St. Kilda. *Proceedings of the Zoological Society of London for 1899*, 77–88.

Barton, N. (1989). Founder effect speciation. In *Speciation and its consequences*, (ed. D. Otte and J. A. Endler), pp. 229–56. Sinauer, Sunderland, MA.

Barton, N. and Charlesworth, B. (1984). Genetic revolutions, founder effects and speciation. *Annual Review of Ecology and Systematics*, **15**, 133–64.

Beirne, B. P. (1952). *The origin and history of the British fauna*. Methuen, London.

Berry, R. J. (1964). Evolution of an island population of the house mouse. *Evolution*, **18**, 468–83.

Berry, R. J. (1972). Genetical approaches to taxonomy. *Proceedings of the Royal Society of Medicine*, **65**, 853–4.

Berry, R. J. (1984). Darwin was astonished. *Biological Journal of the Linnean Society*, **21**

Berry, R. J. (1986). Genetics of insular populations of mammals, with particular refere differentiation and founder effects in British small mammals. *Biological Journal of the Linnean Society*, **28**, 205–30.

Berry, R. J. (1987). Where biology meets; or how science advances. *Biological Journal of the Linnean Society*, **30**, 257–74.

Berry, R. J. (1996*a*). Small mammal differentiation on islands. *Philosophical Transactions of the Royal Society of London* B, **351**, 753–64.

Berry, R. J. (ed.) (1996*b*). Environmental stress and evolutionary adaptation. In *Stress: evolutionary, biosocial and clinical perspectives*, (ed. A. Bittles), pp. 24–40. Macmillan, Basingstoke.

Berry, R. J. (1997). The history and importance of conservation genetics: one person's perspective. In *The role of genetics in conserving small populations*, (ed. T. E. Tew, T. J. Crawford, J. Spencer, D. Stevens, M. B. Usher, and J. Warren), pp. 26–32. J.N.C.C., Peterborough.

Berry, R. J. and Bradshaw, A. D. (1991). Genes in the real world. In *Genes in ecology*, (ed. R. J. Berry, T. J. Crawford, and G. M. Hewitt), pp. 431–9. Blackwell Scientific, Oxford.

Berry, R. J. and Bronson, F. H. (1992). Life history and bioeconomy of the house mouse. *Biological Reviews*, **67**, 519–50.

Berry, R. J. and Tricker, B. J. K. (1969). Competition and extinction: the mice of Foula, with notes on those of Fair Isle and St. Kilda. *Journal of Zoology (London)*, **158**, 247–65.

Berry, R. J., Cuthbert, A. and Peters, J. (1982). Colonization by house mice: an experiment. *Journal of Zoology (London)*, **198**, 329–36.

Berry, R. J., Triggs, G. S., King P., Nash, H. R., and Noble, L. R. (1991). Hybridisation and gene flow in house mice introduced into an existing population on an island. *Journal of Zoology (London)*, **225**, 615–32.

Berry, R. J., Berry, A. J., Anderson, T. J. C., and Scriven, P. (1992). The house mice of Faray. *Journal of Zoology (London)*, **228**, 233–46.

Bonhomme, F., Niyashita, N., Boursot, P., Catalan, J., and Moriwaki, K. (1989). Genetical variation and polyphyletic origin in Japanese *Mus musculus. Heredity*, **63**, 299–308.

Brakefield, P. M. (1991). Genetics and the conservation of invertebrates. In *The scientific management of temperate communities for conservation*, (ed. I. F. Spellerberg, F. B. Goldsmith, and M. G. Morris), pp. 45–79. Blackwell Scientific, Oxford.

Bryant, E. H., Meffert, L. M., and McCommas, S. A. (1990). Fitness rebound in serially bottlenecked populations of the house fly. *American Naturalist*, **136**, 542–9.

Ceballos, G. and Brown, J. H. (1995). Global patterns of mammalian diversity, endemism and endangerment. *Conservation Biology*, **9**, 559–68.

Clarke, W. E. (1904). On some forms of *Mus musculus*, Linn., with description of a new subspecies from the Faeroe Islands. *Proceedings of the Royal Physical Society of Edinburgh*, **15**, 160–7.

Corbet, G. B. (1961). Origin of the British insular races of small mammals and of the 'Lusitanian' fauna. *Nature*, **191**, 1037–40.

Crowcroft, W. P. (1966). *Mice all over*. Foulis, London.

Crowcroft, W. P. and Rowe, F. P. (1961). The weights of wild house mice (*Mus musculus* L.) living in confined colonies. *Proceedings of the Zoological Society of London*, **136**, 177–85.

Crowell, K. L. (1986). A comparison of relict versus equilibrium models for insular mammals of the Gulf of Maine. *Biological Journal of the Linnean Society*, **28**, 37–64.

Darwin, C. R. (1837). Remarks upon the habits of the genera *Geospiza, Camarhynchus, Cactornis* and *Certhidea* of Gould. *Proceedings of the Zoological Society of London*, **5**, 47.

Darwin, C. R. (1871). *The descent of man, and selection in relation to sex*. John Murray, London.

Darwin, C. R. (1872). *The origin of species* (6th edn). John Murray, London.

Davis, S. J. M. (1983). Morphometric variation of populations of house mice *Mus domesticus* in Britain and Faroe. *Journal of Zoology (London)*, **199**, 521–34.

Dueser, R. D. and Brown, W. C. (1980). Ecological correlates of insular rodent diversity. *Ecology*, **61**, 50–6.

Dueser, R. D. and Porter, J. H. (1986). Habitat use by insular small mammals: relative effects of competition and habitat structure. *Ecology*, **67**, 195–201.

Elton, C. S. (1930). *Animal ecology and evolution*. Oxford University Press.

Endler, J. A. (1986). *Natural selection in the wild*. Princeton University Press.

Ford, E. B. (1975). *Ecological genetics*, (4th edn). Chapman and Hall, London.

Foster, J. B. (1964). Evolution of mammals on islands. *Nature*, **202**, 234–5.

Frankham, R. (1997). Do island populations have less genetic variation than mainland populations? *Heredity*, **78**, 311–27.

Godfrey, J. (1958). The origin of sexual isolation between bank voles. *Proceedings of the Royal Physical Society of Edinburgh*, **27**, 47–55.

Gould, S. J. and Lewontin, R. C. (1979). The spandrels of San Marco and the Panglossian paradigm: a critique of the adaptationist programme. *Proceedings of the Royal Society of London* B, **205**, 581–98.

Grant, P. R. (1972). Interspecific competition among rodents. *Annual Review of Ecology and Systematics*, **3**, 79–106.

Grant, P. R. (1974). Reproductive compatibility of voles from separate continents (Mammalia: *Clethrionomys*). *Journal of Zoology (London)*, **174**, 245–54.

Grant, P. R. (1978). Competition between species of small mammals. In *Populations of small mammals under natural conditions: a review and analysis of the contribution of long term experimental and descriptive studies*, (ed. D. Snyder), pp. 38–51. Proceedings of Symposium in Ecology No. 5 at the Pymatuning Laboratory, University of Pittsburgh.

Halkka, O., Raatikainen, M., and Halkka, L. (1974). The founder principle, founder selection, and evolutionary divergence and convergence in natural populations of *Philaenus*. *Hereditas*, **78**, 73–84.

Hanski, I. (1992). Inferences from ecological incidence functions. *American Naturalist*, **139**, 657–62.

Hanski, I. (1993). Dynamics of small mammals on islands. *Ecography*, **16**, 372–5.

Hanski, I. and Gilpin, M. (1991). Metapopulation dynamics: brief history and conceptual domain. *Biological Journal of the Linnean Society*, **42**, 3–16.

Hanski, I. and Kuitunen, J. (1986). Shrews on small islands: epigenetic variation elucidates population stability. *Holarctic Ecology*, **9,** 193–204.

Heaney, L. R. and Patterson, B. D. (eds.) (1986). *Island biogeography of mammals*. Academic Press, London.

Jones, C. S., Noble, L. R., Jones, J. S., Tegelström, H., Triggs, G. S., and Berry, R. J. (1995). Differential male genetic success determines gene flow in an experimentally manipulated mouse population. *Proceedings of the Royal Society of London* B, **260**, 251–6.

Kilpatrick, C. W. (1981). Genetic structure of insular populations. In *Mammalian population genetics*, (ed. M. H. Smith and J. Joule), pp. 28–59. University of Georgia Press, Athens, GA.

King, C. M. (ed.) (1990). *The handbook of New Zealand mammals*. Oxford University Press, Auckland.

Leberg, P. L. (1992). Effects of population bottlenecks on genetic diversity as measured by allozyme electrophoresis. *Evolution*, **46**, 477–94.

Lever, C. (1985). *Naturalized mammals of the world*. Longman, London.

Lidicker, W. Z. (1966). Ecological observations on a feral house mouse population declining to extinction. *Ecological Monographs*, **36**, 27–50.

Lidicker, W. Z. (1976). Social behaviour and density regulation in house mice living in large enclosures. *Journal of Animal Ecology*, **45**, 677–97.

Lomolino, M. V. (1993). Winter filtering, immigrant selection and species composition of insular mammals of Lake Huron. *Ecography*, **16**, 25–30.

Lynch, C. B. (1992). Clinal variation in cold adaptation in *Mus domesticus*: verification of predictions from laboratory populations. *American Naturalist*, **139**, 1219–36.

Lynch, C. B. (1994). Evolutionary inferences from genetic analyses of cold adaptation in laboratory and wild populations of the house mouse. In *Quantitative genetic studies of behavioural evolution*, (ed. C. R. B. Boake), pp. 278–301. University of Chicago Press.

MacArthur, R. H. and Wilson, E. O. (1963). An equilibrium theory of insular biogeography. *Evolution*, **17**, 373–87.

MacArthur, R. H. and Wilson, E. O. (1967). *The theory of island biogeography*. Princeton University Press.

Mayr, E. (1942). *Systematics and the origin of species*. Columbia University Press, New York.

Mayr, E. (1954). Change of genetic environment and evolution. In *Evolution as a process*, (ed. J. Huxley, A. C. Hardy, and E. B. Ford), pp. 157–80. Allen and Unwin, London.

Mayr, E. (1963). *Animal species and evolution*. Harvard University Press, Cambridge, MA.

Mayr, E. (1967). The challenge of island faunas. *Australian Natural History*, **15**, 359–74.

Nash, H. R. , Brooker, P. C., and Davis, S. J. M. (1983). The Robertsonian translocation house-mouse populations of north east Scotland: a study of their origin and evolution. *Heredity*, **50**, 303–10.

Patterson, B. D. and Atmar, W. (1986). Nested subsets and the structure of insular mammalian faunas and archipelagos. *Biological Journal of the Linnean Society*, **28**, 65–82.

Patton, J. L. (1984). Genetical processes in the Galapagos. *Biological Journal of the Linnean Society*, **21**, 29–59.

Patton, J. L., Yang, S.-Y., and Myers, P. (1975). Genetic and morphologic divergence among introduced rat populations (*Rattus rattus*) of the Galapagos Archipelago, Ecuador. *Systematic Zoology* **24**, 296–310.

Pettonen, A. and Hanski, I. (1991). Patterns of island occupancy explained by colonization and extinction rates in shrews. *Ecology*, **72**, 1698–708.

Petras, M. L. and Topping, J. C. (1983). The maintenance of polymorphisms at two loci in house mouse (*Mus musculus*) populations. *Canadian Journal of Genetics and Cytology*, **25**, 190–201.

Provine, W. B. (1989). Founder effects and genetic revolutions in microevolution and speciation: an historical perspective. In *Genetics, speciation and the founder principle*, (ed. L. V. Giddings, K. Y. Kaneshiro, and W. W. Anderson), pp. 43–76. Oxford University Press.

Reid, W. V. and Miller, K. R. (1989). *Keeping options alive: the scientific basis for conserving biodiversity*. World Resources Institute, Washington, D. C.

Sarich, V. M. (1977). Rates, sample sizes and the neutrality hypothesis in evolutionary studies. *Nature*, **265**, 24–8.

Scriven, P. (1992). Robertsonian translocations introduced into an island population of mice. *Journal of Zoology (London)*, **227**, 493–503.

Searle, J. B. (1991). A hybrid zone comprising staggered chromosomal clines in the house mouse (*Mus musculus domesticus*). *Proceedings of the Royal Society of London* B, **246**, 47–52.

Searle, J. B. (1993). Chromosomal hybrids in eutherian mammals. In *Hybrid zones and the evolutionary process*, (ed. R. G. Harrison), pp. 309–53. Oxford University Press.

Sheail, J. (1987). *Seventy-five years in ecology: the British ecological society*. Blackwell Scientific, Oxford.

Skelton, P. (ed.) (1993). *Evolution*. Addison-Wesley, Wokingham.

Smith, A. T. (1993). The natural history of inbreeding and outbreeding in small mammals. In *The natural history of inbreeding and outbreeding*, (ed. N. W. Thornhill), pp. 329–51. University of Chicago Press.

Stewart, D. T. and Baker, A. J. (1992). Genetic differentiation and biogeography of the masked shrew in Atlantic Canada. *Canadian Journal of Zoology*, **70**, 106–14.

Sulloway, F. J. (1982). Darwin and his finches: the evolution of a legend. *Journal of the History of Biology*, **15**, 1–53.

Waddington, C. H. (1957). *The strategy of the genes*. Allen and Unwin, London.

Wallace, M. E. (1981). The breeding, inbreeding and management of wild mice. *Symposium of the Zoological Society of London*, No. 47, 183–204.

Williamson, M. H. (1981). *Island populations*. Oxford University Press.

Williamson, M. H. (1984). Sir Joseph Hooker's lecture on insular floras. *Biological Journal of the Linnean Society*, **22**, 55–77.

Wilson, E. O. (1969). The species equilibrium. *Brookhaven Symposia in Biology* No. 22, 38–47.

Winking. H. (1986). Some aspects of Robertsonian karyotype variation in European wild mice. *Current Topics in Microbiology and Immunology*, No. 127, 68–74.

Wright, S. (1942). Statistical genetics and evolution. *Bulletin of the American Mathematical Association*, **48**, 223–46.

Zimmermann, K. (1959). Uber eine Kreuzung von Unterarten der Feldmaus *Microtus arvalis*. *Zoologische Jahrbuch (Systematiks)*, **87**, 1–12.

4

The maintenance of genetic polymorphism in small island populations: large mammals in the Hebrides

Josephine M. Pemberton, Judith A. Smith, Tim N. Coulson, Tristan C. Marshall, Jon Slate, Steve Paterson, Steve Albon, and Tim Clutton-Brock

4.1 Introduction

In this chapter we seek to emphasize a particular virtue of island populations: they can be used to evaluate the population genetic processes responsible for changing allele and genotype frequencies, and hence levels of genetic variation. This virtue is distinct from the more commonly discussed merits of island populations for investigating speciation, which occupy most of the other contributions to this book, and for investigating loss of genetic variaton in island-hopping founder events, exemplified by Gilbert *et al.*'s (1990) investigation of California Channel Island fox (*Urocyon littoralis*) populations.

Island populations have numerous practical advantages for investigating certain population genetic processes. First, immigration and emigration are often minimal or absent for practical purposes, leaving selection and genetic drift as the only processes likely to affect levels of genetic variation. Second, the finite population size and often single management authority responsible for whole islands makes island populations exceptionally tractable for detailed studies. It is no coincidence that many of the classic long-term, individual-based studies which have contributed to our knowledge of population dynamics, behavioural ecology, and natural selection are based on islands (e.g. red deer (*Cervus elaphus*, Fig. 4.1; see Clutton-Brock *et al.* 1982), Soay sheep (*Ovis aries*, Fig. 4.1; see Jewell *et al.* 1974; Clutton-Brock *et al.* 1991), Darwin's finches (*Geospiza* spp; see Grant 1986), song sparrows (*Melospiza melodia*; see Smith and Dhondt 1980), collared flycatchers (*Ficedula albicollis*; see Gustafsson 1986). Finally and perhaps most importantly, island populations are frequently subject to changes in population size, either due to external factors (e.g. weather: Darwin's finches (Grant 1986) and song sparrows (Keller *et al.* 1994)) or due to intrinsic instability (e.g. Soay sheep; see

Fig. 4.1 (a) The view south down Kilmory Glen, Rum, part of the red deer study area, in October. The glen bottom and coastal regions of the study area contain favoured grazing, allowing frequent observation of most resident individuals. Photo T. C. Marshall. (b) A three-year-old female red deer in the Rum study area. At birth she was weighed, sampled for genetic analysis, tagged, and fitted with an expanding plastic collar for identification. Photo J. M. Pemberton. (c) Village Bay, Hirta, the Soay sheep study area, in November. The former agricultural land (now favoured by Soay sheep) is enclosed by the circular head dyke, within which lie the row of cottages built in 1860 and the current army base. Numerous old dry stone habitations and storage buildings, called cleits, can also be seen; they are used extensively by the sheep for shelter. Photo T. C. Marshall. (d) A Soay ewe suckles a four-month-old ram lamb in August. In November he will compete for matings and, depending on population demography, may well obtain one or more paternities. Photo J. M. Pemberton.

Grenfell *et al.* 1992), offering unusual opportunities to investigate the relation between population dynamics and genetic variation.

It is the conventional wisdom of population genetics theory that fluctuations in population size, and specifically reductions in population size, will result in loss of genetic variation. This is because, after a population decline, the individuals that survive to breed are a random but small sample of the preceding population. Two assumptions underpin this wisdom. First, it assumes that selection is weak or absent during population reduction. Whereas this is likely to be true in deliberately designed laboratory experiments or in some natural catastrophes, it is not necessarily so for all kinds of population reduction. For example, any population reduction which includes intraspecific competition for scarce resources might involve selection, and there are at least two processes through which this could retard loss of genetic variation, namely through selection for heterozygotes or through selection for alternative alleles during periods of population decline and recovery. In general, theoretical studies have predicted that selection that fluctuates in direction, though perhaps slowing loss of genetic variation, cannot maintain it. Recently, however, Ellner and Hairston (1994) have shown that with overlapping generations, genetic variation can be maintained by selection that fluctuates in direction. The second common assumption is that the distribution of male mating success remains constant despite changing population size and demography. There have been few investigations of this point from the population genetic viewpoint, even though we know that in some vertebrates mating systems can vary dramatically with population demography (e.g. fallow deer; see Langbein and Thirgood 1989)

Here, we investigate these issues in two individually monitored island populations of ruminants, red deer living on the Isle of Rum, Inner Hebrides, and Soay sheep living on Hirta in the St Kilda archipelago, Outer Hebrides. In the course of each study, the population has experienced marked changes in population size, allowing us to investigate whether there are systematic associations between fitness measures and genetic factors through population fluctuations, and whether there are systematic changes in the distribution of male mating success associated with population fluctuations.

4.2 The study populations

Red deer

The Isle of Rum (10 600 ha) is one of the Inner Hebrides, lying about 20 km west of the Scottish mainland. After extirpation in the eighteenth century, red deer were reintroduced to Rum from 1845 onwards, from at least four source populations elsewhere in the UK. In recent years, the whole island deer population has ranged between 1200 and 1700.

Red deer living in the North Block of the Isle of Rum have been the subject of intense, individual-based monitoring since 1971 (Clutton-Brock *et al.* 1982; Clutton-Brock and Albon 1989). Over this time approximately 1700 individuals have been monitored from birth, through all reproductive attempts, to death. In general, females adopt a home range similar to their mother's, whereas males emigrate from their natal area at about 2 years of age. There are no physical barriers between the study area and the rest of the island, and through male emigration and immigration, the study population is panmictic with that of the rest of the island. At any time the study population constitutes about one-quarter of the total island population.

Culling ceased in the study area at the start of the study, and this has led to changes in the density and demography of the population. Over the first 8 years the female population rose to about four times its original number and it now fluctuates around the apparent carrying capacity of the study area (Fig. 4.2a). In the same period, the male population has fluctuated and, if anything, declined slightly (Fig. 4.2a) with the

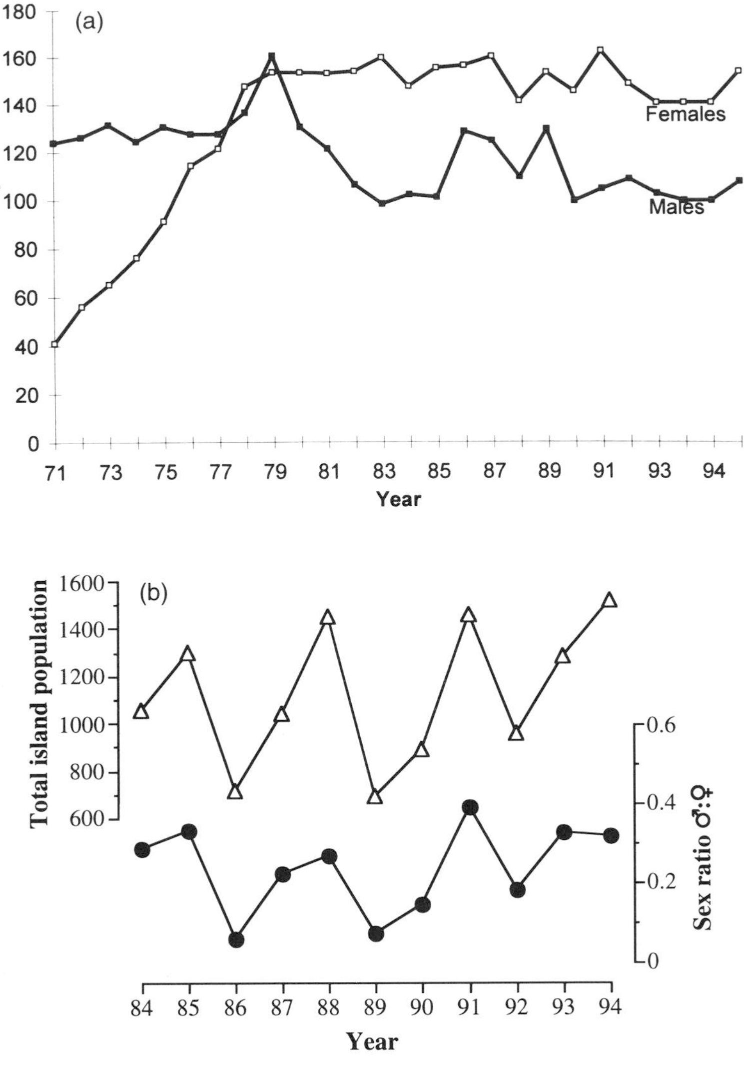

Fig. 4.2 (a) Number of male and female red deer (older than one year) regularly using the Isle of Rum study area (open squares: females; filled squares: males). Since cessation of culling in 1971 female numbers (and hence overall density) have quadrupled, whereas male numbers have declined slightly. (b) Population size on the whole island of Hirta since 1984, and sex ratio of individuals greater than 1 year, showing correlated fluctuations of these two measures.

consequence that what was initially an area with a male-biased population now holds a strongly female-biased population. Rising deer density has been accompanied by density-dependent responses in fitness components, especially in juvenile survival and female fecundity (Clutton-Brock *et al.* 1985, 1987). In common with other sexually dimorphic vertebrates, juvenile survival is strongly female biased, accounting for much of the female bias in the current adult population (Clutton-Brock *et al.* 1985).

Soay sheep

The St Kilda archipelago lies some 200 km west of the Scottish mainland. Soay sheep are primitive domestic sheep which have existed unmanaged on the St Kildan island of Soay (99 ha) for centuries if not millennia. In 1932, 107 sheep were introduced to the neighbouring island of Hirta (638 ha).

Soay sheep have been intensively monitored within the Village Bay area of Hirta (approximately 175 ha) since 1985 (Clutton-Brock *et al.* 1991, 1992), and again about 1700 individuals have been followed over this time. The monitored population constitutes about one-third of the whole island populaton, with which it is panmictic. The study was initiated because annual censuses of the whole island and a previous period of intensive study (Jewell *et al.* 1974) indicated unusual, unstable population dynamics. At intervals of 3–5 years, the population experiences overwinter population crashes in which up to 70% of the population may die (Clutton-Brock *et al.* 1991, 1992; Grenfell *et al.* 1992) (Fig. 4.2b). The sheep population dynamics are a consequence of the generally high level of fecundity in Soay sheep, which allows them, in a single summer, to increase to well above the overwinter carrying capacity of the herbage (Clutton-Brock *et al.* 1997). Parasitism by gastrointestinal nematodes appears to contribute to the severity of the crashes (Gulland 1992; Gulland and Fox 1992; Gulland *et al.* 1993).

Over population crashes, survival is again female biased, probably because males go into the winter in poor condition following the November rut (Stevenson and Bancroft 1995). In addition, mature animals survive crashes better than yearlings, and particularly lambs (Clutton-Brock *et al.* 1992). In consequence, both the sex ratio (Fig. 4.2b) and the age structure (not illustrated) fluctuate systematically with population size, offering the opportunity to study their effects on the distribution of male mating success.

A striking feature of Soay sheep, which precipitated some of the investigations reported here, is that they show two obvious phenotypic polymorphisms. First, there is polymorphism for pelage colour and pattern, with the principal morphs controlled by two loci each having two alleles (Doney *et al.* 1974). Second, Soays have variable horn phenotype, ranging from no horns (polled), through small, mis-shapen horns knowns as scurs, to normal horns. This variation is heritable (Stevenson, unpublished data), but complicated by different expression in the sexes: a greater proportion of males have normal horns.

4.3 Molecular investigations

Since 1982 and 1985 respectively, samples have been collected from all study area deer and sheep when they were handled or found dead. The samples have been subjected to various molecular techniques, including protein electrophoresis, DNA fingerprinting, and more recently microsatellite DNA profiling (Pemberton *et al.* 1988, 1991, 1992; Gulland *et al.* 1993; Bancroft *et al.* 1995*a, b*). These studies have confirmed that despite their history, Soay sheep retain substantial heterozygosity; for example, across 34 protein loci average heterozygosity is 7.78%, high for a mammal (Bancroft *et al.* 1995*a*). The molecular data for known individuals in both populations have been used to investigate whether associations exist between alleles or genotypes and fitness (Pemberton *et al.* 1988, 1991; Gulland *et al.* 1993; Bancroft *et al.* 1995*b*; Illius *et al.* 1995) and to identify paternity (Pemberton *et al.* 1992). The results reported in this chapter constitute a mix of previously reported and new results derived from this work. We now address the two issues raised in Section 4.1 in turn.

4.4 Retarding the loss of variation: competitive effects

Where population fluctuations involve varying intraspecific competition, there may be systematic selective effects which act to retard loss of genetic variation.

The rising density of deer in the Rum study area, and the population crashes and recoveries seen on St Kilda, involve individuals facing changing levels of competition for food. In each case, this leads to increased mortality, particularly affecting younger age classes and males, and in the case of deer, there are also marked changes in fecundity. In a series of previous investigations, we have shown that for some polymorphic systems, mortality and/or fecundity is not random with respect to phenotype, genotype, or alleles (Pemberton *et al.* 1988, 1991; Gulland *et al.* 1993; Bancroft *et al.* 1995*b*; Illius *et al.* 1995; Moorcroft *et al.* 1996). Here, we review patterns which suggest to us that these associations, combined with fluctuating population dynamics, could slow the rate of loss of genetic variation or maintain it.

Isocitrate dehydrogenase (Idh-2) in deer

At the protein level, two alleles are detectable at Idh-2. Remarkably, across all genotyped individuals ($N = 993$) the two allele frequencies are 0.502 and 0.498. Among calves born into the study population, heterozygotes are significantly more likely to survive to 2 years of age than homozygotes (Pemberton *et al.* 1988). This association is particularly strong in females (Pemberton *et al.* 1988, 1991). In common with other associations discussed below, no mechanism for this association is known and it is possible that Idh-2 is marking the effects of a linked locus.

This example serves to illustrate the possibility that one consequence of intraspecific competition may be selection for heterozygous individuals. The idea of selection for heterozygotes has a long history (see review by Allendorf and Leary 1986), and various mechanisms could be responsible for heterozygote advantage. One possibility is that selection favours outbred individuals, as has been documented for a weather-driven population crash in song sparrows on Mandarte Island, British Columbia (Keller *et al.* 1994). However, another possible explanation is that there could be a selective advantage to having more than one version of an enzyme molecule, perhaps due to greater kinetic adaptability. Either way, a consequence of such selection will be to maintain genetic variation in the population at the locus concerned.

The other examples we cite below illustrate an alternative mechanism which may prevent loss of genetic variation in fluctuating populations, namely that fluctuating levels of intraspecific competition may result in temporal variation in the direction of selection.

Mannose phosphate isomerase (Mpi) in deer

At the protein level, two alleles are detectable at Mpi. In the total study population sample, allele s is at approximate frequency 0.85 and allele f at 0.15. Among calves born into the study population, individuals with an f allele are significantly more likely to die before reaching 2 years of age (Pemberton *et al.* 1988, 1991).

Conversely, in adult female fertility (pregnancy) and fecundity (production of calves), there appear to be significant advantages to having an Mpi f allele. Study area females with an f allele breed at an earlier age and tend to be more fecund over their life span (Pemberton *et al.* 1991), and in shot females from non-study area parts of Rum, Mpi f-carriers are more likely to be pregnant (Pemberton *et al.* 1991).

Set in the context of changing population density, these observations would have the following consequences. At low density, when juvenile mortality is negligible, the Mpi f allele should increase through its association with higher fertility and fecundity in females. At high density, when juvenile mortality becomes a more important component of fitness variation (Clutton-Brock *et al.* 1985, 1987), the Mpi s allele will be favoured and its frequency will increase, as illustrated in the study area deer (Pemberton *et al.* 1988). It seems likely that fluctuating density would retain both alleles in the population.

Changes in the study area population density on Rum were dictated by a change in human management strategy and occured relatively slowly, but conceptually similar examples are apparent within the Soay sheep population on St Kilda, in which relatively short-term population fluctuations occur without human intervention (Fig. 4.2b).

Horn morphs in Soay sheep

As outlined above, there is heritable horn variation in the Soay population, with frequencies for normal, scurred and polled individuals being 87, 11 and 2% respectively in rams and 36, 21 and 35% respectively in ewes. Male sheep with normal horns are behaviourally dominant to polled and scurred individuals, and almost certainly have a mating advantage, because they are able to sequester females in exclusive mating consorts (Stevenson, personal communication). One might therefore predict that sexual selection would have fixed normal horns in the population, or at least in males (in many ruminant species, weapons are confined to males and all females are polled). However, individuals of both sexes with scurred horns consistently survive population crashes better than individuals that are polled or have normal horns (Moorcroft *et al.* 1996). The mechanism for this superior survival is unknown and is currently being investigated. Whatever the mechanism, this superior survival helps to explain why horn polymorphism persists in the population.

Adenosine deaminase (Ada) in Soay sheep

At the protein level, two alleles are detectable at Ada. In the total study population sample, allele s is at approximate frequency 0.75 and allele f at 0.25. Over three successive population crashes, ff individuals consistently survived worst, and in the

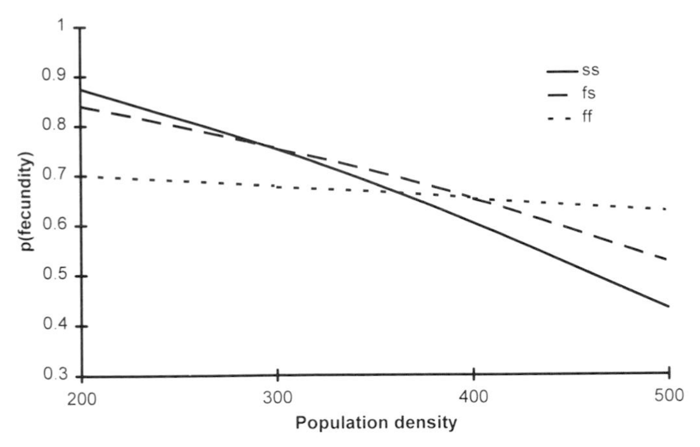

Fig. 4.3 Interaction between the probability that a female Soay sheep has a lamb (*p*(fecundity)), Ada genotype (ss: full line; sf: long-dashed line; ff: short-dashed line) and population density on Hirta (Ada + Ada.population size, $\chi^2 = 14$, d.f. = 4, $P < 0.05$). The fitted ordinal logistic model also included the effects of female age and breeding experience and interactions between these terms and population density.

most extreme crash, heterozygotes survived the best (Gulland *et al.* 1993). Ada ff sheep also have the highest gastrointenstinal nematode burdens at certain times of year, suggesting the survival effect is mediated through parasite resistance (Gulland *et al.* 1993).

Recent investigations of variation in female fecundity on St Kilda have revealed an interaction between Ada genotype, population size, and fecundity (Ada+Ada. population size $\chi^2 = 14$, d.f. = 4, $P < 0.05$; Fig. 4.3). The fecundity of Ada ff females is the least responsive to rising density. It is possible that both the higher worm burdens and greater crash mortality observed are a consequence of these individuals conceiving at high density. Though it may appear from these observations that the f allele should decline, this does not take into account the value of lambs produced at different population sizes. As we will show below, lambs born immediately following a population crash are particularly valuable, broadly because they have a clear 2–4 years in which to breed before another crash occurs. Ada f-carrying individuals, by not responding to rising population size, have a greater probability of producing one of these high-value lambs than ss individuals. It therefore seems possible that the maintenance of the Ada polymorphism in the Soay population is again connected with the different responses of the genotypes to changing population density.

4.5 Retarding the loss of genetic variation: mating success

Where population fluctuations alter the demography of populations, there may be systematic changes in the distribution of male mating success which retard loss of genetic variation.

Calculations of effective population size for polygynous populations often assume that matings are divided between the mature males present in the population regardless of the population's size or demography. Thus, after a population decline leaving a female-biased population, it is assumed that surviving adult males obtain large numbers of the available matings. However, in Soays, behavioural observations show that ram lambs (aged 7 months) participate in the rut to an extent which varies from year to year (Stevenson and Bancroft 1995). They participate most in post-crash ruts, when the older ram population is small and the population is strongly female-biased. In this section, we use paternity data attributed by molecular techniques to investigate the genetic consequences of this and related aspects of the mating system. We concentrate on the population in the years 1986–93. As shown in Fig. 4.2b, over this period, the study population varied substantially in size and sex ratio.

Paternity analysis in Soay sheep

The Soay sheep mating system has been described by Grubb (1974). Soay sheep are highly seasonal and mate in November. Both sexes are promiscuous. Ewes come into oestrus for 24–36 h, and mate repeatedly during this time. Oestrous ewes are often detected by small, young, or scurred rams and chased and mated by several of them. More frequently, an exclusive consort is formed with a single mature ram for hours at a time, resulting in repeated mating by the same pair. For most of the years indicated, rut consort information was collected by censusing, but in tests using DNA fingerprinting or single locus profiling, consort information was found to be a poor predictor of paternity (Bancroft 1993). This is probably because the census-based approach misses many chases and changes in consort partnerships. To pursue the identification of paternity and the success of ram lambs in gaining paternities, we employed an entirely genetical approach to identifying paternity.

We used locus-specific protein and microsatellite markers to investigate paternity for 921 lambs with sampled mothers born into the study area in the years following the ruts 1986–93. Most individuals have been genotyped at 15 to 17 loci (for identity of loci see Bancroft *et al.* (1995*a*) and Smith (1996)), and the candidate fathers for each lamb consisted of all the tagged rams known to be alive in the preceding rut, which varied from 68 to 227 males in different years. We used a computer program (Marshall *et al.*, submitted) written in dBASE IV to conduct all comparisons, and log-likelihood ratios, which can often statistically discriminate two or more candidate males even though they both match a putative offspring at the same set of loci (i.e. have identical probability of non-exclusion). The analyses described below are based on 663 paternities (of 921 cases investigated) attributed to the most likely male with a log-likelihood ratio over 2.0. Simulations which incorporated appropropriate biological and genetic parameters for the population suggest an overall confidence of 80% for these paternity inferences. Similar simulations were used to derive log-likelihood ratio thresholds giving 95% confidence in a paternity identification being correct. Of the 663 cases, 321 (48%) fell into this category. Although the larger sample of paternity identifications is less secure, importantly, it allows us to estimate the number of males which obtained no paternities in each year. Most lambs with no assigned father were probably sired by immigrant, unsampled rams.

Mating success of young rams

As anticipated from rut census data (Stevenson and Bancroft 1995), ram lambs (7 months old at the rut) obtain more paternities in those years when the older rams are few in number (Fig. 4.4a). Similarly, yearling rams obtain more paternities when older rams are few in number (Fig. 4.4b) with the exception that in those ruts when the yearling class has been decimated by a population crash (ruts 1986 and 1989), there are few yearlings around to obtain paternities. Thus, in precisely those years (following a crash) when the effective number of males in the population might be expected to be low, young males, which were *in utero* during the crash, are particularly successful in obtaining paternities.

As a consequence of their success when young, and continued absence of older age classes as they grow older, one might predict that cohorts of rams born following a population crash would have greater mean individual lifetime reproductive success (LRS) than other cohorts. Preliminary data (some relevant animals are still alive at the time of writing) confirm that the two ram cohorts studied which were born after a crash (1986 and 1989) have been more successful than the cohorts that succeeded them.

Variation in sibship size

The distribution of the number of rams obtaining different numbers of paternities in each year varies systematically with population demography. The pattern is that in ruts that occur at high density, paternity is relatively evenly spread, with most rams getting zero, one, or two paternities. Using a multinomial regression model, the probability of a ram siring multiple offspring is significantly associated with population size (χ^2 = 37.92, d.f. = 1, $P < 0.001$; Fig. 4.5a). Thus, despite the depredations of young rams (above), the low-density ruts which follow population crashes do allow surviving older rams to get multiple paternities. Nevertheless, young rams have an important impact on these ruts, obtaining about half the paternities in the one- and two-lamb categories.

Variation in the distribution of male success between years has an important effect on the size of the paternal half-sibships subsequently born. Thus, most lambs conceived in high-density ruts are the sole representative of their father in their cohort. By contrast, most lambs conceived in low-density ruts have paternal half sibs in the same cohort (Fig. 4.5b). Interestingly, therefore, the lambs which are *in utero* during a population crash, which, if they survive, go on to enjoy high LRS, are a particularly representative sample of the pre-crash male population.

4.6 Discussion

The purpose of this chapter has been to outline two general processes, selection and variation in the distribution of male mating success, which may buffer loss of genetic

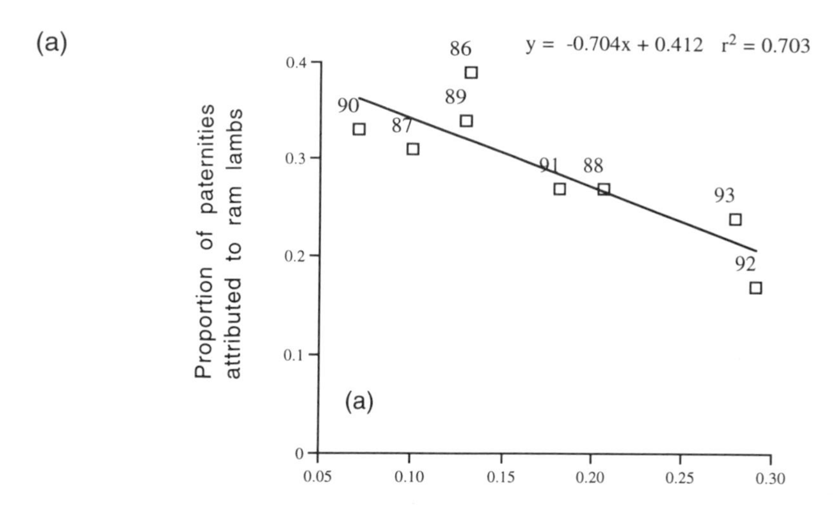

Sex ratio of rams 2 years and older : ewes

(b)

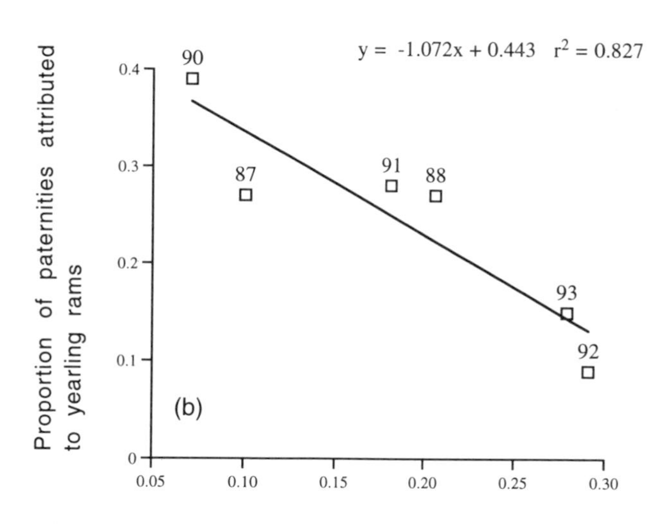

Sex ratio of rams 2 years and older : ewes

Fig. 4.4 Proportion of paternities attributed to young rams in relation to (on horizontal axes) sex ratio of rams 2 years and older:ewes. The latter attempts to measure the extent to which older rams can gain exclusive access to oestrous ewes. (a) Ram lambs (aged 7 months at the rut). (b) Yearling rams (aged 19 months at the rut). Note that the ruts for 1986 and 1989 are missing from this graph since in each case only three rams survived the relevant crash to become yearlings, and as a result their cohort had very low paternity success.

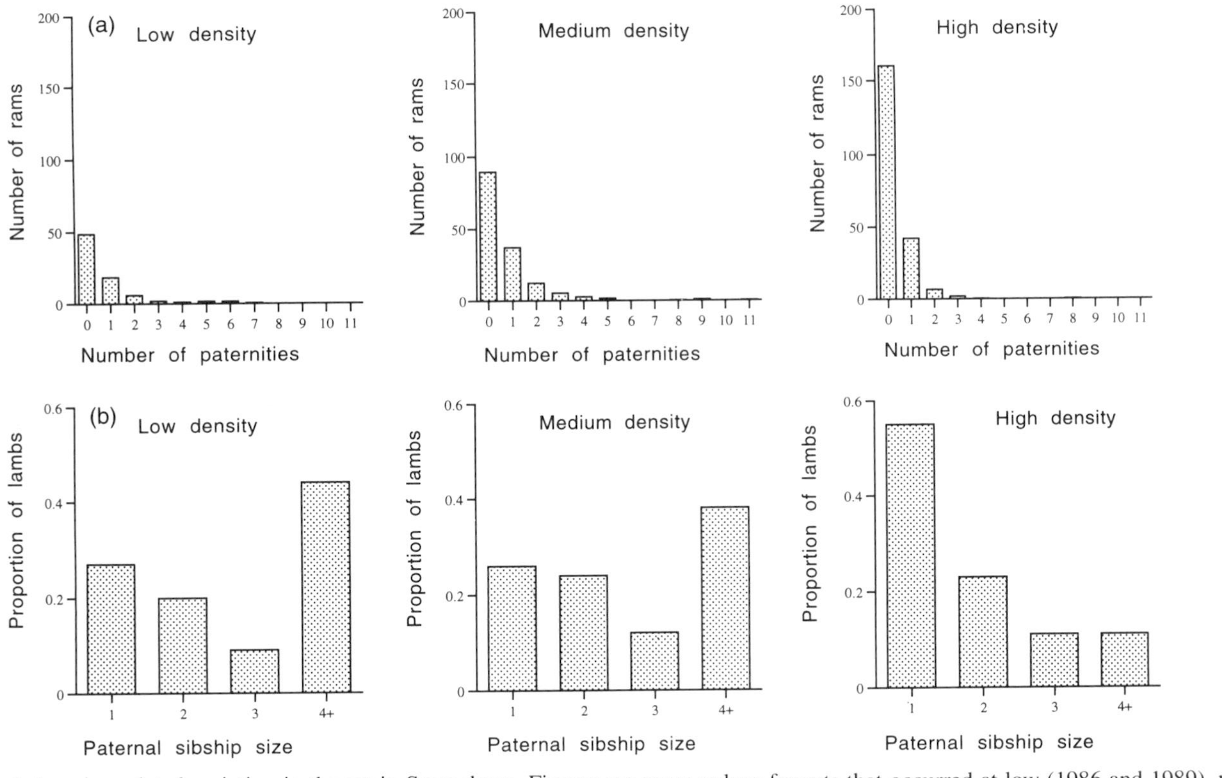

Fig. 4.5 Population size-related variation in the rut in Soay sheep. Figures are mean values for ruts that occurred at low (1986 and 1989), medium (1987, 1990, 1992) and high (1988, 1991, 1993) population density. (a) Distribution of the number of rams with different numbers of paternities. In a multinomial regression model applied to the whole dataset 1986–93, the probability of a male siring multiple offspring is significantly associated with population size ($\chi^2 = 37.92$, d.f. = 1, $P < 0.001$). In low-density ruts, individual older rams are able to obtain multiple paternities, despite the efforts of young rams (Fig. 4.3), which account for about half the cases in the one- and two-lamb categories. (b) Distribution of paternal half-sibship size. Note that in ruts occurring at high density (immediately before a crash) most lambs conceived are the only offspring of their father in their cohort. By contrast, in ruts occurring at low density, most lambs have paternal half-sibs in their cohort.

variation by genetic drift during population fluctuations in nature. In this section we discuss how widespread these processes are likely to be, and some practical implications.

Many documented population reductions have undoubtedly resulted in loss of genetic variation. One of the best examples, among large mammals, is the northern elephant seal (*Mirounga angustirostris*) (Bonnell and Selander 1974; Hoelzel *et al.* 1993). In this case, in which the decline was caused by hunting, it seems likely that mortality was random with respect to genotype and thus adhered to neutral population genetic expectations. However, the overwhelming number of vertebrate population size changes, including declines, are probably caused by habitat change, and, depending on their severity, may involve selection. Few field studies have the data resolution to investigate selection in relation to population change, but all those that do have shown selective processes likely to retain variation. Apart from the red deer and Soay sheep studies reported here, these include Darwin's finches in the Galápagos (Grant 1986) and song sparrows on Mandarte Island (Keller *et al.* 1994). More studies are required to confirm the generality of selection during population size change.

Fluctuations in the distribution of male mating success with changes in population size, density, or demography have not been extensively documented. Again it should be no surprise that individuals have variable involvement in mate competition according to ambient levels of competition. Again, there are few studies which have explicitly addressed this topic. However, in a recent analysis of 20 years of male rut behaviour in red deer, which is strongly correlated with true success as measured by DNA fingerprinting (Pemberton *et al.* 1992), Clutton-Brock *et al.* (submitted) have shown that in this species also, young males have variable mating success. A major task for the future is to assess the strength of these processes in retaining genetic variation. In particular, estimating effective population size in populations with overlapping generations, fluctuating population size, and varying distribution of male mating success is an urgent requirement.

Population genetic investigations in island populations form important practical models for species conservation. It is axiomatic to modern conservation biology that genetic variation must be preserved in endangered species, to avoid inbreeding depression and to allow future genetic adaptation, but at present the models employed make relatively simple assumptions about the process of genetic drift during population size fluctuations. The studies reported here suggest processes which should be investigated further and taken into account in species conservation studies.

4.7 Summary

Conventionally, small populations living on islands are expected to lose genetic variation by drift. Fluctuations in population size, combined with polygynous mating systems, are expected to contribute to the process by increasing sampling effects on genetic variation. However, in individually monitored populations of red deer on Rum and Soay sheep on St Kilda, which experience fluctuations in population size, two processes have been identified which mitigate loss of genetic variation. First, in a number of examples, population reductions are associated with selection. Selection may be in favour of heterozygotes, or, as we have documented in several cases, it may fluctuate in direction temporally. Second, in Soay sheep, in which mortality over population crashes is male-biased, ostensibly leading to low effective numbers of males, molecular studies show that there are systematic changes in the reproductive success of young males, and in variance in male success, that broaden genetic representation compared with expectation.

Acknowledgements

We would like to thank the authorities that allow and help us to work on Rum and St Kilda: Scottish Natural Heritage, The National Trust for Scotland, the Royal Artillery and the Royal Corps of Transport; the financial supporters of our research, the Biotechnology and Biological Sciences Research Council and the Natural Environment Reseach Council; and the numerous collaborators, field assistants, and volunteers who have been involved in the studies over many years.

References

Allendorf, F. W. and Leary, R. F. (1986). Heterozygosity and fitness in natural populations of animals. In *Conservation biology: the science of scarcity and diversity*, (ed. M. E. Soule), pp. 57–76. Sinauer, Sunderland, MA.

Bancroft, D. R. (1993). Genetic variation and fitness in Soay sheep. Unpubl. Ph.D thesis, University of Cambridge.

Bancroft, D. R., Pemberton, J. M., and King, P. (1995*a*). Extensive protein and microsatellite variability in an isolated, cyclic ungulate population. *Heredity*, **74**, 326–36.

Bancroft, D. R., Pemberton, J. M., Albon, S. D., Robertson, A., MacColl, A. D. C., Smith, J. A., Stevenson, I. R., and Clutton-Brock, T. H. (1995*b*). Molecular genetic variation and individual survival during population crashes of an unmanaged ungulate population. *Philosophical Transactions of the Royal Society of London* B, **347**, 263–73.

Bonnell, M. L. and Selander, R. K. (1974). Elephant seals: genetic variation and near extinction. *Science*, **184**, 908–9.

Clutton-Brock, T. H. and Albon, S. D. (1989). *Red deer in the highlands*. Blackwell Scientific Publications, Oxford.

Clutton-Brock, T. H., Albon, S. D., and Guinness, F. E. (1982). *Red deer: behavior and ecology of two sexes*. University of Chicago Press.

Clutton-Brock, T. H., Major, M., and Guinness, F. E. (1985). Population regulation in male and female red deer. *Journal of Animal Ecology*, **54**, 831–46.

Clutton-Brock, T. H., Major, M., Albon, S. D., and Guinness, F. E. (1987). Early development and population dynamics in red deer. I. Density-dependent effects on juvenile survival. *Journal of Animal Ecology*, **56**, 53–67.

Clutton-Brock, T. H., Price, O. F., Albon, S. D., and Jewell, P. A. (1991). Persistent instability and population regulation in Soay sheep. *Journal of Animal Ecology*, **60**, 593–608.

Clutton-Brock, T. H., Price, O. F., Albon, S. D., and Jewell, P. A. (1992). Early development and population fluctuations in Soay sheep. *Journal of Animal Ecology*, **61**, 381–96.

Clutton-Brock, T. H., Illius, A., Wilson, K., Grenfell, B. T., MacColl, A., and Albon, S. D. (1997). Stability and instability in ungulate populations: an empirical analysis. *American Naturalist*, **149**, 195–219.

Clutton-Brock, T. H., Rose K. E., and Guinness, F. E. Density-related changes in sexual selection in red deer. *Evolution*. (Submitted.)

Doney, J. M., Ryder, M. L., Gunn, R. G., and Grubb, P. (1974). Colour, conformation, affinities, fleece and patterns of inheritance of the Soay sheep. In *Island survivors: the ecology of the Soay sheep of St Kilda*, (ed. P. A Jewell, C. Milner, and J. Morton Boyd), pp. 88–125. Athlone Press, London.

Ellner, S. and Hairston, N. G. (1994). Role of overlapping generations in maintaining genetic variation in a fluctuating environment. *American Naturalist*, **143**, 403–17.

Gilbert, D. A., Lehman, N., O'Brien, S. J., and Wayne, R. K. (1990). Genetic fingerprinting reflects population differentiation in the California Channel Island Fox. *Nature,* **344**, 764–7.

Grant, P. R. (1986). *Ecology and evoluton of Darwin's finches*. Princeton University Press.

Grenfell, B. T., Price, O. F., Albon, S. D., and Clutton-Brock, T. H. (1992). Overcompensation and population cycles in an ungulate. *Nature,* **355**, 823–6.

Grubb, P. (1974). The rut and behaviour of Soay rams. In *Island survivors: the ecology of the Soay sheep of St Kilda*, (ed. P. A. Jewell, C. Milner, and J. Morton Boyd), pp. 195–223. Athlone Press, London.

Gulland, F. M. D. (1992). The role of nematode parasites in Soay sheep (*Ovis aries* L.) mortality during a population crash. *Parasitology*, **105**, 493–503.

Gulland, F. M. D. and Fox, M. (1992). Epidemiology of nematode infections of Soay sheep (*Ovis aries* L.) on St Kilda. *Parasitology*, **105**, 481–92.

Gulland, F. M. D., Albon, S. D., Pemberton, J. M., Moorcroft, P., and Clutton-Brock, T. H. (1993). Parasite-associated polymorphism in a cyclic ungulate population. *Proceedings of the Royal Society of London* B, **254**, 7–13.

Gustafsson, L. (1986). Lifetime reproductive success and heritability: empirical support for Fisher's fundamental theorem. *American Naturalist,* **128**, 761–4.

Hoelzel, A. R, Halley, J., O'Brien, S. J., Campagna, C, Arnbom, T., Le Boeuf, B., Ralls, K., and Dover, G. A. (1993). Elephant seal genetic variation and the use of simulation models to investigate historical population bottlenecks. *Journal of Heredity*, **84**, 443–9.

Illius, A. W., Albon, S. D., Pemberton, J. M., Gordon, I. J., and Clutton-Brock, T. H. (1995). Selection for foraging efficiency during a population crash in Soay sheep. *Journal of Animal Ecology*, **64**, 481–92.

Jewell, P. A., Milner, C., and Morton Boyd, J. (eds.). (1974). *Island survivors: the ecology of the Soay sheep of St Kilda*. Athlone Press, London.

Keller, L. F., Arcese, P., Smith, J. N. M., Hochachka, W. M., and Stearns, S. C. (1994). Selection against inbred song sparrows during a natural population bottleneck. *Nature,* **372**, 356–7.

Langbein, J. and Thirgood, S. J. (1989). Variation in the mating systems of fallow deer (*Dama dama*) in relation to ecology. *Ethology*, **83**, 195–214.

Marshall, T.C., Slate, J., Kruuk, L.E., and Pemberton, J.M. Statistical confidence for likelihood-based paternity inference in natural populations. *Genetics*. (Submitted.)

Moorcroft, P. R., Albon, S. D., Pemberton, J. M., Stevenson, I. R., and Clutton-Brock, T. H. (1996). Density-dependent selection in a fluctuating ungulate population. *Proceedings of the Royal Society of London* B, **263**, 31–8.

Pemberton, J. M., Albon, S. D., Guinness, F. E., Clutton-Brock, T. H., and Berry, R. J. (1988). Genetic variation and juvenile survival in red deer. *Evolution*, **42**, 921–34.

Pemberton, J. M., Albon, S. D., Guinness, F. E., and Clutton-Brock, T. H. (1991). Countervailing selection in different fitness components in female red deer. *Evolution*, **45**, 93–103.

Pemberton, J. M., Albon, S. D., Guinness, F. E., Clutton-Brock, T. H., and Dover, G. A. (1992). Behavioural estimates of male mating success tested by DNA fingerprinting in a polygynous mammal. *Behavioral Ecology*, **3**, 66–75.

Smith, J. A. (1996). Polymorphism, parasites and fitness in Soay sheep. Unpubl. Ph.D thesis, University of Cambridge.

Smith, J. N. M. and Dhondt, A. A. (1980). Experimental confirmation of heritable morphological variation in a natural population of song sparrows. *Evolution*, **34**, 1155–8.

Stevenson, I. R. and Bancroft, D. R. (1995). Fluctuating trade-offs favour precocial maturity in male Soay sheep. *Proceedings of the Royal Society of London* B, **262**, 267–75.

5

Molecular and morphological evolution within small islands

Roger S. Thorpe and Anita Malhotra

5.1 Introduction

Many studies of evolution on islands, from Darwin's (1859) time onwards, have tended to emphasize inter-island differences. However, inter-island studies (although very popular) are generally not in a position to contribute much to an understanding of relative importance of historical processes and selection on their own. There have been some efforts to distinguish between the roles of historical factors and current selection pressures (Snell *et al.* 1984; Gardner 1986), but they do not use the appropriate methodology to make much headway with this problem.

Some of the difficulties involved are exemplified by the inter-island differences in the endemic western Canary Island lacertid *Gallotia galloti* (Fig. 5.1). The distinct differences in scalation, colour-pattern, and body dimensions could be due to historical factors like founder effects and drift, or to adaptation to the current ecological conditions that differ among islands (Thorpe 1996). Molecular data, such as mtDNA sequence and RFLPs (which are hopefully minimally confounded by selection effects) can be used to reconstruct a phylogeny (Thorpe *et al.* 1994*a*). The western Canaries have not been joined to one another or the mainland (Carracedo 1979 and references therein). Consequently, any organism naturally distributed across them must have undergone inter-island dispersal. Given this geological background, a rigorous set of rules can be used to convert this phylogeny into a colonization sequence (Thorpe *et al.* 1994*a*; Juan *et al.* 1995). *G. galloti* appears to have arisen on the oldest western island, and colonized the younger islands further to the west. The colonization time for each island inferred from the DNA divergence is appropriately less than the geological time of origin of each island (Thorpe *et al.* 1994*a*).

Once a quantifiable perspective of the historical relationships (with minimal selection effects) is obtained, one can test for adaptation taking into account these historical relationships using partial regression based matrix correspondence tests (see below). Historical relationships may be represented in a variety of ways, including as a matrix of patristic distances among taxa along the branches of a molecular phylogeny (Thorpe 1996; Thorpe *et al.* 1995, 1996; but see also Douglas and Matthews 1992). When there are only a few islands, as in the western Canaries, and each island is generally

Fig. 5.1 Endemic lizard (*Gallotia galloti*) from Tenerife.

being treated as an evolutionary entity, one can only obtain a preliminary and limited answer to the roles played by various evolutionary processes (Thorpe 1996). For example, the presence or absence of blue leg spots in western Canarian lacertids appears to reflect phylogenetic relationships, irrespective of selection for environmental/climatic conditions; yellow dorsal bars appear to be associated with selection for wet climates, irrespective of history or other ecological conditions; and body size appears to associated with how depauperate the environment is, irrespective of history or climatic factors (Thorpe 1996).

Although inter-island studies will remain popular and important, there are inescapable limitations to using a few islands as single entities when trying to investigate the relative contribution of current natural selection and historical factors. More progress can be made by considering islands with habitat differentiation or zonation within them as the heterogeneous entities that they are, and investigating population differentiation among numerous local populations (Thorpe and Brown 1989; Brown *et al.* 1991; Malhotra and Thorpe 1991*a*; Castellano *et al.* 1994; Prentice *et al.* 1995; Thorpe *et al.* 1996). Small islands may offer distinct advantages for these studies of microgeographic variation. Biotic and physical factors pertinent to natural selection may vary substantially over very short geographic distances, which gives considerable logistic advantages over mainland systems where considerable distances may be involved. The general trend is to have a depauperate fauna with few species, which may mean that individual species are found in densities far greater than in mainland systems. This facilitates sampling and may allow exhaustive coverage of the species range over all habitat types.

This chapter surveys our studies of within-island microgeographic variation that use lizards as model organisms. This work has primarily been carried out in the Canarian and Lesser Antillean archipelagos and can be considered under three rubrics: (1) matrix correlation tests for association between observed patterns and patterns generated by putative causal factors; (2) investigations of parallel patterns of variation; and (3) large-scale field experiments on natural selection.

5.2 Matrix correspondence (mantel) tests and molecular studies

Tests of a null hypothesis of no association between an observed pattern of geographic variation and a pattern generated from one or more causal hypotheses are useful in that they enable hypotheses to be rejected. When dealing with geographic patterns, both patterns and hypotheses can conveniently (and in some cases must) be represented as a matrix of dissimilarities between entities (e.g. local demes). The correspondence between the matrices can be measured by a statistic such as a correlation or regression, but the probability of the null hypotheses cannot be tabulated because the elements of a matrix are not independent and the degrees of freedom are unknown. With large matrices, which cannot be exhaustively permutated, the rows and columns of one of the matrices (Manly 1986*a*, *b*, 1991) can be randomized and the statistic recomputed. This is repeated a large number of times (e.g. 10 000 fold in Thorpe *et al.* (1996)) to give a distribution of the statistic so that the probability of the null hypothesis of no association can be found. Where there are several hypotheses, the patterns generated by them may be intercorrelated. To overcome this, partial correlation, or partial regression extensions of the test are used (Smouse *et al.* 1986) where the observed pattern, (e.g. in morphology) is taken as the dependent variable and patterns generated from the hypotheses are taken as the independent variables (Manly 1986*b*; Thorpe and Baez 1993).

These tests are making an important contribution to studies at the intraspecific level and have recently had a high profile (Brown *et al.* 1991; Sokal *et al.* 1991; Waddle 1994; Daltry *et al.* 1996; see also a review by Smouse and Long 1992). Two examples of their application to within-island geographic variation are given here; Tenerife lacertids and Dominican anoles.

Geographic variation of the lacertid G. galloti *within Tenerife*

The colour pattern of sexually mature male lacertids varies markedly across Tenerife. The variation in six colour pattern characters can be treated individually or combined by multivariate analysis (Thorpe and Brown 1989; Thorpe *et al.* 1994*b*). Thorpe *et al.* (1994*b*) used partial regression matrix correspondence (PRMC) methods to test the association of the pattern of geographic variation in the colour pattern, across 67 localities, against several hypotheses including (i) historical separation of ancient precursor islands, (ii) a cloud layer around Teide inducing separation of high-altitude populations from low-altitude populations, (iii) adaptation to altitude, (iv) adaptation to two latitudinal climatic/vegetational biotopes which meet along a sharp ecotone and (vi) geographic proximity representing the opportunity for gene flow and unspecified geographical components. All hypotheses except for adaptation to the climatic/vegetational biotopes can be rejected. It appears that the colour pattern of sexually mature males may be a balance between crypsis to avoid predation (overhead avian predators see the dorsum with its disruptive yellow bars in the north) and sexual selection for lateral, blue, display markings (for laterally positioned conspecifics).

Later studies of the molecular affinities of these Tenerife populations (across largely the same set of localities) by Thorpe *et al.* (1996), using cytochrome b sequence data, revealed three main haplotypes which, when subjected to an outgroup rooted phylogenetic analysis, revealed eastern and western lineages. Historical relationships, whether derived from molecular or other data (Sokal *et al.* 1991; Waddle 1994), can also be tested against alternative historical scenarios. In this case the historical relationships are represented by the molecular phylogenetic affinities (patristic distances among populations on the molecular phylogenetic tree). These were treated as the observed (dependent) pattern. Several alternative historical hypotheses can be tested. These hypotheses are primarily based on the concept of populations existing on the ancient precursor islands that formed Tenerife (Ancochea *et al.* 1990). Patterns generated from these hypotheses are represented as independent variables in a PRMC test. A set of three similar patterns represented hypotheses involving a single western precursor and an eastern precursor, a fourth pattern represented separate north-western and south-western precursors, a fifth pattern represented three separate precursors, one on each of the three ancient areas, a sixth pattern representing cloud induced vicariance, with a seventh matrix representing geographic proximity. A series of pairwise matrix correspondence tests followed by a PRMC tests allows one to reject all hypotheses other than that the pattern was formed by expansion from an ancient western and an ancient eastern precursor island (Thorpe *et al.* 1996).

The earlier colour pattern studies made no allowance for different molecular phylogenetic lineages being present on the island. When this is allowed for by (1) testing colour pattern against biotope and geographic proximity within the eastern and western lineage separately, and (2) testing colour pattern against climatic biotopes, geographic proximity, and phylogenetic patristic distances in a PRMC test (see Thorpe *et al.* 1996; but also Thorpe *et al.* 1995; Daltry *et al.* 1996; Thorpe 1996) then there is still an association between colour pattern of sexually mature males and biotope.

The eastern and western lineages revealed in the mtDNA appear to have introgressed completely as there is no indication of reproductive isolation between these east–west lineages in previous quantitative morphological studies of scalation, shape (Thorpe and Baez 1987), size (Thorpe and Brown 1991), or colour (Thorpe and Brown 1989). Indeed, natural selection of the colour pattern for current biotopes appears to have largely eradicated historical effects. The molecular lineages have an east–west pattern while colour pattern has a marked north–south pattern irrespective of lineage (Fig. 5.2). DNA times (which are compatible with geological times (Thorpe *et al.* 1996)) suggest this introgression occurred after about 700 000 years separation.

Dominican anole, Anolis oculatus

Anolis oculatus is the only anoline lizard on Dominica, a young volcanic island in the Lesser Antillean island chain fringing the eastern Caribbean. Dominica possesses a diverse set of climatic and vegetational regimes with littoral woodland on the Atlantic coast, xeric woodland on the Caribbean coast, and montane rain forest and cloud forest

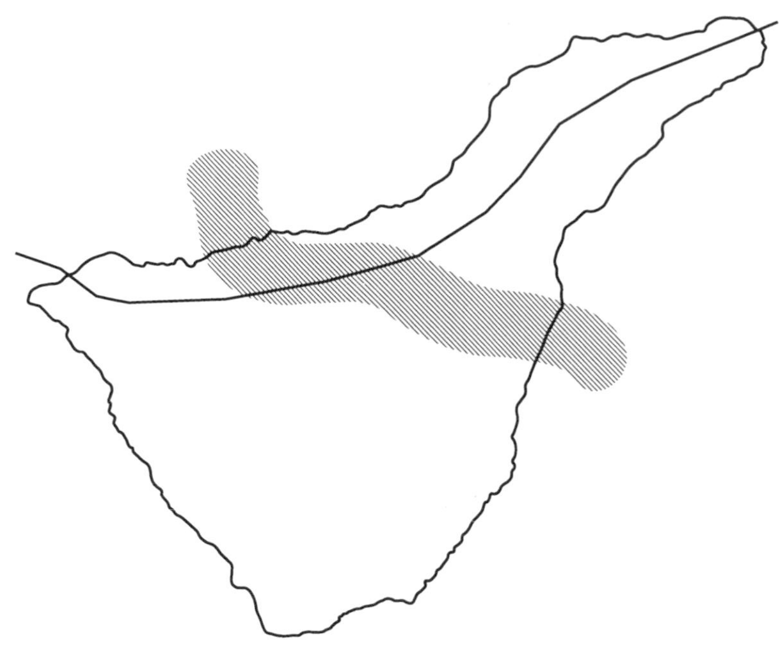

Fig. 5.2 Colour pattern variation and DNA lineages of *Gallotia galloti* on Tenerife. The line indicates the transition between northern and southern colour morphs (Thorpe and Brown 1989) which is associated with an ecotone between different climatic/vegetational biotopes. The shaded band indicates the distribution of the western and north-eastern mtDNA lineages, bearing in mind that some populations contain different haplotypes (Thorpe *et al.* 1996).

in the extremely mountainous centre. Anoles were sampled from 33 localities across the island and showed pronounced geographic variation in the 47 morphological characters (body shape, scalation, and colour pattern) studied (Fig. 5.3). The 'overall' similarity among samples of females using 47 morphological characters is strongly associated ($P < 0.0001$) with the overall ecological similarity among localities (using altitude, rainfall, temperature, and vegetation (Malhotra and Thorpe 1991*a*)) when tested with a partial matrix correspondence test which also includes geographic proximity.

A more detailed picture can be obtained by using PRMC tests to test multivariate character sets such as body shape, scalation, and colour pattern (as dependent variables) against a series of independent variables, i.e. geographic proximity, altitude, temperature, rainfall, and vegetation type. This shows that, using locality means, general body shape and colour are related to vegetation, and general scalation is related to rainfall (Thorpe *et al.* 1994*b*; Malhotra and Thorpe 1997*a*, *b*; Malhotra and Thorpe unpublished data). When individual characters are tested they generally conform to the multivariate sets, but with some exceptions. For example, relative size of enlarged lateral scales is associated with vegetation type, not rainfall (Malhotra and Thorpe 1997*a*), while the cyan element of body hues is related to rainfall, not vegetation type (Malhotra and Thorpe in preparation; see also Thorpe *et al.* 1994*b*).

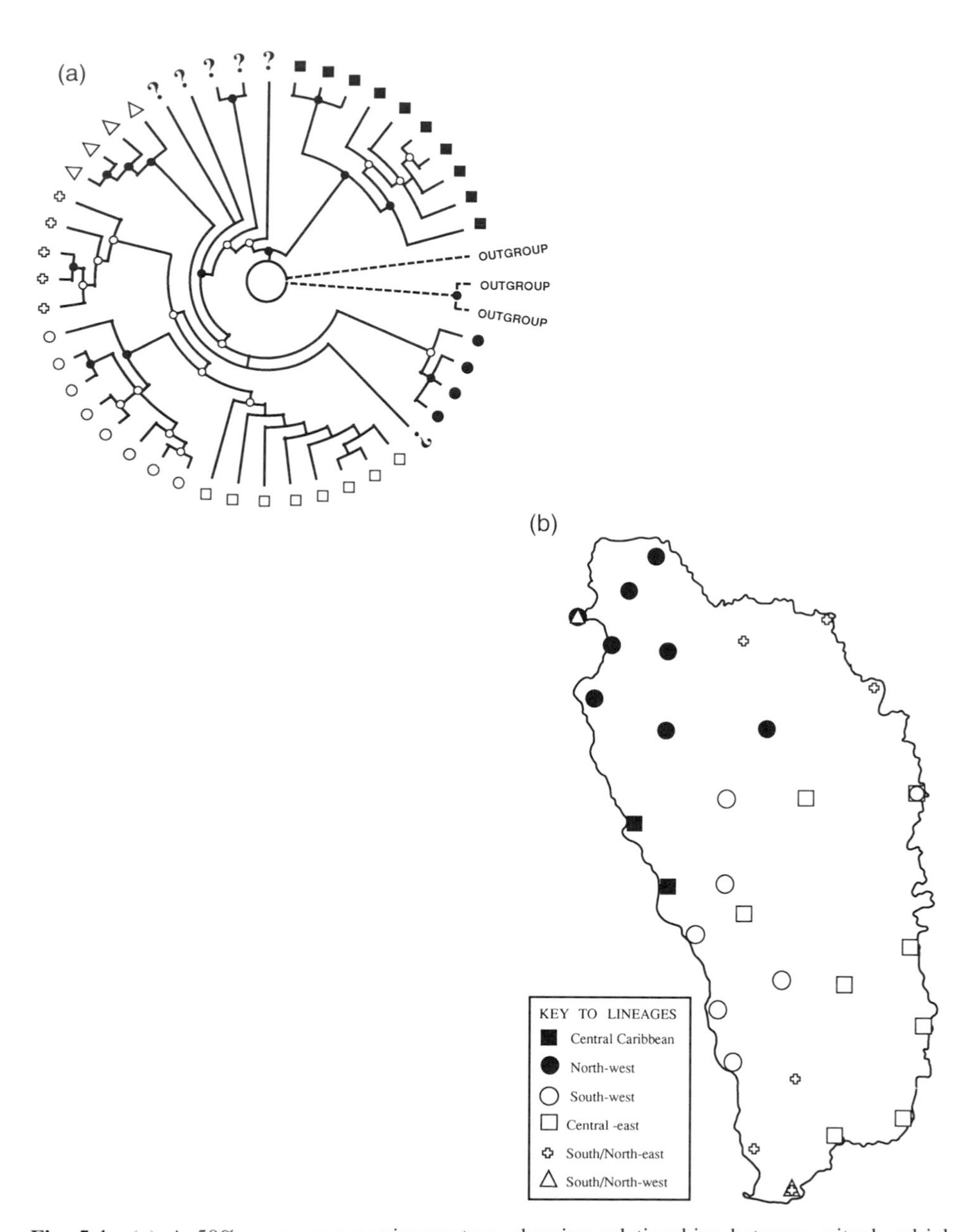

Fig. 5.4 (a) A 50% consensus parsimony tree, showing relationships between mitochondrial cytochrome b haplotypes of *Anolis oculatus*. Although there are a very large number of equally parsimonious trees, most of the nodes are well supported. The level of support for the different nodes is indicated by a black (100%) or white dot (>80%). Nodes shown as resolved but without a symbol are represented in between 50 and 80% of trees. (b) Map of Dominica illustrating the geographic position of the haplotype lineages shown in (a). Individual haplotypes not clearly belonging to any major group (indicated by ? in (a)) are not plotted. Filled squares indicate central Caribbean lineage; filled circles indicate northwest lineage; open circles indicate south-west lineage; open squares indicate central-east lineage; open crosses indicate south/north-east lineage; and open triangles indicate south/north-west lineage.

Generally there is rather limited congruence in geographic variation in individual characters and the extent of morphological difference is commensurate with the extent of ecological difference. However, along the Caribbean coast the morphological change is marked while the ecological change is subtle. Moreover, there is a higher degree of congruence among characters in this area. This raises the question as to whether there is a phylogenetic/historical component to this geographic variation along the Caribbean coast.

A 267bp section of the mtDNA cytochrome b gene was sequenced which showed 22% variability across the 33 localities. A phylogenetic tree (Fig. 5.4) was reconstructed using both parsimony and distance-based methods, which gave very similar results. Because of the high variability, there were many equally parsimonious reconstructions, but a 50% consensus tree was resolved to a surprisingly high degree (Fig. 5.4a), and revealed the presence of several lineages. The most basal split is the best supported (appearing in 100% of all trees) and would have occurred 4 Ma BP (using a rate of 2.5% Ma^{-1}). Populations belonging to this lineage are found in the central part of the Caribbean coast (Fig 5.4b) and the southern edge of its range may be close to the transition between north and south Caribbean ecotypes. Other lineages are more closely related, but also have a degree of geographical coherence when mapped. Nevertheless, the phylogenetic relationships do not adequately explain the morphological difference between north and south Caribbean coast populations. The geographical distribution of lineages shows no relationship to current or known past barriers to gene flow (i.e. position of lava flows). The most divergent populations in terms of mitochondrial sequence are completely introgressed morphologically with populations further north. Although the morphologically differentiated southern Caribbean coast populations do belong to a single lineage, it appears to be relatively recently derived, and its only peculiar feature is that it has a lower haplotype diversity compared with other lineages, but this may simply be a consequence of the somewhat lower population densities in this part of the island.

Moreover, when a ‘phylogenetic’ distance matrix derived from this sequence information was added to partial matrix correspondence tests comparing generalized morphological (47 characters of females) with generalized ecology and geographic proximity, then morphology remains associated with ecology ($P < 0.0001$ with both the inclusion and exclusion of phylogeny), whereas generalized morphology is not closely associated with phylogeny ($P = 0.0168$), or proximity ($P = 0.2385$). The null hypotheses of no association has to be accepted for the last two variables after Bonferroni correction across the three independent variables ($0.05/3 = 0.0167$). Each character system is associated with an aspect of the ecology. This association is not changed when a partial matrix correspondence test is used to test each system in turn against the pertinent ecological factor together with geographic proximity and DNA phylogeny, i.e. colour pattern and body proportions are still associated with vegetation ($P < 0.0001$, $P < 0.0011$ respectively), and scalation is associated with rainfall ($P < 0.0001$). These tests show no association between the DNA phylogeny and scalation ($P = 0.72$), or body proportions ($P = 0.72$), although there is a significant (but slight) association between DNA phylogeny and colour pattern ($P = 0.0151$) even after Bonferroni correction.

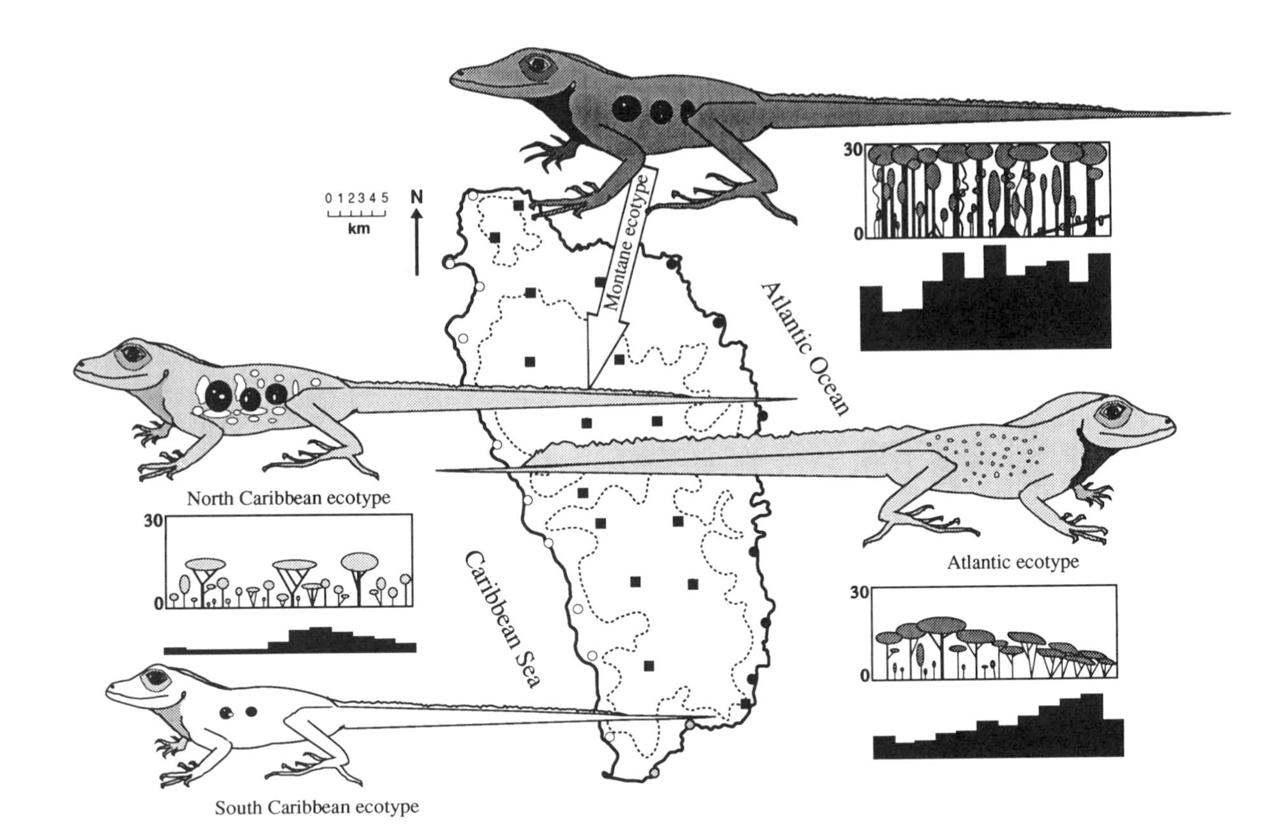

Fig. 5.3 Map of Dominica, indicating the sites from which *Anolis oculatus* morphology and DNA were sampled. The type of vegetation at these sites is illustrated by the following symbols: a filled square indicates evergreen rainforest, a filled circle indicates evergreen littoral woodland, an empty circle indicates seasonal xeric woodland and a grey circle indicates a transition between the latter two. The broken line on the map indicates the 300 m contour. Around the map some of the visually obvious geographic variation in the size, shape, and colour pattern of the lizards is illustrated; there is also significant variation in other characters such as number of scales around the body. A diagrammatic representation of the canopy height (in metres) and structure in the different vegetation zones associated with these ecotypes is given and barcharts illustrate the relative amount and seasonal distribution of rainfall (Lang 1967).

Consequently, the population phylogeny may contribute something to the understanding of the morphological differentiation, but the overwhelming factor appears to be natural selection for current ecological conditions.

5.3 Parallels

Several species on the same island may show similar patterns of geographic variation. For example in Tenerife the lacertid, *G. galloti* (Thorpe and Brown 1989, 1991), the gecko, *Tarentola delallandi* (Thorpe 1991), and the skink, *Chalcides viridanus* (Brown *et al.* 1993) show similar latitudinal patterns. In Dominica the anole, *Anolis oculatus* (Malhotra and Thorpe 1991), the iguana, *Iguana delicatissima* (Day unpublished data), and the ground lizard, *Ameiva fuscata* (Malhotra and Thorpe 1995) show, to varying degrees, a longitudinal pattern of geographic variation with differences between the Caribbean and Atlantic coasts. Similarly, on Gran Canaría the lacertid, *G. stehlini* (Thorpe and Baez 1993), the gecko, *Tarentola boettgeri* (Thorpe unpublished data) and the skink, *C. sexlineatus* (Brown and Thorpe 1991*a*, *b*) all show a latitudinal pattern of geographic variation.

These similar patterns within a single island are, on their own, of limited utility in differentiating between historical vicariance and selection for current ecological conditions. This is because organisms on a common island may have been subject to both the same historical/geological processes at a given time, and the same ecological differentiation. However, when there are islands within archipelagos which have independent histories but common ecological zonation, then parallel patterns of geographic variation and directions of character state change, among different species on different islands, argue for natural selection for adaptation to this zonation because it is the ecological zonation they have in common, not their internal history. Two examples have been elucidated using lizards, the Canary Island skinks and the Lesser Antillean anoles.

Canary Island skinks

The high-elevation islands of Tenerife and Gran Canaria, in the Canarian archipelago, both receive wind-borne rain predominantly from the north. This results in similar ecological zonation in both islands, with a lush, warm, humid habitat on the north-facing slopes (below the 1500 m inversion level in Tenerife) and a barren, hot, arid habitat in the south. The skink, *C. sexlineatus*, shows very pronounced geographic variation across 47 localities within Gran Canaria, in its scalation (Brown and Thorpe 1991*a*), body shape and size (Brown and Thorpe 1991*a*), and colour pattern (Brown and Thorpe 1991*b*). Similarly, the skink, *C. viridanus*, shows geographic variation across 17 localities within Tenerife (Brown *et al.* 1993). The colour pattern variation is particularly noticeable, and in both islands the pattern of geographic variation is shown only to be associated with these climatic biotopes when PRMC tests are employed (Brown *et al.* 1991). In both islands, skinks (both males and females) from the north have brown tails, while those in the south have bright blue tails. Tail autotomy is known to be an antipredator mechanism in lizards, which may be more effective when the tail is conspicuously coloured (Cooper and Vitt 1986). This parallel change in character state, in concert with parallel change in ecological conditions on two independent islands, provides support for the role of natural selection in adapting tail colour for different antipredator strategies in the different habitat types.

Lesser Antillean anoles

The central Lesser Antilles are a series of high-altitude islands with parallel ecological zonation that offers opportunities to investigate parallel patterns in their endemic anoles. Parallel patterns in the morphology of *Anolis oculatus* on Dominica and *Anolis marmoratus* on the neighbouring island of Basse Terre (Guadeloupe) were investigated. Basse Terre belongs to the same period of orogenesis as Dominica and is very similar in topography, climate, and vegetation. In essence, it is a mirror image of Dominica, as the highest mountain is in the south of Basse Terre, but in the north of Dominica. Thus the rain shadow effect results in the south-west coast of Basse Terre and the north-west coast of Dominica being the driest regions of the respective islands. As well as being closely related, these anoles are ecologically similar. Both are solitary species which are widely distributed in a number of different habitats and show a wide range of morphological variation.

The 17 morphological characters (colour pattern, scalation, and body dimensions) recorded were selected from an analysis of morphological variation in *A. oculatus* on the basis of high between-locality *F*-ratios and large squared multiple correlations with ecological variables. Finally, characters showing clear homologies with *A. oculatus* were favoured. This was particularly relevant for colour pattern characters. Males from 25 localities in Basse Terre were investigated and morphological distance matrices representing multivariate generalizations of the three character systems were derived. These dependent variable matrices were each compared with several independent variable matrices representing environmental variation (as described above) using a PRMC test.

The relationship between morphology and environmental variation found in *A. marmoratus* was similar to those found in *A. oculatus*. In both species, generalized scalation is correlated with rainfall, and generalized colour pattern with vegetation type ($P < 0.0005$). There are also parallel state changes in individual characters; body size is significantly correlated with both altitude and rainfall, the number of body scales is significantly correlated with rainfall and altitude, and the number of spots with the occurrence of dry scrub woodland (Malhotra and Thorpe 1994).

The parallel variation strongly suggests that natural selection is responsible for determining morphological geographic variation in these anoles. It may also give us some insight into the cause of differentiation of southern Caribbean coast populations of *A. oculatus* in Dominica, as a corresponding parallel cline is observed on the Caribbean coast of Basse Terre (Malhotra and Thorpe 1994). This suggests a non-historical cause common to both species, and raises the possibility that an ecological factor that is important to the lizard, but is not obvious to humans, does vary along the Caribbean coast of both islands. An intriguing twist to this parallel variation is added by the parallel variation in cytochrome b sequence of these two species along the

Caribbean coast (Malhotra and Thorpe 1994). The sequence variation is congruent with some morphological and ecological clines. At first sight this conflicts with the conventional interpretation of mtDNA variation as reflecting historical changes rather than selection effects. However, a fuller understanding will require a detailed study of cytochrome b variation across both islands.

Other islands in the Lesser Antilles present the opportunity for further tests of parallel variation. A series of islands of independent origin, sharing the same climatic patterns, and having a single endemic species of anole, are present. However, the orogenic history is more complex for some (e.g. St Lucia, Martinique) and the contrast between habitat types may be less pronounced for smaller, less mountainous islands (e.g. Montserrat). Nevertheless, preliminary results from studies in progress on these islands indicate that there are some parallels in common across all these species. For example, male *Anolis luciae* (St Lucia) show a parallel association between patterns of variation in generalized colour pattern and vegetation ($P < 0.0100$) as do both sexes of *Anolis lividus* (Montserrat) ($P < 0.0052$ males, $P < 0.0004$ females). The latter species also shows a parallel association between scalation and moisture levels in females ($P < 0.0090$).

5.4 Field experiments on selection

Relatively few rigorously tested, direct demonstrations of current selection in natural populations exist (Endler 1986; but see Endler 1980; Halkka and Raatikainen 1975; Knights 1979; Price *et al.* 1984). The aim of this manipulative field experiment was to provide such a demonstration of the action of natural selection on morphological variation in *A. oculatus*.

Anolis oculatus, while being a relatively *k*-selected anole (Andrews 1979), has a relatively short generation time, with juveniles reaching sexual maturity in under a year. Other features of its population structure that make it especially suitable for such a study are its extremely high population density, territorial behaviour, and striking degree of phenotypic variation. Since the geographic variation in morphology relates to the four ecological zones the extremes of the continuum are referred to as ecotypes. The relative ecological difference between the habitats indicates the north and south Caribbean coast habitats are very similar, and the Atlantic coast habitat is somewhat intermediate between the former and that of the montane habitat.

In this experiment, large-scale lizard-proof enclosures were constructed in two different habitats. Four enclosures were constructed (Malhotra and Thorpe 1993) in xeric woodland on the northern Caribbean (west) coast, and samples of four source populations (representing the four ecotypes) were translocated into these enclosures. One enclosure contained a 'resident' control, which was subjected to the same procedures as translocated 'foreign' ecotypes. Similarly, two enclosures were constructed on the Atlantic (east) coast containing one resident control population and one translocated population from the Caribbean coast, thus providing a partial reciprocal experiment. Before marking and releasing into the appropriate enclosure, 10 morphological characters (Malhotra and Thorpe 1991*b*) were recorded from each lizard (which was individually marked by toe clipping). This multivariate phenotypic profile was later used to compare morphology of survivors and non-survivors (lizards were not re-measured). The west-coast enclosures (1 to 4) were stocked in June/July 1990 (at the start of the wet season) and monitored in September 1990. The two east-coast enclosures (5 and 6) were stocked in September 1990 and monitored in February 1991.

Multivariate morphological differences between survivors and non-survivors were tested. As the morphological differences between sexes and between ecotypes also needed to be taken into account, a three-way multivariate analysis of variance (MANOVA) was used initially for the west-coast experiment. The model included interactions between sex (male or female), survival (survivor or non-survivor), and ecotype (north Caribbean, south Caribbean, montane, Atlantic). The interaction between survival and ecotype reveals whether the magnitude of morphological difference between survivors and non-survivors is greater in some enclosures than others. A canonical variate analysis was then performed on all groups for each experiment (4 ecotypes × 2 sexes × survival/non-survival = 16 groups for the west-coast experiment and 8 groups for the east-coast experiment). Since the ecotypes differ considerably in size, a possible bias may be introduced into the analysis. Although canonical analysis takes into account the intercorrelating effects of size, this was checked by size-adjusting the linear measurements prior to the analysis, and repeating this with SVL included and excluded. In all cases, the results were unaltered (Thorpe and Malhotra 1992).

West-coast experiment

The results of a three-way MANOVA show a highly significant interaction exists between enclosure and survival versus non-survival ($P < 0.001$). This implies the existence of varying selection intensity between the ecotypes. The multivariate distance (Mahalonobis D^2) between the morphology of survivors and non-survivors of each ecotype was obtained from the canonical analysis. After only two months, the montane population was already showing significant differences between morphology of survivors and non-survivors ($P < 0.01$ for males and $P < 0.001$ for females) (Malhotra and Thorpe 1991*b*).

In order to examine the relationship between selection intensity (represented by the extent of morphological separation between survivors and non-survivors of each ecotype) and the extent of ecological change experienced by the translocated populations more rigorously, the ecological dissimilarity was plotted against the morphological dissimilarity (D^2) averaged across the sexes (see also Malhotra and Thorpe 1991*b*). A curve of best fit to the four data points gave a correlation of 1.0, which is significant ($P < 0.01$) even with the one degree of freedom left by adopting this curvilinear model (Thorpe and Malhotra 1992). This suggests that the intensity of selection on the different population was strongly dependent on the magnitude of ecological change experienced.

East-coast experiment

Few animals in the transferred population survived until the first monitoring session. Even so, in males, there is a significant difference between the morphology of survivors and non-survivors of the translocated north Caribbean ecotype ($D^2 = 8.46$, $P < 0.05$), but not in the control littoral woodland ecotype ($D^2 = 1.20$, $P > 0.05$) (Malhotra 1992).

There is much discussion of the role, mode of action, and rate of natural selection in evolution (Endler 1989). This experiment, designed to run over a long-term period, unexpectedly demonstrated that significant mortality selection can occur over a very short time scale within single generations of perturbed populations. Two points argue strongly for the differences among ecotypes being maintained by natural selection for current ecological conditions. First, both the west- and east-coast experiments indicate a significant difference in morphology of survivors and non-survivors of critical translocated ecotypes but no difference in the control ecotypes. Second, the west-coast experiment indicates a correlation between the extent of morphological difference between survivors and non-survivors and the extent of ecological difference between the enclosure habitat and the habitat from where they were translocated.

5.5 Conclusions

Although explicit rules can be used to hypothesize an inter-island colonization sequence from a molecular phylogeny, and statistical methods can be used to assist in partitioning historical factors from current selective factors, this latter process has limited efficacy where there are a limited number of islands in inter-island studies. Studying numerous populations within small, but heterogeneous, islands allows a better understanding of the factors causing geographic variation.

This overview reveals that even on small islands a species may show pronounced morphological differentiation and distinct, sometimes deep, molecular phylogenetic divisions. A more complete understanding of nature of the geographic variation comes from a combination of molecular and morphological studies. However, patterns may only be revealed by sampling at numerous localities across the entire island. Sampling restricted areas (Malhotra and Thorpe 1994) may not be capable of fully revealing the pattern or underlying process. Coarse sampling, or using conventional subspecies as operational entities is also unlikely to give a sufficiently detailed picture to be of much value and may be positively misleading.

Three main approaches have been used to study within-island geographic variation using lizards on Canarian and Lesser Antillean archipelagos: (1) matrix correspondence tests and their partial regression/correlation extensions on morphological and molecular data; (2) identification of within island patterns of morphological geographic variation paralleled on independent islands; and (3) large-scale field experiments on selection. These studies reveal that, even on small islands, 'island populations' may not be homogeneous in morphology, or molecular phylogeny, and that natural selection for current ecological conditions appears to be a primary force influencing morphological population differentiation, irrespective of phylogenetic history.

5.6 Summary

The sequence of inter-island colonization events can be hypothesized from a molecular phylogeny with the aid of explicit rules. Statistical methods can be used to assist in partitioning historical factors from current selective factors, but they are of limited efficacy when the number of islands is small. Studying numerous populations within small, but heterogeneous, islands allows a better understanding of the factors causing geographic variation. Three main approaches have been used to study within-island geographic variation in lizards on Canarian and Lesser Antillean archipelagos: (1) matrix correspondence tests and their partial regression/correlation extensions on morphological and molecular data; (2) identification of within-island patterns of morphological geographic variation in parallel on independent islands; and (3) large-scale field experiments on selection. These studies reveal that, even on small islands, 'island populations' may not be homogeneous in morphology, or molecular phylogeny, and that adaptation to current ecological conditions appears to be a primary force influencing morphological population differentiation, irrespective of phylogenetic history.

Acknowledgements

We thank R. P. Brown, H. Black, M. Day, N. Giannasi, and M. Harris for information and NERC, SERC/BBSRC, Royal Society, Carnegie, Bonhote, and Percy Sladen trusts for financial support.

References

Ancochea, E., Fuster, J. M., Ibarrola, E., Cendrero, A., Coello, J., Hernan, F., Canatgrei, J. M., and Jamond, C. (1990). Volcanic eruption of the island of Tenerife (Canary Islands) in the light of new K-Ar data. *Journal of Volcanology and Geothermal Research*, **44**, 231–49.

Andrews, R. M. (1979). Evolution of life histories: a comparison of *Anolis* lizards from matched islands and mainland habitats. *Breviora*, **454**, 1–51.

Brown, R. P. and Thorpe, R. S. (1991*b*). Within-island microgeographic variation in the colour pattern of the skink, *Chalcides sexlineatus*: pattern and cause. *Journal of Evolutionary Biology*, **4**, 557–74.

Brown, R. P., and Thorpe, R. S. (1991*a*). Within-island microgeographic variation in body dimensions and scalation of the skink *Chalcides sexlineatus*, with testing of causal hypotheses. *Biological Journal of the Linnean Society*, **44**, 47–64.

Brown, R. P., Thorpe, R. S., and Baez, M. (1991). Lizards on neighbouring islands show parallel within-island micro-evolution. *Nature*, **352**, 60–2.

Brown, R. P., Thorpe, R. S., and Baez, M. (1993). Patterns and causes of morphological population differentiation in the Tenerife skink, *Chalcides viridanus*. *Biological Journal of the Linnean Society*, **50**, 313–28.

Carracedo, J. R. (1979). *Paleomagnetismo e historia volcanica de Tenerife*. Aula de Cultura de Tenerife, Santa Cruz de Tenerife.

Castellano, S., Malhotra, A., and Thorpe, R. S. (1994). Within-island geographic variation in the dangerous Taiwanese snake *Trimeresurus stejnegeri*, in relation to ecology. *Biological Journal of the Linnean Society*, **52**, 365–75.
Cooper, W. E. and Vitt, L. J. (1986). Blue tails and autotomy: enhancement of predator avoidance in juvenile skinks. *Zeitschrift für Tierpsychologie*, **70**, 265–76.
Daltry, J., Wüster, W., and Thorpe R. S. (1996). Diet and snake venom evolution. *Nature*, **379**, 537–40.
Darwin, C. (1859). *On the origin of species by means of natural selection*. John Murray, London.
Douglas, M. E. and Matthews W. J. (1992). Does morphology predict ecology? Hypothesis testing within a freshwater stream fish assemblage. *Oikos*, **65**, 213–24.
Endler, J. A. (1980). Natural selection on colour patterns in *Poecilia reticulata*. *Evolution*, **34**, 76–91.
Endler, J. A. (1986). *Natural selection in the wild*. Princeton University Press.
Endler, J. A. (1989). Problems in speciation. In *Speciation and its consequences*, (ed. D. Otte and J. A. Endler), pp. 625–48. Sinauer, Sunderland, MA.
Gardner, A. S. (1986). Morphological evolution in the day gecko *Phelsuma sundbergi* in the Seychelles—a multivariate study. *Biological Journal of the Linnean Society*, **29**, 223–44.
Halkka, O. and Raatikainen, M. (1975). Transfer of individuals as a means of investigating natural selection in action. *Hereditas*, **80**, 27–34.
Juan C., Oromi, P., and Hewitt, G. M. (1995). Mitochondrial DNA phylogeny and sequential colonization of Canary Islands by darkling beetles of the genus *Pimelia* (Tenebrionidae). *Proceedings of the Royal Society of London* B, **261**, 173–80.
Knights, R. W. (1979). Experimental evidence for selection on shell size in *Cepaea hortensis* (Mull.). *Genetica*, **50**, 51–60.
Lang, D. M. (1967). *Soil and land use surveys, No. 21, Dominica*. University of the West Indies, Trinidad.
Malhotra, A. (1992). What causes geographic variation: A case study of *Anolis oculatus*. Unpubl. Ph.D Thesis, University of Aberdeen.
Malhotra, A. and Thorpe, R. S. (1991*a*). Microgeographic variation in *Anolis oculatus* on the island of Dominica, West Indies. *Journal of Evolutionary Biology*, **4**, 321–35.
Malhotra, A. and Thorpe, R. S. (1991*b*). Experimental detection of rapid evolutionary response in natural lizard populations. *Nature*, **353**, 347–8.
Malhotra, A. and Thorpe, R. S. (1993). An experimental field study of a eurytopic anole, *Anolis oculatus*. *Journal of Zoology (London)*, **229**, 163–70.
Malhotra, A. and Thorpe, R. S. (1994). Parallels between island lizards suggests selection on mitochondrial DNA and morphology. *Proceedings of the Royal Society of London* B, **257**, 37–42.
Malhotra, A. and Thorpe, R.S. (1995). *Ameiva fuscata*. *Catalogue of American Amphibians and Reptiles*, No. 606, 1–3.
Malhotra, A. and Thorpe, R. S. (1997*a*). Microgeographic variation in scalation of *Anolis oculatus* (Dominica, West Indies): A multivariate analysis. *Herpetologica*, **53**, 49–62.
Malhotra, A. and Thorpe, R. S. (1997*b*). Size and shape variation in a Lesser Antillean anole, *Anolis oculatus* (Sauria: Iguanidae) in relation to habitat. *Biological Journal of the Linnean Society*, **60**, 53–72.
Manly, B. F. J. (1986*a*). *Multivariate statistical methods: a primer*. Chapman and Hall, London.
Manly, B. F. J. (1986*b*). Randomization and regression methods for testing associations with geographical, environmental and biological distances between populations. *Researches on Population Ecology*, **28**, 201–18.
Manly, B. F. J. (1991). *Randomization and Monte Carlo methods in biology*. Chapman and Hall, London.

Prentice, H. C., Lonn, M., Lefkovitch, L. P., and Runyeon, H. (1995). Associations between allele frequencies in *Festuca ovina* and habitat variation in the Alvar grassland on the Baltic island of Öland. *Journal of Ecology*, **83**, 391–402.

Price, T. D., Grant, P. R., Gibbs, H. L., and Boag, P. T. (1984). Recurrent patterns of natural selection in a population of Darwin's finches. *Nature,* **309**, 787–9.

Smouse, P. E. and Long, J. C. (1992). Matrix correlation analysis in anthropology and genetics. *Yearbook of Physical Anthropology*, **35**, 187–213.

Smouse, P. E., Long, J., and Sokal, R. R. (1986). Multiple regressions and correlation extensions of the Mantel test of matrix correspondence. *Systematic Zoology*, **35**, 627–32.

Snell, H. L., Snell, H. M., and Tracy, C. R. (1984). Variation among populations of Galápagos land iguanas (*Conolophus*); contrasts of phylogeny and ecology. *Biological Journal of the Linnean Society*, **21**, 185–207.

Sokal, R. R., Oden, N. L., and Wilson, C. (1991). Genetic evidence for the spread of agriculture in Europe by demic diffusion. *Nature*, **351**, 143–5.

Thorpe, R. S. (1991). Clines and cause: microgeographic variation in the Tenerife gecko *Tarentola delalandii. Systematic Zoology*, **40**, 172–87.

Thorpe, R. S. (1996). The use of DNA divergence to help determine the correlates of evolution of morphological characters. *Evolution*, **50**, 524–31.

Thorpe, R. S. and Baez, M. (1987). Geographic variation within an island: univariate and multivariate contouring of scalation, size and shape of the lizard *Gallotia galloti. Evolution*, **41**, 256–68.

Thorpe, R. S. and Baez, M. (1993). Geographic variation in scalation of the lizard *Gallotia stehlini* within the island of Gran Canaria. *Biological Journal of the Linnean Society*, **48**, 75–87.

Thorpe, R. S. and Brown, R. P. (1989). Microgeographic variation in the colour pattern of the lizard *Gallotia galloti* within the island of Tenerife: distribution, pattern and hypothesis testing. *Biological Journal of the Linnean Society*, **38**, 303–22.

Thorpe, R. S. and Brown, R. P. (1991). Microgeographic clines in the size of mature male *Gallotia galloti* (Squamata: Lacertidae) on Tenerife: causal hypotheses. *Herpetologica*, **47**, 28–37.

Thorpe, R. S. and Malhotra, A. (1992). Are *Anolis* lizards evolving? *Nature*, **355**, 506.

Thorpe, R. S., McGregor, D. P., Cumming, A. M., and Jordan, W. C. (1994*a*). DNA evolution and colonization sequence of island lizards in relation to geological history: mtDNA RFLP, cytochrome b, cytochrome oxidase, 12s RRNA sequence, and nuclear RAPD analysis. *Evolution*, **48**, 230–40.

Thorpe, R. S., Brown, R. P., Day, M. L., MacGregor, D. M., Malhotra, A., and Wüster, W. (1994*b*). Testing ecological and phylogenetic hypotheses in microevolutionary studies: An overview. In *Phylogenetics and ecology*, (ed. P. Eggleton and R. Vane-Wright), pp. 189–206. Academic Press, London.

Thorpe, R. S., Malhotra, A., Black, H., Daltry, J. C., and Wüster, W. (1995). Relating geographic pattern to phylogenetic process. *Philosophical Transactions of the Royal Society of London* B, **349**, 61–8.

Thorpe, R. S., Black, H., and Malhotra, A. (1996). Matrix correspondence tests on the DNA phylogeny of the Tenerife Lacertid elucidates both historical causes and morphological adaptation. *Systematic Biology*, **45**, 335–43.

Waddle, D.M. (1994). Matrix correlation tests support a single origin for modern humans. *Nature*, **368**, 452–4.

6

Speciation

Peter R. Grant

6.1 Introduction

The biology of islands is the biology of isolation, and isolation is conducive to speciation. How does speciation occur? This is one of the most enduring questions in evolutionary biology, perhaps *the* most enduring, because it is an historical question and because it is complex. Its elements are how populations diverge under selection and drift, where this happens, how the diverging populations become reproductively isolated from each other, and how and when this happens. In other words there are components of geography and history, of genetics, behaviour, and ecology.

Studies of organisms on islands have contributed to an understanding of all of these components. They are strongly influenced by the ideas of Sewall Wright on the dynamics of genes in subdivided populations, that is in insular demes loosely connected by occasional migration. The present chapter offers a discussion and integration of the major issues raised in the following five chapters. Those chapters develop the theoretical ideas and empirical aspects of speciation on islands in more detail. All authors adopt the biological species concept without being concerned to draw a precise line between populations of the same species and populations of different species that occasionally interbreed. Emphases of the chapters differ.

Since genetic processes are at the core of the speciation process, and genetic processes provide the link with the preceding chapters on microevolution, I start with the theory that is most explicit about genetics, the theory of founder effect speciation. A few words on the historical context will help place it in perspective. The idea of founder effects traces back ultimately to Wright's shifting balance theory of evolution (Provine 1989), and his use of an adaptive landscape metaphor to depict variation in fitness among genotypes (gene combinations) in a population. In Wright's famous diagrams the landscape is three-dimensional; a plane of gene combinations is contoured with multiple peaks and valleys of fitness (Wright 1932; discussed extensively by Provine 1986). A population is held by stabilizing selection on one peak, perhaps for a very long time, then shifts to an unoccupied and higher neighbouring peak through accidental crossing of the intervening valley by random drift. This is a model of evolution, not of speciation. Nevertheless founder effect theories of speciation incorporate the central notion of random drift producing new combinations of interacting genes which are then operated on by selective agents.

For speciation to occur an important question (considered in Chapter 7) is how reproductive isolation arises between an ancestral population, which can be thought of as remaining on peak 1, and the derived and newly founded population on peak 2. The relevant genetic changes are those that prevent an exchange of genes, through either pre-mating or post-mating mechanisms. Genetic differences between closely related species in those mechanisms give us the best indication of what those changes must have been during speciation (e.g. see Palopoli and Wu 1994; Coyne 1996).

6.2 Founder effect speciation: background and theory

The basic idea of founder speciation as developed by Mayr (1942, 1954, 1992) is that an island is colonized by a small number of individuals of a species, and the population rapidly undergoes a genetic reorganization with the result that it becomes reproductively isolated from its parental population. The key features of the process are a reduction in genetic variation under the starting conditions of low population size, a further reduction by drift and by selection against deleterious recessive alleles in homozygotes due to inbreeding, and, as a result of these losses, the formation of new combinations of (coadapted) genes that are subject to selection (see also Chapter 3). Thus the theory combines geographical isolation, which permits divergent evolution, with the effects of very small population size, which promotes divergent evolution.

The basic idea has been modified in two ways. In Carson's (1968, 1975) flush–crash–founder variant natural selection is relaxed as a population rises in density (the flush) in a permissive environment. New combinations of strongly epistatic genes are produced by recombination, combinations that would be selected against in the ancestral population but persist here. The population then falls to near extinction (the crash). A small number of the new recombinants (the founders) survive the crash by chance, and these are preserved in high frequency as the population recovers and expands (see also Slatkin 1996). Selection on this newly constituted genetic variation operates as the population approaches a new stable density, favouring new sets of coadapted genes. The whole cycle of events may be repeated many times. Thus speciation occurs through a series of catastrophic, stochastic, genetic events that lead eventually to the formation and retention under natural selection of new coadapted gene complexes. This is the same endpoint as in Mayr's theory, reached by a different route and without an appreciable loss of genetic variation.

In the genetic transilience model of Templeton (1980, 1996) a key role in the alteration of the population's genetic constitution is played by a small number of genes (or segregating blocks of genes) of large individual effect and by their epistatic modifiers. Frequencies of alleles are abruptly changed in the founding event, resulting in strong selection at the modifier loci, and this is effective because genetic variation has not been reduced to a large extent (unlike in Mayr's scheme). The model is less general than Mayr's because it only applies to species with a certain genetic architecture. Unlike Mayr's, it does not involve a large-scale genome-wide reorganization.

Chapter 8 by Hollocher discusses and compares these models in more detail. All three models lay stress on a break-up of coadapted gene complexes in the founding event or subsequent population bottlenecks (see also Slatkin 1996). Since the reorganization takes place fairly rapidly it is likely to be completed by the time additional colonizations take place, hence it will not be impeded by gene flow. Experimental founding events created with laboratory populations of flies have provided some support for the idea of genetic reorganization (e.g. Powell 1978). For some quantitative morphological traits (though not all) additive genetic variation has been shown to increase through a bottleneck as a consequence of changes in epistatic variance (Bryant *et al.* 1986; Goodnight 1988; Carson and Wisotzkey 1989; Bryant and Meffert 1996; Cheverud and Routman 1996). This is relevant to speciation because a comparison of closely related *Drosophila* species suggests that the evolution of post-zygotic isolation involves epistatic effects (Palopoli and Wu 1994; Wu and Palopoli 1995), as anticipated by Dobzhansky (1937) and Muller (1940). Indeed it must involve epistasis (or dominance) if the negative effects of alleles are expressed in matings between, but not within, populations (Johnson and Wade 1996; see also Orr and Orr 1996). Thus although loss of neutral or near neutral alleles is a repeated feature of island populations (see Sections 2.4 and 3.3), and is in accord with Mayr's reasoning, evolutionarily important additive genetic variance is not necessarily lost and may even be gained (see also Chapter 4).

The idea that speciation occurs as a result of randomly induced genetic changes occurring during or soon after a colonizing event has been controversial for several reasons, both theoretical and empirical (Barton and Charlesworth 1984; Carson and Templeton 1984; Provine 1989; Rice and Hostert 1993; Templeton 1996). One difficulty is in deciding whether the hypothesized genetic reconstruction takes place, by whatever particular mechanism, and another is whether the reconstruction is sufficient by itself to produce reproductive isolation, or whether it simply initiates a process of divergence that culminates in reproductive isolation.

In Chapter 7 Barton examines the evolution of post-zygotic reproductive isolation in two Wright-inspired models, one with distinct peaks separated by a valley or trough in a fitness landscape, and the other with a ridge that connects two peaks. Conditions are found for the evolution of reproductive isolation as a result of founder effects where drift overrides selection, but in general the conditions in these models are restrictive, and are more favourable to selective mechanisms of divergence. It seems to me that the conditions would be less restrictive if the island population that experienced a founder effect had time to diverge further before encountering members of the ancestral population, or even members of another derived one. This would allow for the interplay of drift and selection that occurs in the models of founder speciation devised by its proponents, especially Carson's model. Also drift in a partly isolated hybrid population might establish a fit recombinant, and this could lead to speciation (Rieseberg 1995). Moreover Gavrilets and Hastings (1996) have successfully modelled founder effect speciation in terms close to the original Mayr conception. Barton's objection is that it is hard to see why changes in genetic background (i.e. epistasis), as opposed to simple changes in allele frequencies, should make the evolution of reproductive isolation (post-zygotic) more likely in this model.

6.3 Founder effect speciation: *Drosophila*

Empirical issues are addressed in two other chapters. The first (Chapter 8) compares speciation of *Drosophila* in the Hawaiian archipelago and in the Caribbean. The second examines the applicability of Mayr's original ideas to speciation of birds (Chapter 9).

Analyses of Hawaiian *Drosophila* have done much to stimulate founder effect speciation theories (Carson and Templeton 1984), so the major patterns are worth recounting. Molecular data and amber fossils give estimates of 20–40 Ma for the age of the Hawaiian Drosophilids (DeSalle 1995). They are monophyletic. No single continental taxon can be designated as the sister group to the Hawaiian lineages (DeSalle 1995). Since the oldest current island inhabited by *Drosophila* (Kaua'i) is little more than 5 Ma old the original colonization and diversification must have taken place on islands now partly or wholly submerged. The phylogenetically oldest species occur on the geologically oldest islands. The sheer number of species is extraordinary. More than 500 species in the family Drosophilidae from the Hawaiian archipelago have been named and described, another 250–300 have been collected but not described, and a further 200–250 are estimated to be awaiting collection and description (Kaneshiro *et al.* 1995).

The idea that speciation is associated with founder events comes from a reconstruction of the evolutionary history of the group (Carson 1990). Reconstruction is made possible by excellent geological knowledge of the islands coupled with unequalled detail of genetics. Giant polytene chromosomes in the cells of the salivary glands of mature *Drosophila* larvae are conspicuously banded, and rearrangements recognized by the banding patterns, principally paracentric inversions, have been used to trace the evolutionary history of groups of closely related species, one step at a time. The reconstructions reveal that most species demonstrate a strong tendency towards single-island endemism, their sister taxa generally occurring on adjacent islands. This geographical pattern has been interpreted as evidence that speciation is associated with the colonization of an island (Carson 1990), possibly early in the history of the island. Sister taxa on different islands tend to be similar ecologically but differ in behavioural traits associated with mate recognition, therefore sexual selection pressures may have been especially important in generating species diversity in this group (see Section 8.3).

How important have founding events been in the evolutionary diversification of Hawaiian *Drosophila*? The species appear to have the appropriate genetic architecture, a suitable (polygamous) mating system involving elaborate courtship behaviour (Giddings and Templeton 1983), and a history of inter-island colonization, yet it is difficult to answer the question for two reasons. First, many species show identical banding patterns in the polytene chromosomes, therefore speciation is not necessarily accompanied by chromosomal rearrangements. The history of these species is not easy to deduce. Second, many species have yet to be described. Nevertheless what is clear is that inter-island colonization has happened numerous times, and so has speciation. In addition some species may have been formed allopatrically within islands, perhaps also involving founder effects.

Caribbean *Drosophila* make an interesting comparison. There are 58 species in nine genera in contrast to 1000 species in just two genera in the Hawaiian archipelago. The maximum age of the islands is similar in the two archipelagos, and the maximum age of a *Drosophila* species resident there (>20 Ma) may be similar too. On the other hand the geological history is much more complex than is the Hawaiian history, the archipelago is much closer to a continent, and the islands are closer to each other. Owing to continental proximity there have been many more invasions and fewer speciation events (or many more, unrecorded, extinction events) in the Caribbean. Several signs suggest a lesser role for founder events in speciation there. Unlike the situation in Hawaii, the flies display clinal variation in morphological features (pigmentation of the abdomen) indicative of natural selection. The linear geographical variation in morphology contrasts with a nonlinear sequence of island colonization; colonization has not always followed a simple stepping-stone progression from one end of the island chain to the other. In general the different traits that distinguish the species do not couple tightly with the pattern of colonization. Although degree of reproductive isolation follows the pattern of colonization more linearly there are inconsistencies and the coupling is, once again, not tight. Finally sexual selection appears to have been less prevalent in the Caribbean than in the Hawaiian archipelago.

6.4 Founder effect speciation: birds

Bird species often vary in morphological features clinally and gradually over large continental regions, yet their island relatives are abruptly and markedly different from them (Mayr (1940, 1942). The theory of founder effect speciation was developed to explain the distinctiveness of island populations in terms of a reorganization of sets of epistatically interacting genes at the founding event (Mayr 1942, 1954, 1992). The founder principle has been found useful in explaining low allelic variation (Baker and Moeed 1987; Degnan 1993), and the evolution of dull plumage including neotenous characters (McDonald and Smith 1990) in small island populations.

A direct test of founder effect speciation in birds would involve genetic analysis of the phenotypic differences between an island species and its continental (or archipelago) relative, and a demonstration that different epistatic systems are responsible for the differences in phenotypes. This test has not been done. It could be accomplished with a captive breeding programme, taking advantage of the fact that speciation in birds on islands involves the development of pre-mating barriers to gene exchange (Section 9.6), which can be circumvented in captivity (e.g. Danforth 1950; Sharpe and Johnsgard 1966; Johnsgard 1973). Instead, what has been done repeatedly is to cross-breed continental species in captivity, for a variety of genetic and behavioural purposes. The results can be used for making inferences about speciation in continental taxa. Summarized, they show that size, plumage, and behavioural traits by which species differ are inherited in a predominantly additive fashion; there is little evidence of dominance and epistasis (Grant and Grant 1997; see also Grant and Grant 1994). In so far as speciation involves the same genetic changes on islands as on continents the results of captive breeding experiments do not support the theory of speciation by founder effects in birds. Chapter 9 considers other indirect evidence and concludes it is unnecessary to invoke founder effects to account for speciation of birds on islands. More traditional theories (see below) are sufficient.

6.5 Genetical and ecological theories of speciation

Founder effect speciation is a genetical theory. Behavioural or ecological factors are not irrelevant, but the prime movers in the divergence of a newly founded population are genetic changes that would take place whether or not ecological and behavioural factors remained unaltered. In contrast to genetical theories are theories that see forces of selection as driving the divergence of populations towards the point at which they are no longer capable of exchanging genes. Genetic factors are not irrelevant, but the prime role is played by natural selection or sexual selection acting on *existing* genetic variation and new variation produced by mutation. Inasmuch as the changes occur as a result of selection on phenotypes external to the genome they are exogenous theories. To dichotomize speciation theories and to characterize the two groups as endogenous (i.e. genetical) and exogenous (ecological and behavioural) is potentially misleading, since it implies a discrete difference where this is lacking (Mayr 1963; Carson 1978), nevertheless the categorization is convenient and useful for organizing a discussion of the relevant factors. The main question for all theories to answer is how differences between species in their phenotypes and genotypes are to be explained. Environmental theories attempt to determine, in addition, whether speciation can occur sympatrically or whether it only occurs partly or entirely in allopatry.

Wright's (1931, 1932) adaptive landscape for gene combinations is a useful concept for ecological and behavioural theories when translated into an adaptive landscape for phenotypic characters (Simpson 1944, 1953). Peaks in the landscape are local optima in phenotypic space created by ecological conditions (Lande 1976). A population occupies part of the topographic surface at or close to a peak, with directional selection taking the population towards the peak and frequency-dependent selection tending to take it slightly away from it (Lande 1976). A valley between two peaks is crossed by random drift. Environmental change can cause a reconfiguration of the surface of an adaptive landscape of phenotypes (Simpson 1944), just as it changes the relative heights of peaks and valleys in the fitness surface for gene combinations (see also Weber 1996).

Thus Wright's concept of a landscape or surface is useful in this metamorphosed form to biologists who have no knowledge of individual genes. This is a large advantage in the study of speciation on islands, where phenotypes and environments are far better known than are genes. Although not translatable into the mathematical theory of population genetics that inspired the concept (Provine 1986), adaptive landscapes of phenotypes are closer to common experience of the real world than are fitness surfaces of gene combinations. Like Wright's original concept, the adaptive landscape of phenotypes was not offered by Simpson as an explanation for speciation but as an explanation for how progressive, adaptive, evolution occurs. Nevertheless, with small modifications such a landscape serves the purpose well (see also Wright 1982).

For the study of speciation three features of the Simpsonian adaptive landscape are worth emphasizing. The first is the property of ridges (e.g. see Section 7.5). This is lacking in Simpson's original conception. Ridges facilitate directional evolution more than do isolated peaks. Divergence along two different ridge systems separated by increasingly deep valleys can lead to the incidental acquisition of reproductive isolating mechanisms, both pre-zygotic and post-zygotic. There is a parallel here with the Dobzhansky–Muller theory for the evolution of post-zygotic isolation (e.g. see Coyne 1992; Orr and Orr 1996).

The second feature is the property of peak mobility. For his landscape Wright (1932) emphasized that the environment, biotic and abiotic, was in a state of continual change. He interpreted this to mean that peaks and valleys were constantly changing elevation, and position to a minor extent, with the result that a species might change erratically from time to time without undergoing any net directional (adaptive) change. Simpson, while embracing the same idea of frequent change, introduced a new idea that is important for the study of speciation. This is the idea that peaks may change in their locations relative to each other, as well as in elevation. They become more or less accessible to a population. Simpson stressed greater accessibility, with a diagram (Simpson 1944, Fig. 13) showing that one peak could approach another so closely that the intervening valley would be occupied and crossed by the phenotypically most marginal members of the population. The result would be selection-driven occupation of the second peak by members of the population, isolated from other members on the first peak: two species formed from one. Simpson was not specific on the details of how the population was supposed to split, whether it would happen sympatrically (as implied) or allopatrically as described below. He was more concerned with the problem of explaining how a new adaptive zone would be entered rather than a small-scale niche shift. The important point is that an unoccupied peak could become occupied not simply as a result of a peak shift in a topography where peaks had fixed locations, but largely as a result of the peaks themselves shifting (changing), becoming closer to each other. Conversely one can imagine that two peaks closely positioned and occupied by a single population could move apart carrying with them increasingly isolated, divergently adapted, populations. The ridges referred to above might similarly change shape and position; ridges are not necessarily rigid.

There are two reasons for the peaks to shift. The first is a correlated effect of evolutionary change in a trait or traits not included in the construction of the phenotypic axes of the landscape, i.e. outside the system. This was Simpson's explanation. The second is a change in the external environment, causing changes in the fitness values associated with combinations of traits. Both mechanisms are potentially capable of explaining peak shifts without random drift (see also Price *et al.* 1993; Whitlock 1995, 1997). A change in the environment could cause peaks to join and separate, allowing a species to become the occupant of two peaks from an original one.

The third feature of the Simpsonian adaptive landscape is inter-island variation in landscape topographies. It is highly unlikely that two islands (or island and mainland) will have identical landscapes, and unlikely that those landscapes will change over time in an identical manner. For example, mainland–island differences in plant population breeding characteristics are a function of the geographical distance between them (Section 2.4), and although this could reflect a distance-dependent gene flow it could also mean that ecological similarity of environments decreases with distance. The degree to which the island environments differ will influence the likelihood that speciation will occur, for ecological reasons, following the colonization of one of them: allopatric speciation.

6.6 Ecology of speciation

Two chapters highlight the way in which environmental factors associated with insular or island-like environments guide the course of divergence in the speciation process. The standard allopatric model of speciation is outlined in Section 9.2 as it applies specifically to Darwin's finches on the Galápagos islands, and in Chapter 10 Schluter assembles the evidence for ecologically driven speciation in sticklebacks in post-glacial lakes.

Allopatric speciation is best visualized with the aid of patterns of morphological variation displayed by a group of species in archipelagos (Lack 1947; Amadon 1950; Bock 1970). Speciation involves divergence of two or more populations in allopatry, followed by the establishment of sympatry through dispersal of members of one population to an island occupied by another population. An important question to answer is this: does reproductive isolation evolve entirely in allopatry, or is it initiated in allopatry and completed in sympatry; and if the latter, are the differences that arise in allopatry reinforced in sympatry by selection that minimizes interbreeding or by selection that minimizes competition for ecological resources?

These questions are addressed in Chapter 9. Experiments with model specimens of Darwin's finches presented to reproductively active birds have simulated the invasion of an island by a differentiated relative of a resident species (Ratcliffe and Grant 1983, 1985). Residents generally showed a lack of discrimination between their own type and the invader, in contrast to the clear discrimination shown between own type and another congeneric resident species. These results demonstrate the potential for interbreeding at the secondary contact phase, and hence for reinforcing selection to occur. However, sympatric species that have been coexisting for, presumably, a very large number of generations still hybridize; moreover they do so with no loss of fitness under some ecological circumstances. These important observations imply that there is no basis for the hypothesized selection regime of reinforcing selection on courtship traits, and at the same time it shows that speciation does not run its full course in allopatry. Rather, speciation in these finches involves (1) allopatric divergence in resource exploiting traits (beaks), (2) further divergence of these traits in sympatry, (3) natural selection, in both allopatry and sympatry, that drives the population close to a local peak in an adaptive landscape determined by food supply (Schluter and Grant 1984; Schluter *et al.* 1985), and towards the peak in a new location when the food supply changes (Boag and Grant 1981; Price *et al.* 1984; Gibbs and Grant 1987; Grant and Grant 1995), and (4) pre-mating isolation that arises partly as a consequence of the adaptive divergence of these traits, and partly as a consequence of divergence in culturally inherited song characteristics.

Bird speciation on other islands around the globe is likely to follow a similar course of events. For example there are several islands populated by two or three species that are most closely related to a single continental species (Mayr 1942). Apparently each of these islands received double or triple invasions of the same ancestral species (Section 9.4). The resulting island species, like coexisting Darwin's finch species, display clear ecological segregation; in feeding niche, habitat, or some complementary combination of the two (Grant 1966). However, island species of birds elsewhere differ from Darwin's finches in one conspicuous respect; colour. Variation in plumage colour and pattern implies a role for sexual selection. Speciation in island birds around the world has involved a combination of sexual selection and natural selection (Grant and Grant 1997). A good example is the brightly coloured and ecologically diverse honey-creeper finches in the Hawaiian archipelago (Section 9.7).

Chapter 10 discusses the role of ecology in the speciation of fish in lakes. Co-existence of closely related, lacustrine stickleback species derived from a common (marine) ancestor is characterized by a high degree of ecological and reproductive isolation. Both types of biological separation have evolved fairly rapidly, though neither is complete. Niche differentiation is pronounced, but there is some degree of overlap. Similarly pre-mating isolation has evolved as a consequence of divergent selection between resource environments, yet some hybridization occurs. The species persist sympatrically despite gene flow between them, and despite the fertility of interspecific hybrids, because the hybrids are selected against under some ecological conditions. All these features are indicative of speciation being driven by ecological factors. In marked contrast to most of the literature on genetical models of speciation, Schluter concludes that knowledge of ecological environments is essential for an understanding of the origin of species in adaptive radiations (see also Littlejohn 1994; Schluter 1996).

Several post-glacial lakes have been colonized by marine sticklebacks of the same stock on two separate occasions (Section 10.2); the parallel with double invasion of islands by birds (Section 9.4) is striking. One invasion has given rise to a benthic form and the other has given rise to a limnetic form. Repeated entry into the same or similar environments should lead to replicated adaptive evolution and perhaps to speciation. Hubbs (1940) noted the phenomenon of parallel speciation in fish in an earlier argument for the importance of ecological factors in speciation, and Schluter and Nagel (1995) have narrowed and refined the concept. Parallel speciation occurs when reproductive isolation evolves between two descendant forms from a common ancestor that adapt to different environments (or equivalently, to the same environment in different ways); descendants adapting to similar environments do not become reproductively isolated (Section 10.3). Thus two species in a lake descended from the same ancestor become reproductively isolated as a consequence of adaptive divergence in niche-exploiting morphology and behaviour (e.g. benthics and limnetics); the same process and pattern is repeated in different lakes, similarly colonized by the same stock; and species so formed that occupy the same niche (e.g. benthic region) in different lakes are not reproductively isolated if and when they ever encounter each other.

A large advantage of studying speciation in these and other fish is their suitability for experimental investigation (Schluter 1996). They can be kept in small, ecologically realistic, re-creations of their natural environment, and manipulated. Predictions from speciation theories are therefore open to experimental test (Section 10.3).

6.7 Speciation without ecology, or without genes?

There are three exceptions to the rule that ecological and genetic factors combine to provide sufficient theories of speciation.

The first exception is a genomic theory of speciation. It leaves out ecology. This is the type of genetical theory in which intra- or inter-genomic interactions provide the basis for the evolution of reproductive isolation wholly independent of the external environment. For example the evolution of reproductive isolation has been attributed to mobile element transpositions (Fontdevila 1992), meiotic drive (Frank 1991; Hurst and Pomiankowski 1991) or to symbionts such as *Wollbachia* (Wade and Stevens 1985; Breeuwer and Werren 1990). Superficially they appear to be irrelevant to speciation on islands. However, there is much to be learned from detailed field studies of island and mainland populations of the same organisms, and their irrelevance may be only apparent, not real. For example the meiotic drive system in feral mice (*Mus musculus*) is sensitive to environmental factors, and especially to demography. Frequencies of the *t*-haplotypes involved in the drive system of mice differ between islands and mainland, perhaps as a result of different demographic processes (Anderson *et al.* 1964; Ardlie 1995).

The second exception is en epigamic theory of speciation in which divergent sexual selection pressures in geographically isolated populations favour divergent courtship signals and behaviour that are the basis of mate recognition systems (e.g. see Lande 1981). The traits are genetically based, unlike the situation in the theory below, and diverge as a result of mutation and recombination. The theory of epigamic speciation differs from ecological speciation in not specifying a role for natural selective pressures arising from the different resources in the separate environments. Divergence in ecologically significant traits may nonetheless occur as a result of random drift.

The third exception leaves out genetics in the crucial matter of mate recognition systems. This is a cultural speciation theory. Culturally transmitted behaviour acts as a barrier to interbreeding. Chapter 9 briefly refers to an example of culturally (not genetically) transmitted song characteristics of birds, and courtship responses to them, which act as a barrier to gene exchange (see also Grant and Grant 1998). In theory, divergence in mate recognition systems by a process of cultural drift could be sufficient to produce two non-interbreeding species from one. Genetic drift could cause divergence of ecologically significant traits, with the result that at secondary contact in sympatry the two populations would be reproductively and ecologically isolated from each other. While these two processes might be feasible in principle, I know of no situation where they are both likely to have occurred. The closest is Darwin's finches, but speciation in these is associated with adaptive divergence of ecologically (and reproductively) significant traits, in addition to culturally inherited patterns of reproductive behaviour (Section 9.6). Cultural traits may be important factors in speciation in the many other birds where sexual imprinting is known to affect mate choice (Ten Cate *et al.* 1993), and perhaps in other organisms too. A possible example is the pre-mating isolation between some *Drosophila* species that arises from being reared on different larval substrates (Brazner and Etges 1993). The possibility depends on whether choice of oviposition site is culturally inherited as a result of habitat imprinting or is genetically inherited.

6.8 Hybridization

When Mayr (1963) wrote his book on speciation in animals it was thought that hybridization in animals was much rarer and less important in the speciation process than in plants. While that contrast is still valid, numerous reports of hybridization in animals in recent years have led to a re-assessment and elevation of its evolutionary importance; for example in echinoderms (Byrne and Anderson 1994), coelenterates (corals) and other marine invertebrate phyla (Veron 1995), cladoceran crustacea (Schwenk and Spaak 1995), butterflies (Sperling 1990; Mallet *et al.* 1998), stick insects (Bullini and Nascetti 1990), snails (Woodruff 1989), fish (Smith 1992), amphibia (Dawley and Bogart 1989; Ptacek *et al.* 1994), lizards (Sites *et al.* 1990), birds (Grant and Grant 1992), and mammals (Gray 1972). As pointed out in Chapter 1, hybridization occurs moderately frequently on islands, which is to be expected given that it is most likely to occur in the early stages of speciation and adaptive radiation, and islands are arenas of both. The next three chapters discuss various aspects of hybridization in island birds (Chapter 9), lacustrine fish (Chapter 10), and island snails (Chapter 11).

Introgressive hybridization is widespread in birds, though rarely common; approximately one in ten of the global species are known to have hybridized in nature (Grant and Grant 1992), more have done so in captivity (Gray 1958). Hybrids and backcrosses are sometimes at little or no intrinsic fitness disadvantage for genetic reasons, but are more likely to suffer a disadvantage for ecological reasons, perhaps episodically (Grant and Grant 1993, 1996), or in competition for mates (Johnson and Johnson 1985). These facts suggest that speciation in birds involves the evolution of pre-mating barriers to gene exchange, with post-mating isolation evolving later, perhaps many millions of years later (Section 9.6). Species which shared a common ancestor more than 20 million years ago retain the capacity to hybridize (Prager and Wilson 1975), and some of the hybrids are capable of backcrossing.

At first sight warm-blooded birds on islands and cold-blooded fish in lakes would seem to have little in common other than their backbones. Nevertheless in their speciation they share many features, including the propensity to hybridize. Speciation of fish in post-glacial lakes of the north temperate zone has been especially well studied in three groups, coregonids (Svärdson 1970), other salmonids (Smith 1992), and sticklebacks (Schluter and McPhail 1992). These hybridize, and so do other groups of freshwater fish, including catastomids, cyprinids, and poeciliids (Hubbs 1955; Dowling *et al.* 1989; Smith 1992; Dowling and De Marais 1993). In sticklebacks hybridization is moderately frequent, the hybrids are fertile, and they are viable to a degree dependent on ecological context; they are selected against for ecological reasons since they are less efficient feeders than either parental species (Section 10.3). As in birds, pre-mating isolation, and niche differentiation that may or may not be directly associated with it, evolve much faster than post-mating isolation. These characteristics are not taxonomically or geographically restricted, as they appear to be true of some mammals (Patton and

Smith 1994), butterflies (Sperling 1990), and some of the Hawaiian drosophilids (Kaneshiro 1990).

How does introgressive hybridization contribute to the speciation process? No simple answer can be given to this question, as there are five different genetic effects of introgression:

1. Introgression of alleles with predominantly additive effects on phenotypes makes two species more similar, both genetically and phenotypically, than they would be in the absence of hybridization (Section 9.6).
2. Introgression elevates levels of heterozygosity, and thus may give rise to interspecific hybrid vigour (e.g. Lewontin and Birch 1966; Nagle and Mettler 1969), whether by true or apparent overdominance.
3. Introgression results in new combinations of genes producing novel phenotypic properties (Chiba 1993), and thus provides the opportunity for new epistatically interacting systems of genes to become established (Grant and Grant 1994).
4. Introgression elevates genetic variances of continuously varying traits, and it either increases or decreases genetic covariances depending upon the degree of similarity or difference in the genetic covariance structures of the hybridizing species (Section 9.5).
5. Introgression may enhance mutation and recombination (e.g. Woodruff 1989; Fontdevila 1992).

Thus introgressive hybridization may retard the process of divergence in the speciation process (1), and may even collapse it if strong enough; the outcome in the latter case is referred to as reticulate evolution. However, it may also lead to a transformation of the genetic characteristics of the interbreeding species (2–5), and depending upon demography and ecological context this could enhance and accelerate the speciation process (Sections 7.5 and 9.6). Perhaps rarest of all, for animals though not for plants (Rick 1963; Rieseberg 1995), hybridization of two rather different species could give rise to an intermediate third species with a distinctive ecology (Svärdson 1970; Sperling and Harrison 1994) and set of courtship traits (Wallace *et al.* 1983). This is equivalent to suggesting that two species on peaks spaced widely apart interbreed and produce hybrids that occupy an intermediate peak but do not interbreed with the parental species. A stronger possibility is that the hybrids have greater access to a fourth peak than do either of the parental species. This could lead to speciation if hybrids formed a moderately large fraction of the colonists of a new island, and on this island had a selective advantage as a result of their proximity to this peak.

It is important to know about the incidence and consequences of introgressive hybridization when attempting to reconstruct the evolutionary diversification of a group of organisms and the speciation that contributed to it. Such knowledge is needed to distinguish between competing hypotheses. Schluter provides a clear example in Section 10.2. Genetic similarity between coexisting species of fish can be explained as the result of either sympatric speciation or gene exchange following a second colonization of a lake (Bernatchez *et al.* (1996) for salmonids; Taylor *et al.* (1997) for sticklebacks). In Chapter 11 Clarke and colleagues consider in detail the effects of hybridization on the reconstruction of the evolution of *Partula* snails on Moorea, one of the Society islands in the south-western Pacific.

In situ (within island) speciation of *Partula* snails has been inferred from their allozyme similarities (Johnson *et al.* 1986, 1994); sister taxa of any one species often occur on the same island, and not like Hawaiian drosophilids on different islands. The interpretation of speciation in these snails is re-examined in the light of new findings implicating introgressive hybridization (Section 11.5). Genetic differences between two species are surprisingly small in the north of Moorea, despite marked morphological and ecological differences, and are much larger in the south. One explanation for the northern genetic similarity is occasional introgressive hybridization of neutral and advantageous alleles. This is plausible because several other pairs of *Partula* species hybridize without losing their morphological identity (integrity). However, there is no direct field evidence of hybridization between the members of this particular pair of species (*P. saturalis* and *P. taeniata*), nor do they hybridize in the laboratory. To add to the puzzle, the two species are of opposite coiling type in the north, and snails of opposite coiling type are known to be less likely to mate with each other than are snails of the same coiling type (Clarke and Murray 1969). They have the same (dextral) coiling type in the south.

The explanation for genetic similarity in terms of hybridization has two important and general implications. The first concerns speciation. Two coexisting species may become progressively more similar at neutral loci, although at a diminishing rate, as they become progressively less able to interbreed through the evolution of characters that reduce their ability or opportunity to mate with each other. This implies that genetic similarity is as much if not more a function of past introgressive hybridization as of recent hybridization. To judge from fossil evidence of freshwater snails, hybridization may continue for a million years or more (Geary 1992).

The second implication follows from the first, and concerns the attempt to reconstruct phylogeny; genetic similarity can mislead us about phylogenetic history in two ways when introgressive hybridization occurs, regardless of whether the reconstruction is based on genetic distances as in Chapter 11 or cladistic treatment of the genetic data. The first problem is with the age of estimation of the time in the past when two taxa diverged from a common ancestor; the point of coalescence backwards through time. Introgression yields genetic distances that are smaller than they would be in the absence of hybridization, so the estimates of the time since the split occurred are too short (unless introgression causes high rates of mutation). This holds even if hybridization resulted in introgression in the first few million years in the life of two sister taxa, and does not occur now or occurs only rarely. The second problem is that variable rates of introgression among related taxa will give variable degrees of genetic similarity attributable to secondary gene exchange, and this will confound attempts to reconstruct the exact sequence of phylogenetic events (e.g. DeSalle and Giddings 1986; Avise 1989, 1996; Patton and Smith 1994; Moore 1995).

Clarke and colleagues discuss the problem of finding phylogenetically useful parts of the genome in hybridizing species. They suggest the most reliable information for reconstructing phylogeny resides in the genes that are kept distinct in hybridizing species by natural selection. One possibility is to use genes that evolved distinctively in the taxa in allopatry, and others closely linked to them, which have remained unaffected by introgression after the subsequent establishment of sympatry. This may be possible with DNA preserved in fossils of some animals in the allopatric phase of their history. Taking the more accessible modern allopatric state (if it exists) as a proxy for historical allopatry is the only alternative, but would be fraught with difficulties of interpretation, because some of the adaptations in allopatry may have occurred recently.

Thus I am not optimistic that such genes (alleles) of pre-sympatry origin can be reliably identified. The challenge remains to devise a means of finding phylogenetically useful and reliable parts of the genome of hybridizing species. The problem is not restricted to the inhabitants of islands and island-like environments. Clearly there is a need to assess the effect of introgressive hybridization on the process of speciation and our reconstruction of phylogenies in all those radiations where hybridization is known to occur, regardless of the degree to which the species are genetically similar. This applies to the examples given in Chapters 8 to 11, others considered later in the book (Chapters 12, 14, and 17), and yet others not treated in the book, including the radiations of plants in Macaronesia (Francisco-Ortega *et al.* 1996) and the snails of Hawaii (Cowie 1995).

References

Amadon, D. (1950). The Hawaiian honeycreepers (Aves: Drepaniidae). *Bulletin of the American Natural History Museum*, **95**, 151–262.

Anderson, P. K, Dunn, L. C., and Beasley, A. B. (1964). Introduction of a lethal allele into a feral house mouse population. *American Naturalist*, **98**, 57–64.

Ardlie, K. G. (1995). The frequency, distribution, and maintenance of *t* haplotypes in natural populations of mice (*Mus musculus domesticus*). Unpublished Ph.D thesis, Princeton University.

Avise, J. C. (1989). Gene trees and organismal trees: a phylogenetic approach to population biology. *Evolution*, **43**, 1192–208.

Avise, J. C. (1996). Three fundamental contributions of molecular genetics to avian ecology and evolution. *Ibis*, **138,** 16–25.

Baker, A. J. and Moeed, A. (1987). Rapid genetic differentiation and founder effect in colonizing populations of Common Mynas (*Acridotheres tristis*). *Evolution*, **41**, 525–38.

Barton, N. and Charlesworth, B. (1984). Genetic revolutions, founder effects, and speciation. *Annual Review of Ecology and Systematics*, **15**, 133–64.

Bernatchez, L., Vuorinen, J. A., Bodaly, R. A., and Dodson, J. J. (1996). Allopatric origin of sympatric populations of lake whitefish (*Coregonus clupeaformis*) as revealed by mitochondrial-DNA restriction analysis. *Evolution*, **44**, 1263–71.

Boag, P. T. and Grant, P. R. (1981). Intense natural selection in a population of Darwin's finches (*Geospizinae*) in the Galápagos. *Science*, **214**, 82–5.

Bock, W. (1970). Microevolutionary sequences as a fundamental concept in macroevolutionary models. *Evolution*, **24**, 704–22.

Brazner, J. C. and Etges, W. J. (1993). Pre-mating isolation is determined by larval rearing substrates in cactophilic *Drosophila mojavensis*. II. Effects of larval substrates on time to copulation, mate choice and mating propensity. *Evolutionary Ecology*, **7**, 605–24.

Breeuwer, J. A. J. and Werren, J. H. (1990). Microorganisms associated with chromosome destruction and reproductive isolation between two insect species. *Nature*, **346**, 558–60.

Bryant, E. H. and Meffert, L. M. (1996). Nonadditive genetic structuring of morphometric variation in relation to a population bottleneck. *Heredity*, **77**, 168–76.

Bryant, E. H., McCommas, S. A., and Combs, L. M. (1986). The effect of an experimental bottleneck upon quantitative genetic variation in the housefly. *Genetics*, **14**, 1191–211.

Bullini, L. and Nascetti, G. (1990). Speciation by hybridization in phasmids and other insects. *Canadian Journal of Zoology*, **68**, 1747–60.

Byrne, M. and Anderson, M. J. (1994). Hybridization of sympatric *Patiriella* species (Echinodermata: Asteroidea) in New South Wales. *Evolution*, **48**, 564–76.

Carson, H. L. (1968). The population flush and its genetic consequences. In *Population biology and evolution*, (ed. R. C. Lewontin), pp. 123–37. Syracuse University Press.

Carson, H. L. (1975). The genetics of speciation at the diploid level. *American Naturalist*, **109**, 83–92.

Carson, H. L. (1978). Speciation and sexual selection in Hawaiian *Drosophila*. In *Ecological genetics: the interface*, (ed. P. F. Brussard), pp. 93–107. Springer, New York.

Carson, H. L. (1990). Evolutionary process as studied in population genetics: clues from phylogeny. *Oxford Surveys in Evolutionary Biology*, **7**, 129–56.

Carson, H. L. and Templeton, A. R. (1984). Genetic revolutions in relation to speciation phenomena: the founding of new populations. *Annual Review of Ecology and Systematics*, **15**, 97–131.

Carson, H. L. and Wisotzkey, R. G. (1989). Increase in genetic variance following a genetic bottleneck. *American Naturalist*, **134**, 668–71.

Cheverud, J. M. and Routman, E. J. (1996). Epistasis as a source of increased additive genetic variance at population bottlenecks. *Evolution*, **50**, 1042–51.

Chiba, S. (1993). Modern and historical evidence for natural hybridization between sympatric species in *Mandarina* (Pulmonata: Camaenidae). *Evolution*, **47**, 1539–56.

Clarke, B. and Murray, J. (1969). Ecological genetics and speciation in land snails of the genus *Partula*. *Biological Journal of the Linnean Society*, **1**, 31–42.

Cowie, R. H. (1995). Variation in species diversity and shell shape in Hawaiian land snails: in situ speciation and ecological relationships. *Evolution*, **49**, 1191–202.

Coyne, J. J. (1992). Genetics and speciation. *Nature*, **355**, 511–15.

Coyne, J. J. (1996). Genetics of differences in pheromonal hydrocarbons between *Drosophila melanogaster* and *D. simulans*. *Genetics*, **143**, 353–64.

Danforth, C. H. (1950). Evolution and plumage traits in pheasant hybrids, *Phasianus*x *Chrysolophus*. *Evolution*, **4**, 301–15.

Dawley, R. M and Bogart, J. P. (eds.) (1989). *Evolution and ecology of unisexual vertebrates*. New York State University Press, Albany, New York.

Degnan, S. (1993). Genetic variability and population differentiation inferred from DNA fingerprinting in silvereyes (Aves: Zosteropidae). *Evolution*, **47**, 1105–17.

DeSalle, R. (1995). Molecular approaches to biogeographic analysis of Hawaiian Drosophilidae. In *Hawaiian biogeography: evolution on a hot-spot archipelago*, (ed. W. L. Wagner and V. A. Funk), pp. 72–89. Smithsonian Institution Press, Washington, D. C.

DeSalle, R. and Giddings, L. V. (1986). Discordance of nuclear and mitochondrial DNA phylogenies in Hawaiian Drosophila. *Proceedings of the National Academy of Sciences USA*, **83**, 6902–6.

Dobzhansky, T. (1937). *Genetics and the origin of species*. Columbia University Press, New York.

Dowling, T. E. and De Marais, B. D. (1993). Evolutionary significance of introgressive hybridization in cyprinid fishes. *Nature*, **362**, 444–6.

Dowling, T. E., Smith, G. R., and Brown, W. M. (1989). Reproductive isolation and introgression between *Notropis cornutus* and *Notropis chrysocephalus* (family Cyprinidae): comparison of morphology, allozymes, and mitochondrial DNA. *Evolution*, **43**, 620–34.

Fontdevila, A. (1992). Genetic instability and rapid speciation: are they coupled? *Genetica*, **86**, 247–58.
Francisco-Ortega, J., Jansen, R. K., and Santos-Guerra, A. (1996). Chloroplast DNA evidence of colonization, adaptive radiation, and hybridization in the evolution of the Macaronesian flora. *Proceedings of the National Academy of Sciences USA*, **93**, 4085–90.
Frank, S. (1991). Divergence of meiotic drive-suppression systems as an explanation for sex-biased hybrid sterility and inviability. *Evolution*, **45**, 262–7.
Gavrilets, S. and Hastings, A. (1996). Founder effect speciation: a theoretical assessment. *American Naturalist*, **147**, 466–91.
Geary, D. H. (1992). An unusual pattern of divergence between two fossil gastropods: ecophenotypy, dimorphism, or hybridization? *Paleobiology*, **18**, 93–109.
Gibbs, H. L. and Grant, P. R. (1987). Oscillating selection on Darwin's finches. *Nature*, **327**, 511–13.
Giddings, L. V. and Templeton, A. R. (1983). Behavioral phylogenies and the direction of evolution. *Science*, **220**, 372–7.
Goodnight, C. J. (1988). Epistasis and the effects of founder events on the additive genetic variance. *Evolution*, **42**, 441–54.
Grant, B. R. and Grant, P. R. (1993). Evolution of Darwin's finches caused by a rare climatic event. *Proceedings of the Royal Society of London* B, **251**, 111–17.
Grant, B. R. and Grant, P. R. (1996). The diet of Darwin's Finch hybrids in relation to bill morphology. *Ecology*, **77**, 500–9.
Grant, B. R. and Grant, P. R. (1998). Hybridization and speciation in Darwin's Finches: the role of sexual imprinting on a culturally transmitted trait. In *Endless forms: species and speciation*, (ed. D. J. Howard and S. L. Berlocher). Oxford University Press.
Grant, P. R. (1966). Ecological compatibility of bird species on islands. *American Naturalist*, **100**, 451–62.
Grant, P. R. and Grant, B. R. (1992). Hybridization of bird species. *Science*, **256**, 193–7.
Grant, P. R. and Grant, B. R. (1994). Phenotypic and genetic effects of hybridization in Darwin's finches. *Evolution*, **48**, 297–316.
Grant, P. R. and Grant, B. R. (1995). Predicting microevolutionary responses to directional selection on heritable variation. *Evolution*, **49**, 241–51.
Grant, P. R. and Grant, B. R. (1997). Genetics and the origin of bird species. *Proceedings of the National Academy of Sciences USA*, **94**, 7768–75.
Gray, A. P. (1958). *Bird hybrids*. Commonwealth Agricultural Bureaux, Farnham Royal, UK.
Gray, A. P. (1972). *Mammal hybrids*. Commonwealth Agricultural Bureaux, Farnham Royal, UK.
Hubbs, C. L. (1940). Speciation of fishes. *American Naturalist*, **74**, 198–210.
Hubbs, C. L. (1955). Hybridization between fish species in nature. *Systematic Zoology*, **4**, 1–20.
Hurst, L. D. and Pomiankowski, A. (1991). Causes of sex ratio bias may account for unisexuality in hybrids: a new explanation of Haldane's rule and related phenomena. *Genetics*, **128**, 781.
Johnsgard, P. A. (1968). *Waterfowl: their biology and natural history.* University of Nebraska Press, Lincoln, NB.
Johnsgard, P. A. (1973). *The grouse and quail of North America.* University of Nebraska Press, Lincoln, NB.
Johnson, M. S., Murray, J., and Clarke, B. (1986). An electrophoretic analysis of phylogeny and evolutionary rates in the genus *Partula* from the Society Islands. *Proceedings of the Royal Society of London* B, **227**, 161–77.
Johnson, M. S., Murray, J., and Clarke, B. (1994). The ecological genetics and adaptive radiation of *Partula* on Moorea. *Oxford Surveys in Evolutionary Biology*, **9**, 167–238.
Johnson, N. A. and Wade, M. J. (1996). Genetic covariances within and between species: indirect selection for hybrid inviability. *Journal of Evolutionary Biology*, **9**, 205–14.

Johnson, N. K. and Johnson, C. B. (1985). Speciation in sapsuckers (*Sphyrapicus*): II. Sympatry, hybridization, and mate preferences in *S. ruber daggetti* and *S. nuchalis*. *Auk*, **100**, 1–15.

Kaneshiro, K. Y. (1990). Natural hybridization in Drosophila, with special reference to species from Hawaii. *Canadian Journal of Zoology*, **68**, 1800–5.

Kaneshiro, K. Y., Gillespie, R. G., and Carson, H. L. (1995). Chromosomes and male genitalia of Hawaiian *Drosophila*. In *Hawaiian biogeography: evolution on a hot-spot archipelago*, (ed. W. L. Wagner and V. A. Funk), pp. 57–71. Smithsonian Institution Press, Washington, D. C.

Lack, D. (1947). *Darwin's finches*. Cambridge University Press.

Lande, R. (1976). Natural selection and random genetic drift in phenotypic evolution. *Evolution*, **30**, 314–34.

Lande, R. (1981). Models of speciation by sexual selection on polygenic traits. *Proceedings of the National Academy of Sciences USA*, **78**, 3721–5.

Lewontin, R. C. and Birch, L. C. (1966). Hybridization as a source of variation for adaptation to new environments. *Evolution*, **20**, 315–36.

Littlejohn, M. J. 1994. Homogamy and speciation: a reappraisal. *Oxford Surveys in Evolutionary Biology*, **9**, 135–65.

Mallet, J., MacMillan, W. O., and Jiggins, C. D. (1998). Mimicry and warning colour at the boundary of microevolution and macroevolution. In *Endless forms: species and speciation*, (ed. D. J. Howard and S. L. Berlocher). Oxford University Press.

Mayr, E. (1940). Speciation phenomena in birds. *American Naturalist*, **74**, 249–78.

Mayr, E. (1942). *Systematics and the origin of species*. Columbia University Press, New York.

Mayr, E. (1954). Change of genetic environment and evolution. In *Evolution as a process*, (ed. J. Huxley, A. C. Hardy, and E. B. Ford), pp. 157–80. Allen and Unwin, London.

Mayr, E. (1963). *Animal species and evolution*. Belknap Press, Cambridge, MA.

Mayr, E. (1992). Controversies in retrospect. *Oxford Surveys in Evolutionary Biology*, **8**, 1–34.

McDonald, M. A. and Smith, M. H. (1990). Speciation, heterochrony, and genetic variation in Hispaniolan Palm-Tanagers. *Auk*, **107**, 707–17.

Moore, W. S. (1995). Inferring phylogenies from mtDNA variation: mitochondrial-gene trees versus nuclear-gene trees. *Evolution*, **49**, 718–26.

Muller, H. J. (1940). Bearing of the *Drosophila* work on systematics. In *The new systematics*, (ed. J. S. Huxley), pp. 185–268. Clarendon Press, Oxford.

Nagle, J. J. and Mettler, L. E. (1969). Relative fitness of introgressed and parental populations of *Drosophila mojavensis* and *D. arizonensis*. *Evolution*, **23**, 519–24.

Orr, H. A. and Orr, L. H. (1996). Waiting for speciation: the effect of population subdivision on the time to speciation. *Evolution* **50**, 1742–9.

Palopoli, M. F. and Wu, C.-I. (1994). Genetics of hybrid male sterility between *Drosophila* sibling species: a complex web of epistasis revealed in interspecific studies. *Genetics*, **138**, 329–41.

Patton, J. L. and Smith, M. F. (1994). Paraphyly, polyphyly, and the nature of species boundaries in pocket gophers (genus *Thomomys*). *Systematic Biology*, **43**, 11–26.

Powell, J. R. (1978). The founder-flush speciation theory: an experimental approach. *Evolution*, **32**, 465–74.

Prager, E. R. and Wilson, A. C. (1975). Slow evolutionary loss of the potential for interspecific hybridization in birds: a manifestation of slow regulatory evolution. *Proceedings of the National Academy of Sciences USA*, **72**, 200–4.

Price, T. D., Grant, P. R., Gibbs, H. L., and Boag, P. T. (1984). Recurrent patterns of natural selection in a population of Darwin's finches. *Nature*, **309**, 787–9.

Price, T. D., Turelli, M., and Slatkin, M. (1993). Peak shifts produced by correlated response to selection. *Evolution*, **47**, 280–90.

Provine, W. B. (1989). Founder effects and genetic revolutions in microevolution and speciation: an historical perspective. In *Genetics, speciation and the founder principle*, (ed. L. V. Giddings, K. Y. Kaneshiro, and W. W. Anderson), pp. 43–76. Oxford University Press.

Provine, W. B. (1986). *Sewall Wright and evolutionary biology*. University of Chicago Press.

Ptacek, M. B., Gerhardt, H. C., and Sage, R. D. (1994). Speciation by polyploidy in treefrogs: multiple origins of the tetraploid, *Hyla versicolor*. *Evolution*, **48**, 898–908.

Ratcliffe, L. M. and Grant, P. R. (1983). Species recognition in Darwin's Finches (*Geospiza*, Gould). II. Geographic variation in mate preference. *Animal Behaviour*, **31**, 1154–65.

Ratcliffe, L. M. and Grant, P. R. (1985). Species recognition in Darwin's Finches (*Geospiza*, Gould). III. Male responses to playback of different song types, dialects and heterospecific songs. *Animal Behaviour*, **33**, 290–307.

Rice, W. R. and Hostert, E. E. (1993). Laboratory experiments on speciation: what have we learned in 40 years? *Evolution*, **47**, 1637–53.

Rick, C. H. (1963). Biosystematic studies on Galápagos tomatoes. *Occasional Papers of the California Academy of Sciences*, **44**, 59–77.

Rieseberg, L. H. (1995). The role of hybridization in evolution: old wine in new skins. *American Journal of Botany*, **82**, 944–53.

Schluter, D. (1996). Ecological causes of adaptive radiation. *American Naturalist*, **148**, supplement, S40-S64.

Schluter, D. and Grant, P. R. (1984). Determinants of morphological patterns in communities of Darwin's Finches. *American Naturalist*, **123**, 175–96.

Schluter, D. and McPhail, J. D. (1992). Ecological character displacement and speciation in sticklebacks. *American Naturalist*, **140**, 85–108.

Schluter, D. and Nagel, L. M. (1995). Parallel speciation by natural selection. *American Naturalist*, **146**, 292–301.

Schluter, D., Price, T. D., and Grant, P. R. (1985). Ecological character displacement in Darwin's Finches. *Science*, **227**, 1056–9.

Schwenk, K. and Spaak, P. (1995). Evolutionary and ecological consequences of interspecific hybridization in cladocerans. *Experientia*, **51**, 465–81.

Sharpe, R. S. and Johnsgard, P. A. (1966). Inheritance of behavioral characters in F_2 Mallard × Pintail (*Anas platyrhynchos* L. × *Anas acuta* L.) hybrids. *Behaviour*, **27**, 259–72.

Sites, J. W., Jr., Peccinini-Seale, D. M., Moritz, C., Wright, J. W., and Brown, W. M. (1990). The evolutionary history of parthenogenetic *Cnemidophorus lemmiscatus* (Sauria, Teiidae). I. Evidence for a hybrid origin. *Evolution*, **44**, 906–21.

Simpson, G. G. (1944). *Tempo and mode in evolution*. Columbia University Press, New York.

Simpson, G. G. (1953). *The major features of evolution*. Columbia University Press, New York.

Smith, G. (1992). Introgression in fishes: significance for paleontology, cladistics, and evolutionary rates. *Systematic Biology*, **41**, 41–57.

Slatkin, M. (1996). In defense of founder-flush theories of speciation. *American Naturalist*, **147**, 493–505.

Sperling, F. A. H. (1990). Natural hybrids of *Papilo* (Insecta: Lepidoptera): poor taxonomy or interesting evolutionary problem? *Canadian Journal of Zoology*, **68**, 1790–9.

Sperling, F. A. H. and Harrison, R. G. (1994). Mitochondrial DNA variation within and between species of the *Papilio machaon* group of swallowtail butterflies. *Evolution*, **48**, 408–22.

Svärdson, G. (1970). In *Biology of coregonid fishes*, (ed. C. C. Lindsay and C. S. Woods), pp. 33–59. University of Manitoba Press, Winnipeg.

Taylor, E. B., McPhail, J. D., and Schluter, D. (1997). History of ecological selection in sticklebacks: uniting experimental and phylogenetic approaches. In *Molecular evolution and adaptive radiation*, (ed. T. J. Givnish and K. J. Sytsma), pp. 511–34. Cambridge University Press.

Templeton, A. R. (1980). The theory of speciation via the founder principle. *Genetics*, **94**, 1011–38.

Templeton, A. R. (1996). Experimental evidence for the genetic-transilience model of speciation. *Evolution*, **50**, 909–15.

Ten Cate, C., Vos, D. R., and Mann, N. (1993). Sexual imprinting and song learning: two of one kind? *Netherlands Journal of Zoology*, **43**, 34–45.

Veron, J. E. N. (1995). *Corals in space and time: the biogeography and evolution of the Scleratinia*. Comstock, Ithaca, NY.

Wade, M. J. and Stevens, L. (1985). Microorganism mediated reproductive isolation in flour beetles (genus *Tribolium*). *Science*, **278**, 527–8.

Wallace, B., Timm, M. W., and Strambi, M. P. P. (1983). The establishment of novel mate-recognition systems in introgressed hybrid *Drosophila* populations. *Evolutionary Biology*, **16**, 467–88.

Weber, K. E. (1996). Large genetic change at small fitness cost in large populations of *Drosophila melanogaster* selected for wind tunnel flight: rethinking fitness surfaces. *Genetics*, **144**, 205–13.

Whitlock, M. C. (1995). Variance-induced peak shifts. *Evolution*, **49**, 252–9.

Whitlock, M. (1997). Founder effects and peak shifts without genetic drift: adaptive peak shifts occur easily when environments fluctuate slightly. *Evolution*, **51**, 1044–8.

Woodruff, D. S. (1989). Genetic anomalies associated with *Cerion* hybrid zones: the origin and maintenance of new electrophoretic variants called hybrizymes. *Biological Journal of the Linnean Society*, **36**, 281–94.

Wright, S. (1931). Evolution in mendelian populations. *Genetics*, **16**, 97–159.

Wright, S. (1932). The roles of mutation, inbreeding, crossbreeding and selection in evolution. *Proceedings of the Sixth International Congress of Genetics*, **1**, 356–66.

Wright, S. (1982). Character change, speciation, and the higher taxa. *Evolution*, **36**, 427–43.

Wu, C.-I. and Palopoli, M. F. (1995). Genetics of postmating reproductive isolation in animals. *Annual Review of Genetics*, **28**, 283–308.

7

Natural selection and random genetic drift as causes of evolution on islands

N. H. Barton

7.1 Introduction

Evolution on islands may differ in many ways from that on the mainland. A novel physical environment, and fewer competing species, may allow unusual adaptive radiations (Schluter 1996). The total number of individuals in the species may be low, giving fewer advantageous mutations, and allowing deleterious mutations to accumulate (Kondrashov 1995; Lynch *et al.* 1995). Finally, populations may be founded by one or a few colonists, giving an immediate burst of random drift (Mayr 1954). I will concentrate on this last aspect of evolution on islands, and in particular, on the idea that genetic drift in founder events may lead to the formation of new species. Because random drift is a well understood and ubiquitous process, its effects are most amenable to theoretical generalities; indeed, most discussions of speciation on islands (verbal and mathematical) have concentrated on drift in founder events. However, this should not lead us to neglect the importance of natural selection: while it is hard to say much in general, its importance in particular cases is clear (e.g. Schluter 1996; Chapters 5, 9–11).

I will begin by describing the apparent obstacles to the evolution of the reproductive isolation which defines biological species, and the way founder effects have been seen as avoiding these difficulties. I will then summarize general arguments which suggest that random drift (whether in established populations or during founder events) is unlikely to lead to strong reproductive isolation. These arguments were set out in detail by Barton and Charlesworth (1984) and Barton (1989). I then discuss recent models which have been suggested as ways of avoiding these arguments, and which would make it likely that strong reproductive isolation could evolve in a single step. Finally, I consider the kinds of evidence which might tell us whether random genetic drift has contributed to the often dramatic radiations of species on islands.

Before getting mired in the details of particular models, it is important to keep in mind the general questions. Do populations typically contain genetic variation that can be reshuffled so as to give strong reproductive isolation? Do new species form when random genetic drift reshuffles this variation? Does this drift occur during drastic population bottlenecks? The answer to the first question, in particular, is relevant to many alternative mechanisms of speciation, as well as to the likely genetic basis of adaptive evolution.

7.2 Obstacles to speciation

The origin of species involves a special difficulty, which does not apply to adaptive evolution in general. Suppose that we take species to be *biological* species—that is, we define a species as a group of individuals which share genetic differences which prevent them exchanging genes with other such species. The evolution of biological species therefore requires the establishment of genotypes which cause some reproductive isolation. The key difficulty is that such genotypes should not be able to invade the population: they would either not be chosen as mates, or if they did mate, they would leave fewer offspring. Thus, models for the evolution of reproductive isolation lead to multiple stable equilibria, so that evolution away from one equilibrium state is resisted by natural selection. This is easiest to understand with post-zygotic isolation, where recombinant or heterozygous genotypes have lower fitness. Selection then tends to increase mean fitness, driving the population towards alternative 'adaptive peaks'. On this view, movement from one peak to another requires passage through an unfit intermediate state, and so is resisted by selection. However, multiple equilibria also emerge in models where non-random mating leads to stable reproductive isolation (e.g. Moore 1979). However, such models usually involve frequency-dependent fitnesses, and hence cannot be described by an 'adaptive landscape'.

The simplest genetic model of reproductive isolation is where heterozygotes at a single locus have reduced fitness; the classic example is where heterozygotes for chromosome rearrangements are partly sterile because of failure in pairing and segregation at meiosis. There are thus two peaks in the graph of mean fitness against allele frequency. Frequency-dependent selection can have a similar effect if it favours the commoner morph. For example, the snail *Partula suturalis* may coil in either the sinistral or dextral direction; coiling direction is determined by a single locus. Because the rarer morph will usually mate with the opposite type, it will both fertilize and be fertilized less efficiently. Hence, just as with underdominant chromosome rearrangements, populations tend to fix one or other morph, and polymorphism is confined to the narrow zones where the alternative morphs meet (Johnson *et al.* 1990; Chapter 11). Other examples of alternative equilibria include the production of colicins in bacteria, where the cost of producing the toxin is outweighed by its advantage in killing competitors only above a threshold frequency (Chao 1979); the spread of *Wolbachia* through *Drosophila simulans*, where a similar frequency-dependent advantage arises because infected males sterilise their uninfected mates (Turelli and Hoffmann 1995); and Müllerian mimicry, where the rarer morph is at a disadvantage because it is not recognized as being distasteful (Chapter 16). These examples have a simple genetic basis; below, we consider whether more complex models of reproductive isolation necessarily sustain alternative stable equilibria, and if so, how populations may diverge into these alternative states. (For a review of these issues in the context of Sewall Wright's 'shifting balance' theory of evolution, see Coyne *et al.* (1997).)

7.3 Defining reproductive isolation

The formation of biological species requires genetic differences that prevent gene exchange—that is, the evolution of reproductive isolation. In order to compare mechanisms of speciation, we need to know how much reproductive isolation they produce. This is not straightforward; indeed, much of the theoretical argument over whether founder events can cause strong reproductive isolation depends on how this isolation is measured. There are two difficulties: first, the effect of genetic differences on gene flow depends on just how the divergent populations meet, and second, their eventual contribution to future reproductive isolation is not in general predictable from their immediate effect on gene flow. I have argued that the degree of isolation is best measured by the mean fitness of the population that forms when two divergent populations hybridize (Barton and Charlesworth 1984; Barton 1989). This gives a common yardstick against which different models can be compared, at least where evolution can be described by an 'adaptive landscape'. However, no one measure is entirely satisfactory: the effect of genetic divergence on current gene exchange depends on just how populations meet, and its effect on future divergence is unpredictable. Here, I summarize this argument, and discuss the difficulties with this and with alternative measures.

Suppose that divergent populations meet along a continuous habitat. Selection on those genetic differences responsible for keeping the populations at their different equilibria will maintain a set of clines, which may persist indefinitely. Unless F_1 hybrids are completely inviable or sterile, genes at other loci can flow from one population to the other. However, this exchange will be impeded by the selection on other loci: genes may be eliminated by selection against introgression of the genes with which they are initially associated before they can recombine onto the other genetic background.

This 'hitch-hiking' effect is reflected directly in the sharp clines in marker alleles seen in hybrid zones. For example, the toads *Bombina bombina* and *B. variegata* are separated by a hybrid zone, in which many quantitative traits, and many genetic markers change at the same place (Fig. 7.1a). This concordance makes it unlikely that each locus is responding directly to selection; indeed, these markers are in strong 'linkage disequilibrium' with each other in hybrid populations, even though they segregate independently in crosses. Rather, the sharp clines reflect the net effects of selection on the rest of the genome. The ratio between the step in these clines (Δu) and the gradient ($\mathrm{d}u/\mathrm{d}x$) on either side, $B = \Delta u/(\mathrm{d}u/\mathrm{d}x)$, gives a measure of the strength of the barrier to gene flow (Fig. 7.1a). This has the dimensions of a distance, and can be thought of as the span of unimpeded habitat which would present the same obstacle to flow of a neutral allele. It applies to any localized barrier, whether genetic or physical, and can be used to find the effect on selected as well as neutral alleles.

A simple model in which linkage disequilibria are generated by the mixing of divergent populations predicts that the gradients in marker alleles should depend primarily on the mean fitness of the population, through the relation $(\mathrm{d}u/\mathrm{d}x) \approx \bar{W}^{-1/r}$ (where r is the harmonic mean recombination rate with the selected loci and $\bar{W}$ is the mean relative fitness of the hybrid population (Barton 1986)). Hence, the ratio between the barrier strength and cline width is $(B/w) = \bar{W}^{1/r}$ (see Fig. 7.1b).

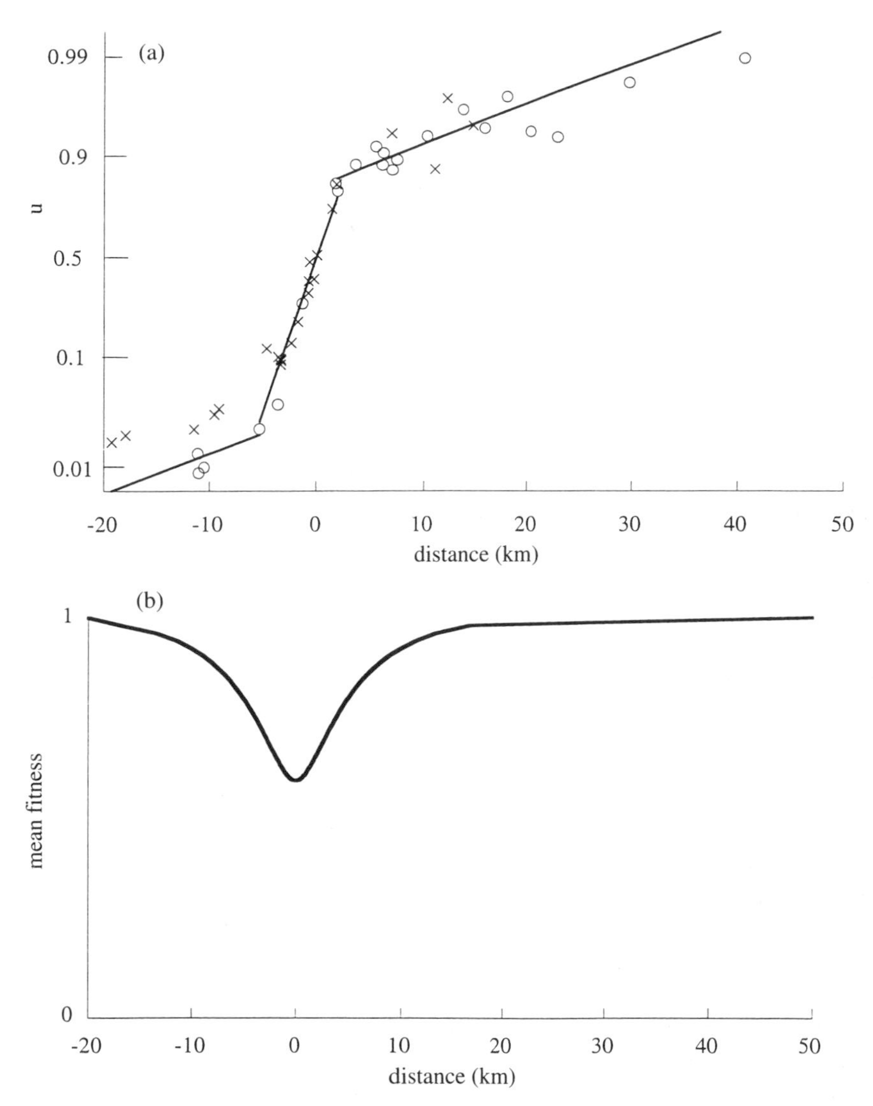

Fig. 7.1 (a) The barrier to gene flow between the toads *Bombina bombina* and *B. variegata*. Symbols show the average frequency, u, of alleles diagnostic for *variegata* at five enzyme loci, across two transects in southern Poland (Szymura and Barton 1991) (open circles: Przemysl; crosses: Krakow). The lines show the best fitting cline, which has width $w = (\mathrm{d}u/\mathrm{d}x)_{max} = 6.0$ km, and corresponds to a barrier $B_v = \Delta u/(\mathrm{d}u/\mathrm{d}x)_v = 51$ km to flow into *variegata*, and $B_b = 260$ km to flow into *bombina*. (b) The reduction in mean fitness needed to account for this stepped cline (harmonic mean recombination rate is $r = 0.25$, and assuming selection against heterozygotes at 55 loci, as estimated from these data).

This relation applies to any model in which genotypes have fixed fitnesses; simulations show that though derived as a weak selection limit, it is accurate up to fairly strong selection (Barton and Gale 1993). However, if selection is strong, the mean fitness of the hybrid population may be much less than that of the ancestral populations connecting the two forms (see below). This is because if hybrids are on average unfit, particular recombinants with high fitness cannot easily be reconstructed. Similarly, if immigrants are rare, the barrier to gene flow depends primarily on the fitness of the first generation hybrids (Bengtsson 1985; Barton and Bengtsson 1986), which may be much lower than that of the hybrids which might form in a continuous habitat, or of the ancestral genotypes. Evidence of this effect is seen in the transition between chromosome races of the grasshopper *Podisma pedestris* . In most places, these meet in a smooth cline, showing no evidence of a barrier. However, across two small streams, gene flow is reduced much more than can be explained by the direct reduction in dispersal. This can be accounted for by a genetic barrier to gene flow due to reduced fitness of F_1 and backcross hybrids (Jackson 1992; Barton and Gale 1993).

Thus, the effect of genetic divergence on the mean fitness of hybrids, and on gene flow, depends on just how those hybrids are formed. A further difficulty is that the contribution of divergence to eventual speciation may not be reflected by its immediate effect on gene exchange. One could argue that speciation may be completed in allopatry, in which case the flow of genes across continuous hybrid zones is irrelevant. Even if speciation is entirely parapatric, gene flow may hardly slow divergence. If selection against hybrids is strong, as is required for biological speciation, the rate of gene flow depends mainly on F_1 or backcross fitness: one could measure progress to full speciation by the contribution to reducing this fitness. If one assumes that shifts in different sets of genes have multiplicative effects, then one might use the reduction in F_1 or backcross fitness as the measure of isolation, even though the immediate effect on gene flow depends more on the mean fitness of a hybrid population and may be much smaller. However, effects may well not be multiplicative; by analogy with deleterious mutations (Kondrashov 1988), hybrid fitness may decrease faster than geometrically with divergence.

The fundamental problem in defining a measure of 'reproductive isolation' is that the effect of any one difference on gene flow depends on what else is different. In the early stages of speciation, hybridization can readily produce a variety of recombinants; the mean fitness of a hybrid population can be increased by selection of the fitter recombinants, and it is this mean fitness that is the most appropriate measure. When hybrids have low fitness, they are likely to be kept at low frequency, and it is the average fitness of these randomly produced hybrids that determines the degree of isolation. Both measures must therefore be considered in judging the plausibility of founder effect speciation.

7.4 Does random genetic drift cause reproductive isolation?

Models of founder effect speciation

If species are maintained by natural selection at stable equilibria, then it is at first hard to see how they could be formed. This difficulty led to the idea that random drift causes speciation, by knocking populations from one equilibrium to another despite the opposition of selection. Mayr (e.g. 1942, 1982) has been the most influential proponent of such ideas. He argues that species cannot change significantly in the main part of their range, because gene flow prevents divergence, and because genes are closely coadapted with each other. Mayr's model of 'peripatric speciation' proposes that species form as a result of founder events in peripheral isolates; these cause a loss of heterozygosity which supposedly changes selection pressures and triggers 'genetic revolutions'. Carson (1968, 1975) has proposed a similar model of 'founder-flush speciation', which includes the effects of relaxed selection during the rapid increase following colonization. Finally, Templeton (1980) has proposed that founder events may trigger a 'genetic transilience' if a fewer major loci interact with many modifiers of small effect (see Chapter 8).

Mayr has argued that founder events have effects distinct from random drift, and that theories of founder effect speciation originated independently of Wright's (1932) theory of a 'shifting balance' between different evolutionary processes. However, Mayr, Carson, and Templeton's theories of founder effect speciation all trace back to Wright's ideas (see Provine 1989). Wright proposed the 'shifting balance' primarily as a more efficient means of adaptation, but recognised that it would also lead to reproductive isolation (Wright 1940, 1982). The 'shifting balance' differs from the other theories in that it involves a network of partially isolated demes, giving many opportunities for peak shifts to occur by drift and founder effects. A new 'adaptive peak' which arises in one of many trials can spread through the whole population, making reproductive isolation more likely. Theories of founder effect speciation assume that the colonizing population spreads over a large area. A population therefore traces its ancestry back through just a few successive bottlenecks, which greatly reduces the chances that strong isolation will evolve.

There are many difficulties with the particular theories outlined above; these are discussed in reviews by Barton and Charlesworth (1984) and Barton (1989). Here, I will concentrate on the general question of whether random genetic drift (especially, in founder events) is likely to lead to significant reproductive isolation.

Populations of steady size

First, consider the effects of drift in a small isolated population which has a steady effective size of N_e diploid individuals. Quite generally, the rate at which drift knocks the population from one adaptive peak to another is proportional to $\bar{W}^{2N_e}$. Here $\bar{W}$ is the mean fitness of the population at the unstable equilibrium which separates the two peaks, relative to the mean fitness at the original equilibrium. This relation was first derived by Wright (1941), but is quite general; it applies whenever the system can be described by an 'adaptive landscape' (Barton and Rouhani 1987). For example, it gives a good approximation to disruptive or stabilizing selection on a quantitative trait (Lande 1985; Barton 1989), selection against heterozygotes (Lande 1979; Hedrick 1981), or Wagner *et al.*'s (1994) model of epistatic selection on two loci (see below).

Now, we have seen that the reproductive isolation which is generated by a peak shift also depends on mean fitness, being proportional to $\bar{W}^{1/r}$. Hence, shifts which lead to strong isolation are necessarily unlikely. One can make the quantitative argument that because the rate of shifts decreases exponentially with $N_e \Delta \log(\bar{W})$ ($\Delta \log(\bar{W})$ being a measure of the depth of the adaptive valley), reproductive isolation will build up mainly through the gradual accumulation of shifts of magnitude $\Delta \log(\bar{W}) \approx 1/N_e$ (Walsh 1982; Barton and Charlesworth 1984).

There is an assumption here, that the mean fitness of the population which forms when divergent populations hybridize is similar to the mean fitness of the 'adaptive valley' which separates the two peaks. The hybrid population may in fact have lower mean fitness, leading to stronger reproductive isolation: since the depth of the 'adaptive valley' is defined by the saddle point in the adaptive landscape, the hybrids cannot have higher mean fitness. Two factors tend to reduce the fitness of the hybrid population: gene flow introduces maladapted genotypes, and linkage impedes the generation of the fitter recombinants. The distinction between the fitness of hybrids, and of the ancestors that link divergent genotypes is crucial to models which allow strong isolation to evolve with high probability; we consider it in detail below.

Founder events

Suppose now that there is a drastic bottleneck, as when a new population is founded. During this brief period, selection is negligible compared with drift, since there will be too little time for it to change allele frequencies appreciably. Therefore, the chance of a shift can be approximated by the chance that the new population will have drifted into the domain of attraction of a new equilibrium. (This approximation overestimates the chance of a shift, because selection will tend to impede divergence.)

The problem can be further simplified, because the exact sequence of population sizes during the bottleneck is irrelevant: under the diffusion approximation, the distribution of allele frequencies afterwards depends on a single parameter, which we can take to be the reduction in genetic variance, $(1-F)$. The mean of an additive quantitative trait will be normally distributed with variance $2FV_g$, where the genetic variance is reduced from V_g to $(1-F)V_g$ (Barton and Charlesworth 1984). Similarly, the distribution of allele frequency after the bottleneck depends primarily on F, and the initial frequency (Fig. 7.2). Thus, the net effect of a bottleneck can be measured by observing the loss of neutral variability: the detailed demography of the bottleneck is irrelevant, provided that drift is concentrated in a short period.

Templeton (1980) has suggested that a bottleneck which involves a severe reduction in 'variance effective size', but little reduction in 'inbreeding effective size', is favourable to speciation, because it allows drift in allele frequencies and yet little loss of variation. There is a fundamental mistake here, since both drift to a new adaptive peak and loss of variation are due to random changes in allele frequency and cannot be

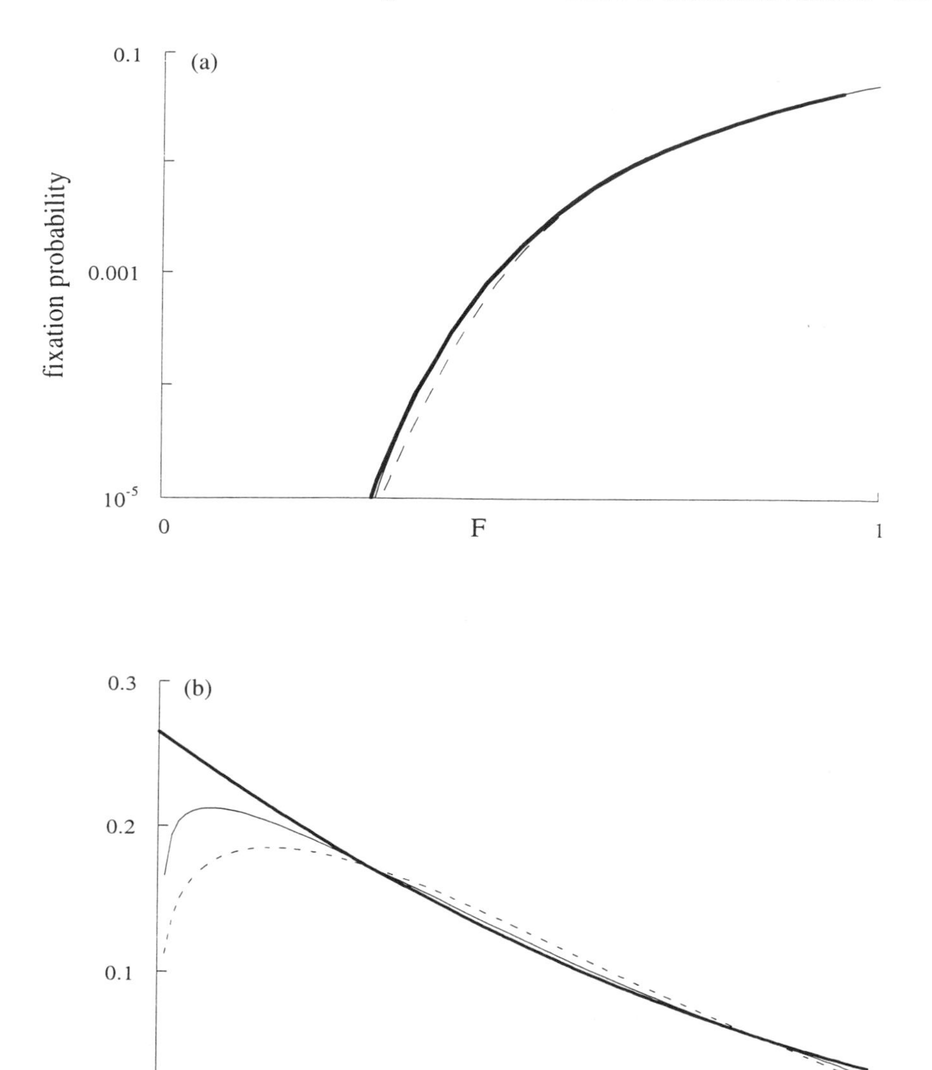

Fig. 7.2 Accuracy of the diffusion approximation for the distribution of allele frequencies after a bottleneck. (a) The probability that an allele initially at frequency $p_0 = 0.05$ will be fixed after a bottleneck which causes inbreeding F. On the diffusion approximation (heavy line), the fixation probability is determined solely by F (Crow and Kimura 1970). The dashed line shows the fixation probability for a population which grows geometrically from two diploid individuals, calculated assuming Wright–Fisher sampling, for a range of growth rates. A light line shows exact results for a population growing from four diploid individuals, but is indistinguishable from the diffusion approximation. (b) The distribution of allele frequencies for $F = 0.555$. The heavy line gives the diffusion approximation, the light line a population increasing by 17.4% per generation, in the sequence 4, 5, 6, 6, 8, 9, 10, 12, ..., and the dashed line growth by 50% per generation, in the sequence 2, 3, 4, 7, ...

disentangled. For example, if a population grows from two diploid individuals sampled from a large pool, the inbreeding effective size is much greater than the variance effective size in the first generation (∞ vs. 2), because the founders are unrelated, but have a large variance in allele frequency. Figure 7.2 shows that for given F, the distribution of allele frequencies, and hence the chance of a shift, is almost independent of whether growth is rapid or slow, and hence of the transient discrepancy between inbreeding and variance effective sizes.

Slatkin (1996) has argued that in a founding population, favourable alleles have an increased probability of fixation, which facilitates founder effect speciation. Founder events increase the probability of fixation of an allele that is present in the initial propagule, for two reasons. First, each individual leaves more offspring when the population is growing, and so is more likely to have descendants in the future. Second, if the allele is recessive, its favourable effects are more likely to be expressed if the population is small. However, Slatkin's (1996) argument rests on the assumption that the allele is lucky enough to be included in the founding population. It is more appropriate to count the chances of fixation of alleles in the base population. Since few alleles can be represented in the founder population, the net chance of a shift is necessarily reduced by a drastic bottleneck (Charlesworth 1997).

These general considerations show that for strong isolation to evolve during a founder event, there must be initial variation in traits or genes that can cause isolation by reducing hybrid fitness. Thus, founder effect speciation is unlikely in models of discrete loci where alleles are initially held at low frequency in a mutation–selection balance. This argument suggests that the most favourable cases are either balanced polymorphisms where epistatically interacting alleles are held at high frequency, or polygenic variation involving many genes. Carson (1968, 1975) has put forward verbal models in which alternative coadapted combinations of genes segregate. The simplest representation of this scheme involves two loci, with selection being maintained by overdominance, and epistasis allowing two alternative equilibria which differ in the sign of linkage disequilibria (Charlesworth and Smith 1982). The expected reproductive isolation produced by this model is proportional to the non-additive variance in fitness in the base population. While there are few direct measurements of the components of fitness variance (Burt 1995), comparisons of the effects of chromosomes extracted from male and from female *Drosophila* (which differ in that only the latter have undergone recombination) show that the non-additive variance in viability and fertility is small (a few per cent: Mukai and Yamaguchi 1974; Charlesworth and Charlesworth 1975; Charlesworth and Barton 1996). Under this model, the amount of reproductive isolation generated by a bottleneck is constrained by the standing variation in fitness associated with epistatically interacting genes. We consider below whether this is a general constraint.

Shifts are more likely to occur if they are based on the random drift of a quantitative trait with high genetic variance. If this variation is additive, we can make a general prediction about the effect of a bottleneck. In generation t, the genetic variance is reduced to $V_{t+1} = V_t(1-1/2N_t)$, and the variance of the mean around its original value increases by V_t/N. The cumulative effect of drift over many generations is to reduce the genetic variance by a factor $1-F = \prod_t(1-1/2N_t)$, and to give a variance of the mean

$2FV_0$. Thus, regardless of the exact sequence of population sizes, the mean will vary in proportion to the net inbreeding. A shift of more than two standard deviations, or $2\sqrt{(2FV_g)}$, is unlikely; thus, even with complete inbreeding, the mean is unlikely to shift more than $2\sqrt{2}$ genetic standard deviations.

This result puts two constraints on the likely level of reproductive isolation. First, the genetic variance must be high enough to include unfit intermediate genotypes if a shift is to be likely; this implies a standing genetic load. Second, if the phenotypic variance is too high, it is unlikely that there will be two distinct 'adaptive peaks': above a critical variance, enough individuals will approach the fittest phenotype that the population will move to that state under selection alone (Kirkpatrick 1982). In the simplest model of disruptive selection on a single additive trait, the genetic variance is constrained to a narrow range: if it is too small, the expected isolation produced in a bottleneck is very small, whilst if it is too large, there is only a single adaptive peak. Within this range, the expected reproductive isolation is smaller than the standing genetic load. In the example shown in Fig. 7.3, disruptive selection favouring optima at ±1 only allows alternative equilibria if the phenotypic variance is less than 0.279; however, if the variance is much smaller than this threshold, the chance of a shift is small, even after an extreme bottleneck. The expected isolation is at most 3.9% of the standing load.

The discussion so far has centred on the simplest model, in which random changes in genetic variance are ignored, and the trait is assumed to be additive. Relaxing these assumptions introduces considerable complications, since the outcome depends on the genetic basis of the trait, rather than on purely phenotypic arguments. First, consider fluctuations in variance. Whitlock (1995) has shown that in a bottleneck, genetic variance may by chance increase substantially. Enough individuals may then approach the fitter optimum to trigger a shift. Whitlock (1995) assumed that the distribution of allelic effects at each locus is Gaussian, and included no mechanism for maintaining the initial variation. However, if variation is maintained by mutation, it is likely that each locus is near fixation, so that variation is due to rare alleles (Turelli 1984). Figure 7.4 shows the evolution of the mean and variance of an additive trait, based on 16 loci, assuming the rare alleles approximation (Barton and Turelli 1987). In a population of steady size, drift is most likely to cause a shift by taking the population across the unstable equilibrium which lies between the alternative equilibria; this involves a substantial increase in genetic variance (from 0.033 to 0.171). However, a founder event is likely to cause less increase in variance (see simulated scatter in Fig. 7.4) This is because a shift is most likely to occur along the shortest route to the domain of attraction of the new state, rather than by making an excursion via the unstable state. Nevertheless, allowing for fluctuations in variance does increase the chance of a shift (Whitlock 1995). For example, using the same parameters as Fig. 7.4, a shift is unlikely for $F<0.4$, and rises to around 1% for larger F (Fig. 7.5). It is an order of magnitude larger than would be the case ignoring fluctuations in variance (curve at lower right of Fig. 7.5). However, allowing for changes in genetic variance also reduces the strength of isolation produced, because a hybrid population can recover fitness by increasing its variance to encompass the fitter phenotypes (Fig. 7.4). The expected reproductive isolation is at most 3.5% of the standing genetic load, similar to the value of 3.9% calculated neglecting fluctuations in variance.

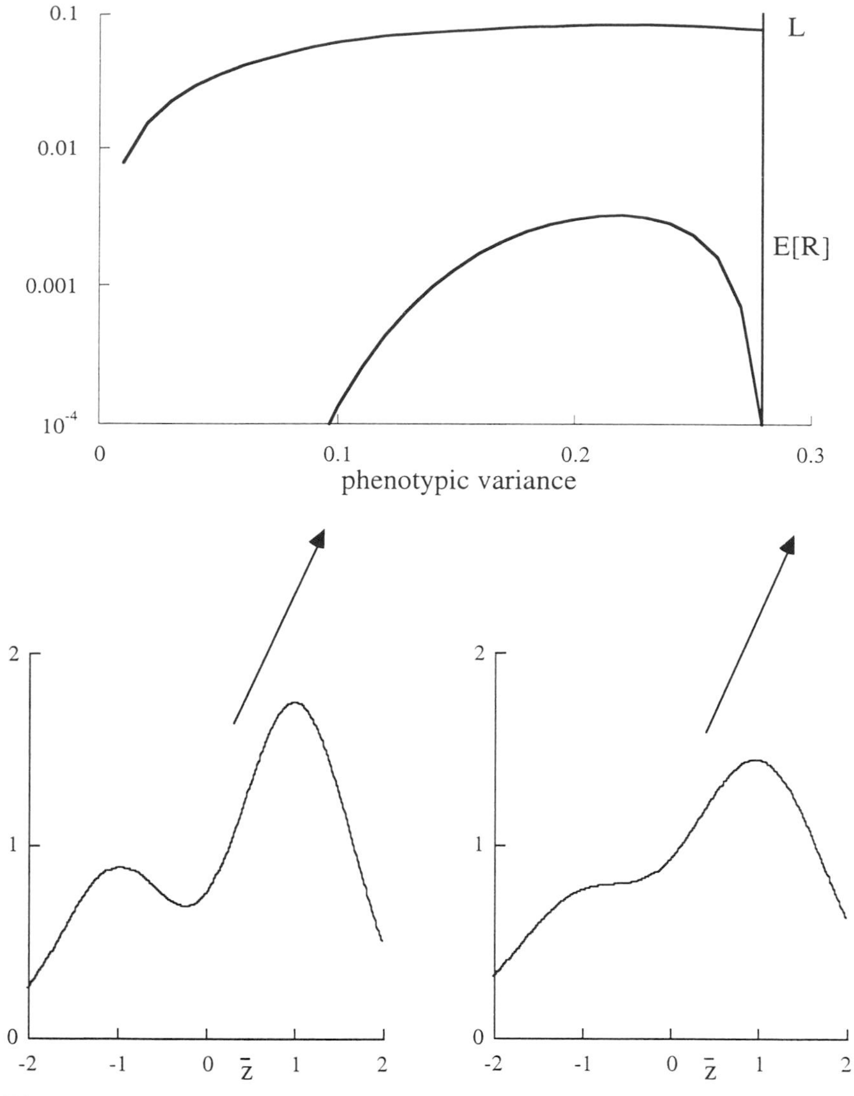

Fig. 7.3 The relation between the expected reproductive isolation, $E[R]$, caused by an extreme bottleneck (F=1), the standing genetic load, L, and the phenotypic variance, V_g+V_e. Disruptive selection acts on an additive trait, z, determined by infinitely many unlinked loci. Individual fitness is $W = A\ \exp[(z-1)^2/2V_s] + \exp[(z+1)^2/2V_s]$, as in Kirkpatrick (1982). There are potentially two peaks of width $\sqrt{V_s}$, one near $z = -1$, and one near $z = +1$ which is higher by a factor A. Environmental and genetic contributions to z are assumed to follow Gaussian distributions, so that mean fitness is given by the above formula, with V_s replaced by $V_s+V_g+V_e$. In this example, $A = 2$, $V_s = 0.3$, and $h^2 = V_g/V_p = 50\%$. Reproductive isolation is defined by the ratio between initial mean fitness and mean fitness in the adaptive valley, $R = -\log_e(\bar{W}_p/\bar{W}_h)$. The standing load is defined by the ratio of initial mean fitness to that in the absence of genetic variation.

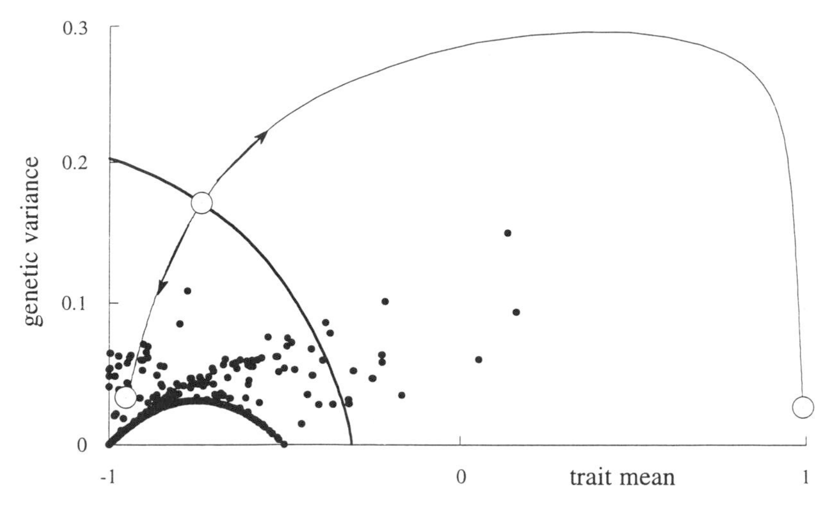

Fig. 7.4 Shifts in an additive trait under disruptive selection. Selection acts as in Fig. 7.3, with $V_s = 0.3$, $A = 2$, $V_e = 0.1$. Mutation at a net rate $U = \Sigma\mu = 0.03$ maintains genetic variance at 16 loci, each segregating for two alleles with effect $a = 0.25$. There are two stable equilibria at ($\bar{z}$ = –0.952, $V_g = 0.0332$, and $\bar{z}$ = +0.990, $V_g = 0.0268$; these are separated by an unstable equilibrium at $\bar{z}$ = –0.737, $V_g = 0.171$ (open circles). The heavy line separates the domains of attraction of the two equilibria, whilst the arrows show paths away from the unstable equilibrium. These are calculated using the rare alleles approximation. This was checked by numerical iterations of the 16 allele frequencies. These showed that the outcome is determined almost entirely by the initial ($\bar{z}$ V_g); the domains calculated in this way are separated by a line close to that shown, but slightly flatter. The points show the outcome of 1000 bottlenecks, in which 50% of genetic variation was lost ($F = 0.5$); the distribution was calculated using the diffusion approximation. Shifts occurred in 13 cases, and resulted in reproductive isolation $R = \log_e(\bar{W}_p/\bar{W}_h) = 0.0531$. This compares with a load due to genetic variation around the initial optimum of $L = 0.0337$; $E[R]/L = 0.020$.

There has been considerable interest in the observation that additive genetic variance can increase substantially following a bottleneck (Bryant *et al.* 1986; Bryant and Meffert 1993, 1995). The simplest explanation is the chance increase of recessive alleles that were rare in the base population (Willis and Orr 1993), though epistasis may also contribute. The key question for founder effect speciation is whether such non-additive inheritance makes it more likely that a founder population will shift into the domain of a new equilibrium. The variance of the mean following a bottleneck is increased above the additive expectation of $2FV_g$ only if the bottleneck is strong (Fig. 7.6). (Note that a moderate bottleneck may, as in some of Bryant *et al.*'s experiments, increase the additive genetic variance; the issue here is the likely shift in the mean, not the variance.) Finally, the most plausible reason why recessives should initially be rare is that they are deleterious; the population would then tend to return to its original state.

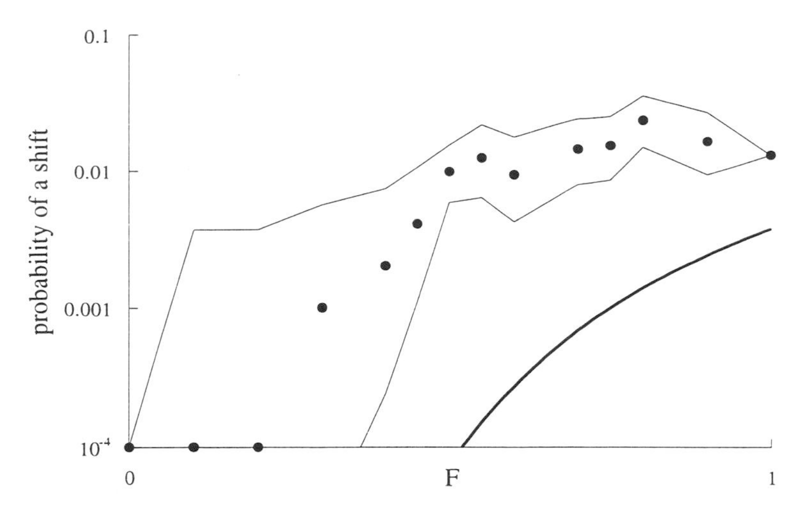

Fig. 7.5 The probability of shifts to a new peak, for an additive trait under disruptive selection. Parameters are as in Fig. 7.4. Each value is calculated from 1000 trials, for varying F (solid circles); the thin lines show 95% confidence intervals. The maximum rate found was 23/1000 for F=0.8; this corresponds to $E[R]/L$ = 0.035. The smooth curve on the lower right shows the prob ability of a shift, calculated by ignoring fluctuations in genetic variance.

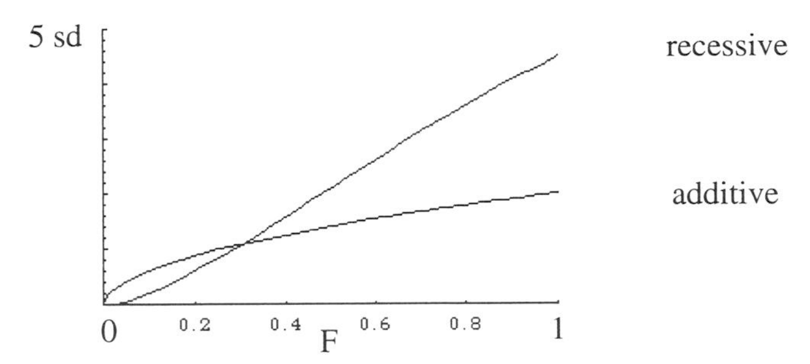

Fig. 7.6 The largest likely deviation (defined as having probability <2.5%) after a bottleneck, measured in phenotypic standard deviations. (h^2 = 50%). The graphs show predictions for additive traits ($2\sqrt{2FV_g}$)) and for traits based on recessive alleles, with initial frequency 10%. (From Barton 1989.)

7.5 Avoiding the obstacles to speciation

As Darwin (1859, Ch. 6). realized, in considering the origin of complex adaptations such as the eye, two distinct morphs may be connected by a relatively fit set of intermediates This possibility aids speciation as well as adaptation, since it allows populations to diverge to incompatible states without suffering low fitness. It can be appreciated by contrasting F_2 or backcross hybrids with the genotypes connecting the parental populations, which (if ancestral alleles have not been lost) will be contained within the set of F_2 genotypes. Randomly produced hybrids have mostly never been tested by selection, and so may have much reduced fitness; the ancestral genotypes make up a small fraction of those hybrids, and may have high fitness. Indeed, recombinant genotypes may be fitter than either parent, and may spawn 'hybrid species' if they can rise to high frequency (Rieseberg 1995). In this section, I consider how far evolution along ridges in the adaptive landscape aids the evolution of strong reproductive isolation, and whether it makes founder effect speciation more or less likely than alternative mechanisms.

Those who have invoked the need for random drift to overcome natural selection have not accepted that extant genotypes are connected by fit intermediates. Wright (1932) argued, in his theory of the 'shifting balance', that populations could not find paths of increasing fitness which would allow them to escape inferior 'adaptive peaks'. Similarly, Mayr (1942) held that 'coadaptation' would prevent significant change in large populations. The idea that obstacles to speciation can be avoided by evolution along ridges in the adaptive landscape has a long pedigree, having been proposed in a formal two-locus model by Dobzhansky (1937), and developed by (among others) Muller (1942), Bengtsson and Christiansen (1983), and Nei *et al.* (1983). It has recently received somewhat more attention; in particular, Gavrilets and Hastings (1996) and Wagner *et al.* (1994) have used it to argue that random drift may readily lead to strong reproductive isolation.

Do ridges in the 'adaptive landscape' favour divergence by drift?

Two difficulties arise. First, while ridges in the 'adaptive landscape' make divergence by random drift easier, they also make divergence by selection alone more likely. This objection applies particularly where slight changes in the parameters can remove the adaptive valley altogether. In the example of Fig. 7.3, founder effects produce the greatest isolation when the phenotypic variance is close to the threshold at which the two peaks merge into one. Thus, slight changes in conditions could cause divergence with no need for random drift to oppose selection (cf. Kirkpatrick 1982). Moreover, the existence of ridges in the fitness surface aids founder effect speciation less than it aids divergence by drift in stable populations, or in response to fluctuating selection. This is because divergence in a founder event is most likely to take a direct path: the population will not have time to seek out a tortuous path of high fitness (for example, compare the ridge connecting the peaks in Fig. 7.4 with the outcome of founder events).

Wagner *et al.* (1994) present an example in which the existence of a path of high fitness aids the evolution of reproductive isolation. They consider stabilizing selection on a quantitative trait, determined by two loci with two alleles. If the trait is additive, the optimal phenotype can be obtained by fixing either of the homozygous genotypes ($U_1U_1V_2V_2$) or ($V_1V_1U_2U_2$); there are two peaks in the 'adaptive landscape' (Fig. 7.7a). Allowing epistasis changes the relative fitnesses of the less fit homozygotes ($U_1U_1U_2U_2$, $V_1V_1V_2V_2$), such that the population can shift to a new adaptive peak without passing through a deep adaptive valley. For example, with epistasis of strength $\beta = -4/3$ (the value analysed by Wagner *et al.*), the F_1 heterozygote has fitness reduced by $(1-16s/9)$,

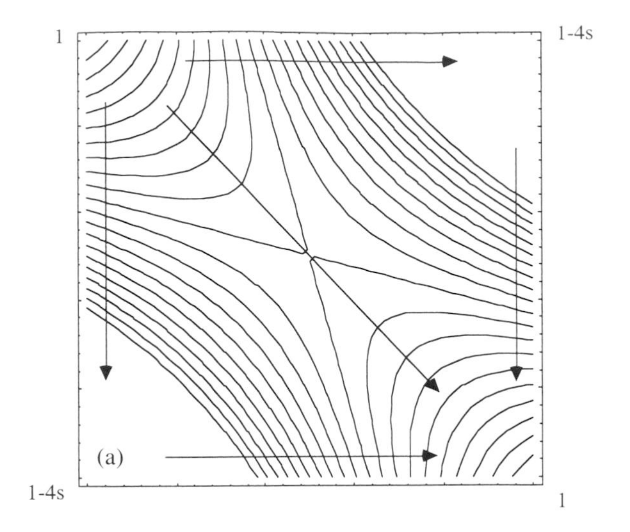

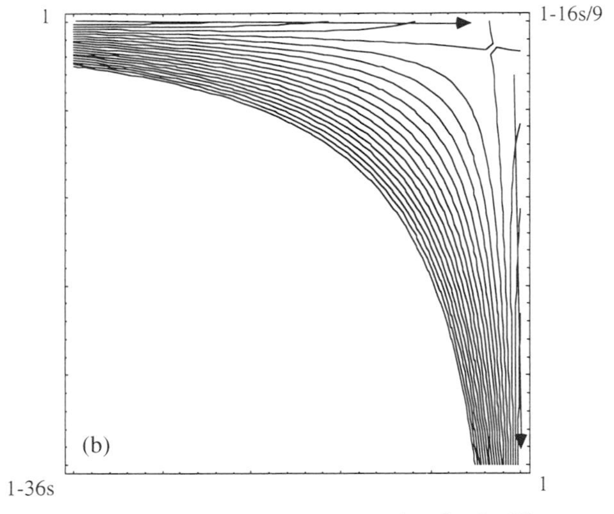

Fig. 7.7 The mean fitness as a function of allele frequencies, for the Wagner *et al.* (1994) model. Contours are spaced at intervals of 0.05. (a) Stabilising selection on an additive trait ($\beta = 0$); the F1 has fitness 1. (b) Stabilising selection on a non-additive trait ($\beta = 0$). The F1 now has fitness $1-16s/9$. Arrows show the alternative paths by which a shift may occur.

but the homozygote ($U_1U_1U_2U_2$) has fitness (1–4*s*/9) where *s* is the strength of stabilizing selection (Fig. 7.7b). The interpretation in terms of stabilizing selection on a non-additive trait is not essential: the model is really one in which the location and height of the ridge can be tuned by the parameter β. With $\beta = -1$, all genotypes homozygous for U_1U_1 or U_2U_2 (or both) have maximum fitness, and there is no 'adaptive valley'—as in the models of Dobzhansky (1937) and Bengtsson and Christiansen (1983).

A small population can readily drift along the ridge, even when the F_1 and F_2 have low fitness (e.g. Fig. 7.7b); in the extreme case, the F_1 could be completely sterile, and drift could cause full speciation. However, when β is close to –1, the 'adaptive valley' is very shallow, and slight changes in selective conditions could also readily lead to isolation. This kind of model therefore favours speciation by any means, rather than by drift in particular. Whitlock (1997) discusses several examples, which show that whenever the adaptive valley is shallow enough to give a reasonable chance of stochastic divergence, slight changes in the parameters can trigger a shift by selection alone. Moreover, this kind of model is relatively insensitive to founder effects: the initial variation is low, being sustained by mutation, and the chance of a shift depends on the domains of attraction of the alternative states, which are independent of epistasis. This is illustrated in Fig. 7.8, which shows the chances that a shift will occur following a founder event in which half the genetic variation is lost ($F = 0.5$), as a function of mutation. Unless mutation is frequent ($\mu/s > 0.1$), it is most likely that one mutation will fix first, followed by evolution up to the new peak (see arrows in Fig. 7.7). (The same is true for drift in a stable population: Wagner *et al.* (1994) found that the rate of shifts scales with μ rather than μ^2). The rate is 4.5 fold higher for the non-additive model; however, this is solely because the selection against one allele is weaker along the ridge, allowing it to become more common in the base population. This contrasts with a

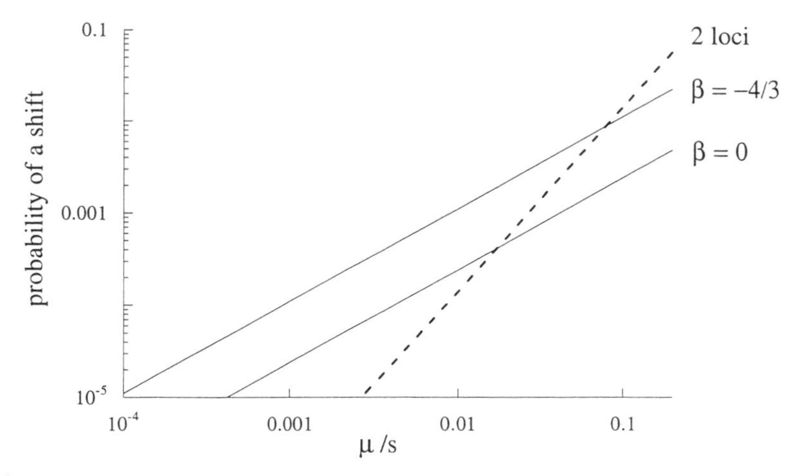

Fig. 7.8 The chance that a founder event ($F = 0.5$) will trigger a peak shift in the Wagner *et al.* (1994) model, plotted against the ratio between mutation and selection (μ/s). The dotted line shows the chance that both loci fix the rare allele; this is the same for $\beta = 0$, $\beta = -4/3$. The two solid lines show the chance that one locus will fix, followed by a shift at the second locus. This is higher for the case $\beta = -4/3$.

steady population, where the existence of a ridge gives an overwhelming advantage to the non-additive model when selection is strong (Wagner *et al.* 1994).

Gavrilets and Hastings (1996) also discuss a variety of models in which drift can lead to strong isolation, as measured from first-generation hybrids. One class of their models relies on ridges in the landscape, and is similar to one developed by Wagner *et al.* (1994; e.g. their Figs. 1, 5). However, they introduce another class of model, in which divergence is due to the accumulation of mutations which arise after the founder event. For example, they suppose that at one locus, polymorphism is maintained by symmetric overdominance. A second locus is fixed for an allele which is favoured when associated with heterozygotes at the first locus. However, if the polymorphism is lost in a founder event, a new allele is favoured at the second locus; if this is fixed, it can lead to strong isolation. This model is close to Mayr's proposal that different alleles can become favourable in the homozygous state that may follow a founder event. However, it is hard to see why the changed genetic background that follows a founder event should be more likely to trigger divergence than changes due to any alteration in gene frequencies. For example, the analogue of Gavrilets and Hastings' (1996; Fig. 7.3b) model would be one where a new allele at the second locus is favoured following a substitution at the first, for whatever reason. These kinds of model allow strong reproductive isolation to be triggered by any change in genetic background, and do not seem especially favourable to founder effect speciation.

Are ridges in the adaptive landscape consistent with strong isolation?

The second difficulty is that the fit intermediates may be reconstructed following hybridization, leading to the collapse of reproductive isolation. This process can be observed directly. For example, F_1 males from the cross between *Chorthippus parallelus parallelus* and *C. p. erythropus* are sterile, yet natural hybrids are fertile (Fig. 7.9). The mechanism by which hybrid fitness can be recovered is understood in the case of chromosome rearrangements. For example, races of the shrew *Sorex araneus* have mainly metacentric karyotypes. Since these involve different chromosome combinations, meiosis is severely disrupted in F_1 hybrids. However, in natural hybrid populations, acrocentric chromosomes are found at high frequency; since simple heterozygotes between fused and unfused chromosomes segregate regularly, there is little loss of fertility, and hence little barrier to gene flow (Searle 1986).

This difficulty can be avoided in two ways. First, ancestral alleles may have been lost: in the *Sorex* example, hybrid fertility is ameliorated by acrocentric chromosomes which are not found elsewhere in the species and might in principle not have been generated by mutation. Second, if divergence has gone so far that the initial hybrids are very unfit, then fit recombinants may not be able to be established: in the limit where F_1 hybrids are completely infertile or inviable, speciation is complete. As discussed above, the outcome depends on how the divergent populations meet. A trickle of gene flow is less likely to allow the establishment of fit recombinants than free hybridization in a continuous hybrid zone; however, even there, strong selection can prevent the establishment of particular genotypes. There is, however, a chance that drift in a partially isolated hybrid population can establish a fit recombinant, perhaps leading to 'hybrid

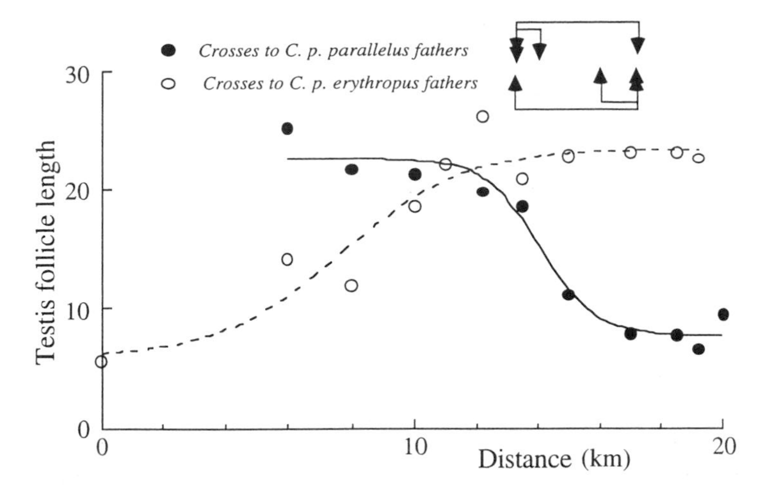

Fig. 7.9 Male fertility in crosses between grasshoppers from the hybrid zone between *Chorthippus parallelus parallelus* and *C. p. erythropus*, as measured by testis follicle length (data from Virdee and Hewitt 1994). Filled circles show the fertility of sons from mothers crossed to *parallelus* fathers, plotted against the position from which the females were collected; open circles are for crosses to *erythropus* fathers. Females from the centre of the hybrid zone give fertile sons for both sires.

speciation' (McCarthy *et al.* 1995; Rieseberg 1995). Since hybrid populations may have an exceptionally high standing load and high non-additive genetic variance, these may be the most favourable sites for founder effect speciation. Such hybrid populations are unlikely to be found on islands (though see Grant and Grant (1995) and Chapter 9), and so this kind of hybrid speciation, aided by random drift, might be most likely to occur within extensive hybrid zones between mainland populations.

7.6 Conclusions

There are strong theoretical constraints on the degree of reproductive isolation which is likely to be produced by a sudden founder event. The genetic variation present in the population must be reshuffled so as to substantially reduce the fitness of hybrids with other populations. Hence, the expected degree of isolation is limited by the initial non-additive variation in fitness, and the standing genetic load. It may be that evolution is facilitated by the existence of sequences of fit genotypes that sustain a ridge on the 'adaptive landscape'. However, these are more favourable to selective mechanisms of divergence, or to drift in a steady population, than to founder effects, in which drift overrides selection. The fit intermediates may also be reconstructed if diverging populations meet before speciation is complete, leading to the breakdown of isolation.

These theoretical arguments suggest what kinds of evidence are relevant. The effect of a bottleneck depends primarily on the fraction of heterozygosity which is lost (F).

Although founder effects can cause a loss of allozyme and DNA variation (e.g. Baker and Moeed 1987; Janson 1987), there is no evidence of a drastic reduction in the classic example of the Hawaiian *Drosophila* (Sene and Carson 1977; DeSalle and Templeton 1988). Ancient polymorphisms that persist through speciation rule out severe bottlenecks (e.g. Klein *et al.* 1993; Richman *et al.* 1996); a systematic survey of their prevalence would be valuable. The way genes interact to determine fitness can be investigated by examining the inheritance of fitness in the base population and in species crosses. Haldane's rule and the large effects of the sex chromosomes indicate that isolation is due to interactions between recessive alleles (Turelli and Orr 1995); the dissection of hybrid fitness in *Drosophila* is just beginning to disentangle the relative fitnesses of recombinant genotypes and to isolate the genes involved (Wu and Palopoli 1994). The strongest evidence, however, is the most straightforward: bottlenecks induce little reproductive isolation in artificial populations of *Drosophila* (Rice and Hostert 1993). Indeed, artificial selection can produce divergence at least as easily as drift (Cohan and Hoffmann 1989; Cohan *et al.* 1989), and the classic examples of speciation on islands involve adaptive radiation in response to a novel physical and biotic environment (Grant 1986; Schluter 1996; Chapters 5, 9–11).

7.7 Summary

The evolutionary processes responsible for adaptation and speciation on islands differ in several ways from those on the mainland. Most attention has been given to the random genetic drift that arises when a population is founded from just a few colonizing genomes. Theoretical obstacles to 'founder effect speciation' are discussed, together with recent proposals for avoiding them. It is argued that while certain kinds of epistasis can facilitate the evolution of strong reproductive isolation, such epistasis favours divergence by selection as much as by random drift.

Acknowledgements

This work was supported by the Darwin Trust of Edinburgh, and by the Biotechnology and Biological Sciences Research Council. Thanks are also due to M. Whitlock for his helpful comments on the manuscript.

References

Baker, A. J. and Moeed, A. (1987). Rapid genetic differentiation and founder effect in colonising populations of common mynahs (*Acridotheres tristus*). *Evolution*, **41**, 525–38.

Barton, N. H. (1986). The effects of linkage and density-dependent regulation on gene flow. *Heredity*, **57**, 415–26.

Barton, N. H. (1989). The divergence of a polygenic system under stabilising selection, mutation and drift. *Genetical Research*, **54**, 59–77.

Barton, N. H. (1989). Founder effect speciation. In *Speciation and its consequences*, (ed. D. Otte and J. A. Endler), pp. 229–56. Sinauer, Sunderland, MA.

Barton, N. H. and Bengtsson, B. O. (1986). The barrier to genetic exchange between hybridising populations. *Heredity*, **57**, 357–76.

Barton, N. H. and Charlesworth, B. (1984). Genetic revolutions, founder effects, and speciation. *Annual Review of Ecology and Systematics*, **15**, 133–64.

Barton, N. H. and Gale, K. S. (1993). Genetic analysis of hybrid zones. In *Hybrid zones and the evolutionary process*, (ed. R. G. Harrison), pp. 13–45. Oxford University Press.

Barton, N. H. and Rouhani, S. (1987). The frequency of shifts between alternative equilibria. *Journal of Theoretical Biology*, **125**, 397–418.

Barton, N. H. and Turelli, M. (1987). Adaptive landscapes, genetic distance, and the evolution of quantitative characters. *Genetical Research*, **49**, 157–74.

Bengtsson, B. O. (1985). The flow of genes through a genetic barrier. In *Evolution: essays in honour of John Maynard Smith*, (ed. J. J. Greenwood, P. H. Harvey, and M. Slatkin), pp. 31–42. Cambridge University Press.

Bengtsson, B. O. and Christiansen, F. B. (1983). A two-locus mutation-selection model and some of its evolutionary implications. *Theoretical Population Biology*, **24**, 59–77.

Bryant, E. H. and Meffert, L. M. (1993). The effect of serial founder flush cycles on quantitative genetic variation in the housefly. *Heredity*, **70**, 122–9.

Bryant, E. H. and Meffert, L. M. (1995). An analysis of selectional response in relation to a population bottleneck. *Evolution*, **49**, 626–34.

Bryant, E. H., McCommas, S. A., and Combs, L. M. (1986). The effect of an experimental bottleneck upon additive genetic variation in the housefly. *Genetics*, **114**, 1191–211.

Burt, A. (1995). The evolution of fitness. *Evolution*, **49**, 1–8.

Carson, H. L. (1968). The population flush and its genetic consequences. In *Population biology and evolution*, (ed. R. C. Lewontin), pp. 123–37. Syracuse University Press.

Carson, H. L. (1975). The genetics of speciation at the diploid level. *American Naturalist*, **109**, 73–92.

Chao, L. (1979). *The population of colicinogenic bacteria: a model for the evolution of allelopathy*. University of Massachussetts Press, Amherst, MA.

Charlesworth, B. (1997). Is founder effect speciation defensible? *American Naturalist*, **149**, 600–3.

Charlesworth, B. and Barton, N. H. (1996). Recombination load associated with selection for increased recombination. *Genetical Research*, **77**, 27–41.

Charlesworth, B. and Charlesworth, D. (1975). An experiment on recombination load in *Drosophila melanogaster*. *Genetical Research*, **25**, 267–74.

Charlesworth, B. and Smith, D. B. (1982). A computer model of speciation by founder effects. *Genetical Research*, **39**, 227–36.

Cohan, F. M. and Hoffmann, A. A. (1989). Uniform selection as a diversifying force in evolution: evidence from Drosophila. *American Naturalist*, **134**, 613–37.

Cohan, F. M., Hoffmann, A. A., and Gayley, T. W. (1989). A test of the role of epistasis in divergence under uniform selection. *Evolution*, **43**, 766–74.

Coyne, J. A., Barton, N. H., and Turelli, M. (1997). A critique of Wright's shifting balance theory of evolution. *Evolution*, **51**, 643–71.

Crow, J. F. and Kimura, M. (1970). *An introduction to population genetics theory*. Harper and Row, New York.

Darwin, C. (1859). *On the origin of species by means of natural selection*. John Murray, London.

DeSalle, R. and Templeton, A. R. (1988). Founder effects and the rate of mitochondrial DNA evolution in Hawaiian Drosophila. *Evolution*, **42**, 1076–84.

Dobzhansky, T. (1937). *Genetics and the origin of species*. Columbia University Press, New York.

Gavrilets, S. and Hastings, A. (1996). Founder effect speciation: a theoretical reassessment. *American Naturalist*, **147**, 466–91.
Grant, P. R. (1986). *Ecolgy and evolution of Darwin's finches*. Princeton University Press.
Grant, P. R. and Grant, B. R. (1995). The founding of a new population of Darwin's finches. *Evolution*, **49**, 229–40.
Hedrick, P. W. (1981). The establishment of chromosomal variants. *Evolution*, **35**, 322–32.
Jackson, K. S. (1992). The population dynamics of a hybrid zone in the alpine grasshopper *Podisma pedestris:* an ecological and genetical investigation: Unpubl. Ph.D thesis, University College London.
Jansson, K. (1987). Genetic drift in small and recently founded populations of the marine snail *Littorina saxatalis*. *Heredity*, **58**, 31–8.
Johnson, M. S., Clarke, B., and Murray, J. (1990). The coil polymorphism in *Partula suturalis* does not favour sympatric speciation. *Evolution*, **44**, 459–64.
Kirkpatrick, M. (1982). Quantum evolution and punctuated equilibrium in continuous genetic characters. *American Naturalist*, **119**, 833–48.
Klein, J., Satta, Y., Takahata, N., and O'Huigin, C. (1993). Trans-specific MHC polymorphism and the origin of species in primates. *Journal of Medicine and Primatology*, **22**, 57.
Kondrashov, A. S. (1988). Deleterious mutations and the evolution of sexual reproduction. *Nature*, **336**, 435–41.
Kondrashov, A. S. (1995). Contamination of the genome by very slightly deleterious mutations: why have we not died 100 times over? *Journal of Theoretical Biology*, **175**, 583–94.
Lande, R. (1979). Effective deme sizes during long-term evolution estimated from rates of chromosomal rearrangement. *Evolution*, **33**, 234–51.
Lande, R. (1985). Expected time for random genetic drift of a population between stable phenotypic states. *Proceedings of the National Academy of Sciences USA*, **82**, 7641–5.
Lynch, M., Conery, J., and Burger, R. (1995). Mutation accumulation and the extinction of small populations. *American Naturalist*, **146**, 489–518.
Mayr, E. (1942). *Systematics and the origin of species*. Columbia University Press, New York.
Mayr, E. (1954). Change of genetic environment and evolution. In *Evolution as a process*, (ed. J. Huxley, A. C. Hardy, and E. B. Ford), pp. 157–80. Allen and Unwin, London.
Mayr, E. (1982).*The growth of biological thought: diversity, evolution and inheritance*. Belknap Press, Cambridge, MA.
McCarthy, E. M., Asmussen, M. A., and Anderson, W. W. (1995). A theoretical assessment of recombinational speciation. *Heredity*, **74**, 502–9.
Moore, W. S. (1979). A single locus mass action model of assortative mating, with comments on the process of speciation. *Heredity*, **42**, 173–86.
Mukai, T. and Yamaguchi, O. (1974). The genetic structure of natural populations of *D. melanogaster.* XI Genetic variability in a local population. *Genetics*, **76**, 339–66.
Muller, H. J. (1942). Isolating mechanisms, evolution and temperature. *Biological Symposia*, **6**, 71–125.
Nei, M., Maruyama, T., and Wu, C.-I. (1983). Models of evolution of reproductive isolation. *Genetics*, **103**, 557–79.
Provine, W. B. (1989). Founder effects and genetic revolutions in microevolution and speciation: an historical perspective. In *Genetics, speciation, and the founder principle*, (ed. L. V. Giddings, K. Y. Kaneshiro, and W. W. Anderson), pp. 43–76. Oxford University Press.
Rice, W. R. and Hostert, E. E. (1993). Laboratory experiments on speciation: what have we learned in forty years? *Evolution*, **47**, 1637–53.
Richman, A. D., Uyenoyama, M. K., and Kohn, J. R. (1996). Allelic diversity and gene genealogy at the self-incompatibility locus in the Solanaceae. *Science*, **273**, 1212–16.
Rieseberg, L. H. (1995). The role of hybridization in evolution: old wine in new skins. *American Journal of Botany*, **82**, 944–53.

Schluter, D. (1996). Ecological causes of adaptive radiation. *American Naturalist*, **148**, supplement, S40–S64.

Searle, J. B. (1986). Factors responsible for a karyotypic polymorphism in the common shrew, *Sorex araneus*. *Proceedings of the Royal Society of London* B, **229**, 277–98.

Sene, F. M. and Carson, H. L. (1977). Genetic variation of Hawaiian Drosophila IV. Allozymic similarity between *D. silvestris* and *D. heteroneura* from the island of Hawaii. *Genetics*, **86**, 187–98.

Slatkin, M. (1996). In defense of founder-flush theories of speciation. *American Naturalist*, **147**, 493–505.

Szymura, J. M. and Barton, N. H. (1991). The genetic structure of the hybrid zone between the fire-bellied toads *Bombina bombina* and *B. variegata*: comparisons between transects and between loci. *Evolution*, **45**, 237–61.

Templeton, A. R. (1980). The theory of speciation via the founder principle. *Genetics*, **94**, 1011–38.

Turelli, M. (1984). Heritable genetic variation via mutation-selection balance: Lerch's zeta meets the abdominal bristle. *Theoretical Population Biology*, **25**, 138–93.

Turelli, M. and Hoffmann, A. A. (1995). Cytoplasmic incompatibility in *Drosophila simulans*: dynamics and parameter estimates from natural populations. *Genetics*, **140**, 1319–38.

Turelli, M. and Orr, H. A. (1995). The dominance theory of Haldane's Rule. *Genetics*, **140**, 389–402.

Virdee, S. R. and Hewitt, G. M. (1994). Clines for hybrid dysfunction in a grasshopper hybrid zone. *Evolution*, **48**, 392–407.

Wagner, A., Wagner, G. P., and Similion, P. (1994). Epistasis can facilitate the evolution of reproductive isolation by peak shifts: a two-locus two-allele model. *Genetics*, **138**, 533–45.

Walsh, J. B. (1982). Rate of accumulation of reproductive isolation by chromosome rearrangements. *American Naturalist*, **120**, 510–32.

Whitlock, M. (1995). Variance-induced peak shifts. *Evolution*, **49**, 252–9.

Whitlock, M. (1997). Founder effects and peak shifts without genetic drift: adaptive peak shifts occur easily when environments fluctuate slightly. *Evolution*, **51**, 1044–8.

Willis, J. H. and Orr, H. A. (1993). Increased heritable variation following population bottlenecks: the role of dominance. *Evolution*, **47**, 949–56.

Wright, S. (1932). The roles of mutation, inbreeding, crossbreeding and selection in evolution. *Proceedings of the Sixth International Congress of Genetics*, **1**, 356–66.

Wright, S. (1940). Breeding structure of populations in relation to speciation. *American Naturalist*, **74**, 232–48.

Wright, S. (1941). On the probability of fixation of reciprocal translocations. *American Naturalist*, **75**, 513–22.

Wright, S. (1982). Character change, speciation, and the higher taxa. *Evolution*, **36**, 427–43.

Wu, C.-I. and Palopoli, M. F. (1994). Genetics of postmating reproductive isolation in animals. *Annual Review of Genetics*, **27**, 283–308.

8

Island hopping in *Drosophila*: genetic patterns and speciation mechanisms

Hope Hollocher

8.1 Introduction

Islands have always been attractive systems for the study of evolution, and in particular speciation, for a wide variety of reasons, the most obvious being that they tend to harbour numerous and splendidly diverse arrays of species not typically found in continental areas. These arrays are characterized by high levels of endemism. Therefore, it follows that islands represent conditions that are extremely conducive for repeated speciation. If we can understand which evolutionary processes are important for speciation on islands, we may be able to define general principles that are applicable to a number of different biological situations.

Islands are also attractive because they define clear geographical boundaries that can be used as a natural framework for looking at patterns of diversification among assemblages of species. The geological formation of the islands themselves suggest hypotheses about the evolution of the species that inhabit them, which adds an extra dimension for testing ideas about specific patterns and processes of speciation. In some cases, information about the geological formation of the islands can provide important corroborative evidence for timing different speciation events. Even if the formation of the islands pre-dates speciation that has occurred on them, they can still serve as a natural sampling design that can be used for testing hypotheses about speciation mechanisms. In order to get the very most out of island studies of speciation, it is important to compare island systems with each other as well as with continental systems in order to develop a broad-based understanding of the relative importance of different evolutionary forces in speciation.

In this chapter I will describe two island systems that have been used to study speciation patterns in *Drosophila*—the Hawaiian islands and the Caribbean islands. The Hawaiian system is probably the best known for studying *Drosophila* evolution. While evolution on the Caribbean islands is not as familiar, it too has been an active area of research for *Drosophila* evolution. Although both island archipelagos began their formation at approximately the same time in history, the two are extremely different with respect to how they were formed, their proximity to the mainland, the number of endemic species that have evolved, and their pattern of speciation. They thus provide a good comparison for evaluating the importance of different speciation mechanisms for the evolution of *Drosophila* on islands in general.

8.2 The Hawaiian Drosophilidae

The study of the evolution of Hawaiian Drosophilidae has had an enormous impact on ideas about speciation on island systems. Begun in 1963 as the Hawaiian Drosophilidae Project by W. S. Stone and D. E. Hardy, the study has involved several dozen researchers who have used a wide range of techniques to understand the biology and evolution of this diverse group. The summary that follows is by no means an exhaustive review of all the work that has been done in the area, but rather serves as an overview of some of the major patterns and processes revealed through this work.

General patterns of speciation

In order to understand the patterns of speciation on the Hawaiian islands, we must first briefly review the geology of the region (Fig. 8.1). The present-day south-east corner of the Big Island of Hawaii sits over a stationary thermal plume of volcanic activity that has given rise to all the islands in the Hawaiian archipelago, past and present (McDougal 1979; reviewed in Carson and Clague 1995). Starting 75 to 80 million years ago (Ma), lava from this plume has built up on the ocean floor forming islands, and then movement of the Pacific plate has carried these newly formed islands to the north-west. Continuous volcanic and plate tectonic activity of this nature has built up the successive islands of the Hawaiian archipelago in a process analogous to a moving conveyor belt, with each new island progressively younger than the one immediately preceding it in the chain. Of the islands existing today, Kauai is the oldest, estimated to have formed 5 to 6 Ma, and Hawaii, the youngest, is estimated to have formed only 0.5 Ma. This pattern of the geological age of the islands decreasing to the south-east holds true not only for the entire island chain, but also within the larger island of Hawaii, in that northern parts of the island are older than southern parts of the island.

The Hawaiian system represents probably the most outstanding example of speciation known to have occurred in the family Drosophilidae. With over 500 extant species already named, and an additional 250 still to be described, estimates for the total number of species of Drosophilidae in the Hawaiian archipelago surpass 1000 (Kaneshiro 1993; Kaneshiro *et al.* 1995). In spite of their morphological diversity, all the species are believed to be closely related phylogenetically, belonging to only two different genera, *Drosophila* and *Scaptomyza* (Throckmorton 1975; Kaneshiro *et al.* 1995). These genera represent sister taxa that began diverging from each other over 24 Ma (Thomas and Hunt 1991; DeSalle 1992, 1995), thus predating the formation of the currently oldest island in the chain, Kauai. This indicates that speciation has been occurring continuously along the Hawaiian archipelago and involves islands that have long since disappeared.

Most details of the patterns of speciation of the Hawaiian *Drosophila* come from studies of the picture-winged species group, which contains 111 species divided into

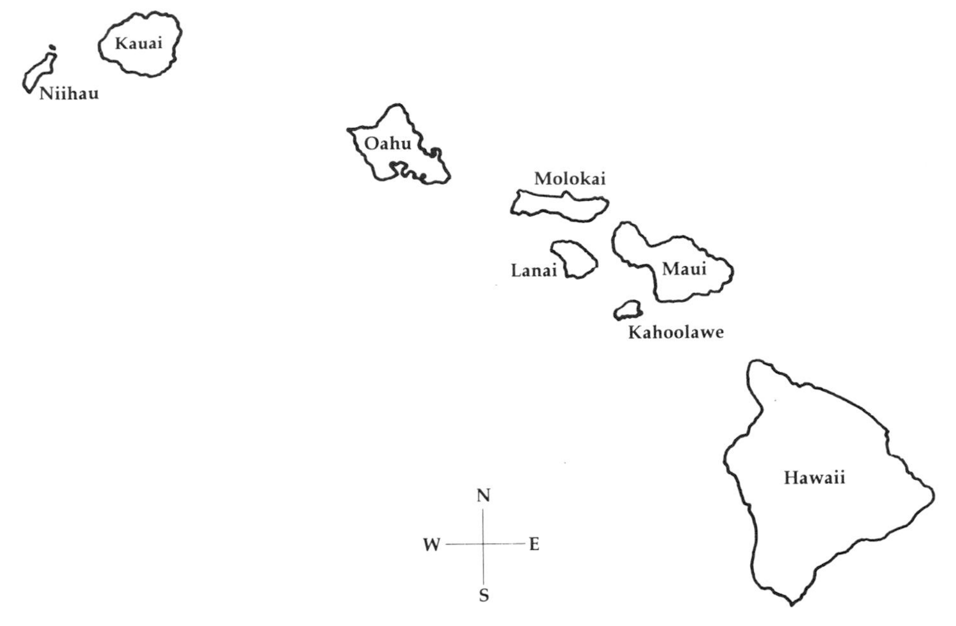

Fig. 8.1 A map of the Hawaiian islands.

approximately a dozen subgroups based on male genitalia (Kaneshiro *et al.* 1995). Initial analysis of polytene chromosomal inversions by Carson (1983) yielded a very detailed genetic phylogeny for the picture-winged subgroups. Because direction cannot be inferred from polytene inversion data alone, Carson (1983, 1987) then used information about the age of the different islands to assign a direction to the network of relationships and to produce a geographical phylogenetic scheme for the flies. In doing so, a general pattern of colonization from older to younger islands was revealed. A more recent re-analysis of the chromosomal data, in which the evolutionary direction was inferred by outgroup comparison and overlaid with the geographical data, showed that the ancestral species for a particular subgroup generally occurred on either Kauai or Oahu, with the oldest species of the entire picture-winged group concentrated on Kauai (Kaneshiro *et al.* 1995). DeSalle (1995) performed sequence and restriction fragment analyses on six subgroups and found evidence that corroborates this same general pattern of speciation from older islands to younger islands and even from older parts to younger parts within the island of Hawaii for different populations of *Drosophila silvestris.*

Together, these data show that the overall pattern of speciation in the Hawaiian picture-winged *Drosophila* is characterized by repeated colonization from older to younger islands. Since most species in this group are single-island endemics and sister taxa tend to occur on adjacent islands, it appears that the formation of new species of Hawaiian *Drosophila* is tightly coupled to colonization between islands. Congruent with this pattern of colonization, shifts in mate recognition systems also appear to have been very important. Sister taxa on different islands tend to be remarkably similar in terms of their ecological niche, yet differ dramatically with respect to morphological and behavioural traits associated with mate recognition (Kaneshiro 1976; Carson and Templeton 1984).

Processes of speciation

The central role of colonization events in the pattern of speciation for the Hawaiian picture-winged *Drosophila* was very influential in the formulation of founder effect speciation theory as elaborated by Carson and Templeton (Templeton 1980, 1981, 1982; Carson and Templeton 1984). The idea of founder effect speciation has its roots in earlier ideas expressed by Mayr. Greatly influenced by the views of Wright, Mayr felt that because organisms were balanced, integrated genetic systems characterized by high levels of fitness epistasis, natural selection would be slow and only marginally effective at bringing about speciation (Mayr 1963; Provine 1989). In expressing his view, Mayr wrote, 'The real problem of speciation is not how to produce difference but rather to escape from the cohesion of the gene complex' (Mayr 1963, p. 518). For Mayr, founder events provided one such escape route. Because response to selection of a particular allele in a population is a function of its complex interactions with alleles at other loci, the response could change as a function of population structure. Mayr saw small founding populations as changing the genetic structure of a population in such a way as to set the stage for selection to act in directions previously blocked by the genetic structuring of the parental population.

Mayr's model of 'genetic revolution', as he originally conceived it, involved high levels of inbreeding which would result in more and more recessive alleles being exposed to natural selection. Inbreeding could change the selective values of different alleles causing some alleles that were favoured in the original population to be lost, and as the population regained equilibrium with natural selection, new integrated gene systems eventually could be stabilized. Because Mayr's model required high levels of homozygosity which in the long run would not actually be conducive to rapid response to selection, Carson (1975) proposed his founder-flush model and Templeton (1980) proposed his genetic transilience model, both of which relaxed the requirement for extended periods of inbreeding which would otherwise lead to the inevitable loss of genetic variability (Carson and Templeton 1984). In their models, genetic restructuring occurs during the initial stages of the founder event, followed by a period of rapid population recovery during which genes will change in response to either greatly relaxed selection in the new ecological environment (Carson 1975) or to selection in the new genetic environment triggered by the founding event (Templeton 1980). These models also differ from Mayr's original proposal in not requiring the genetic restructuring to involve the entire genome and, in fact, they predict that different genetic systems will respond differently to the founder event. For example, neutral nuclear genes are not expected to change much at all under these models. On the other hand, because they are haploid and maternally inherited, mitochondrial genes are expected to be more sensitive to the founder event and show reduced variability. This predicted discrepancy between nuclear and mitochondrial genes has been confirmed in studies by DeSalle and Giddings (1986) and Templeton *et al.* (1987).

Just as neutral traits are expected to be more or less sensitive to founder events depending on which part of the genome is considered, selected traits with different underlying genetic architectures are also expected to be differentially susceptible to changes brought about by founder events. Templeton (1980, 1982) continually stressed that the genetic architecture underlying a trait, i.e. the number and types of genes and their interaction, is very important in determining how that particular trait will respond to founder events. He suggested that traits governed by a few major genes with many epistatic modifiers were more likely to experience a genetic transilience than additive polygenic traits. Initial evidence supporting the importance of a few major genes with epistatic modifiers in speciation comes from genetic analysis of the *abnormal abdomen* system which was found to be responsible for isolation in artificial bottlenecks in *D. mercatorum* (Templeton and Rankin 1978; Templeton 1979*a*, 1982; Hollocher, *et al.* 1992), and from the analysis of head shape differences between two picture-winged species, *D. heteroneura* and *D. silvestris* (Templeton 1977; Val 1977).

Although the idea of founder effect speciation has been very well received, it has not been universally accepted (see Chapter 7). The crux of the controversy involves the relative importance of random genetic drift versus natural selection in promoting speciation and how exactly these two evolutionary forces manifest their effects in founder populations. This controversy surrounding founder effect speciation is not new, and represents but one phase of a 'persistent controversy' that was started by Fisher and Wright back in the 1930s (Provine 1989) and will mostly likely persist until empirical studies provide new bases for useful model building. Opponents of founder effect speciation favour Fisher's genetic view over Wright's, and dismiss the idea that speciation is impeded by genetic cohesion of species. Natural selection operating in moderately sized populations is favoured as being most effective for promoting rapid speciation (Barton and Charlesworth 1984; Barton 1989; Provine 1989). These arguments are based on models that evaluate the effects of founder events on speciation by assessing the probability that genetic drift alone will cause a shift from one adaptive equilibrium or peak to a new adaptive peak separated from the first by a valley of lower average fitness. Generally, these models show that the probability a population will shift to a new adaptive peak during a founder event is relatively low and that such a shift will not generate very high levels of reproductive isolation by itself. However, the models fail to reflect accurately all the processes described in founder effect speciation, which involve not only drift, but also emphasize changes in non-additive effects between genes that could shift during a founder event as well as natural selection following the initial founder event. Such dynamic effects are never suitably analysed in these models, since they only deal in one phenotypic dimension (Carson and Templeton 1984; Templeton, personal communication).

Other models which come closer to representing the genetic events that could be occurring during founder effect speciation have been proposed by Goodnight (1987, 1988) and Cheverud and Routman (1996), who have demonstrated that founder events can trigger epistatic genetic variance to be converted to additive genetic variance, thus allowing a renewed response to selection after a founder event which could result in a shift to a new adaptive peak. Also, Bryant and Meffert (1988, 1990, 1996) have shown an increase in additive genetic variance affecting morphological traits to result from a bottleneck, indicating that complex genetic interactions between traits can be shifted during a founder event, promoting evolution along new trajectories. By incorporating changes in epistasis into their models, these researchers have concluded that founder events may be important in promoting speciation, contradicting the results of models that largely dismiss the importance of epistatic effects.

More recently Wagner *et al.* (1994) have taken Barton's (1989) original model and incorporated epistasis in a way which allows for gene effects to be context dependent rather than one dimensional and have found that this can increase the probability of the peak shift and result in much greater levels of reproductive isolation than the original additive models. Although Wagner *et al.* (1994) were modelling shifts in moderately sized populations rather than in extreme founder events, their results illustrate the large impact that different assumptions about genetic architecture can have on the outcome of the same evolutionary model. Similarly, models incorporating epistasis explored by Gavrilets and Hastings (1996) also show that genetic changes induced by founder events can lead to rapid peak shifts. Interestingly, the main effect of adding epistasis in these models is to change the path that a population can follow as it shifts from one fitness optimum to another, allowing it to avoid having to traverse the point of lowest fitness, making peak shifts via drift more probable (Wagner *et al.* 1994; Gavrilets and Hastings 1996). Thus, there is ample theoretical evidence that founder events can trigger genetic changes that could result in the evolution of new fitness equilibria. There is also a growing body of experimental evidence that has demonstrated founder events can facilitate the evolution of reproductive isolation (Ringo *et al.* 1985; Templeton 1989; Powell 1989; Meffert and Bryant 1991; Templeton 1996; for a contrasting view see Rice and Hostert 1993) as well as morphological divergence (Bryant and Meffert 1988, 1990, 1996).

8.3 The Caribbean Drosophilidae

Because so many of the specific characteristics displayed by the Hawaiian *Drosophila* went into formulating the model of founder effect speciation itself, it is important to move away from this system in order to evaluate the different components of the model for their relative importance in speciation. Speciation of Drosophilidae in the Caribbean pales in comparison to that of the Hawaiian islands. Although the numbers of species are decidedly less impressive, the Caribbean island system nonetheless possesses some unique characteristics that make it particularly suitable for comparative studies of speciation with the Hawaiian islands. As a point of comparison, the Caribbean is intermediate between the intense isolation of the Hawaiian islands and the continuous distribution of species on continents. Because the Caribbean islands show an intermediate level of isolation from the mainland, we may be able to tease apart the effect of the actual founder event in promoting genetic reorganization from the extreme geographical isolation that invariably is associated with the founder event in the Hawaiian islands. The Caribbean Drosophilidae also have very closely related continental groups that the Hawaiian Drosophilidae lack and these can be used for comparison with the insular forms. For closely related Hawaiian species, only island versus island comparisons can be made; however, with Caribbean species, comparisons of mating behaviour, reproductive isolation, and morphological traits can all be made between continental and insular forms, giving more insight into what types of evolutionary changes may distinguish island species from mainland species. By investigating these two distinct island systems for the same taxonomic group of organisms, we can begin to distinguish the relative role of different evolutionary processes in speciation on islands.

Patterns of speciation

Unlike the rather simple geological formation of the Hawaiian islands, the formation of the Caribbean islands (Fig. 8.2) is complex and still controversial (reviewed in Donnelly 1988). Although situated rather close to continental areas, none of the Antillean islands is thought to be a continental fragment. The Greater Antilles were formed beginning about 80 Ma from an independent Cretaceous island arc that was stretched and fragmented into a discontinuous series of islands during the Cenozoic period as the Caribbean plate moved into its present location about 55 Ma. The formation of the Greater Antilles was not smooth, and at various points in its history parts of different islands were pushed together just to be later pulled apart. Early in the history of the Greater Antilles connections to the Central American mainland may have existed between Cuba and the Yucatán and between Honduras and Jamaica, although it is not known how important these connections were for the present biogeography. It is also not clear which island areas were submerged during the middle of the Cenozoic, although it is rather likely that Jamaica was completely under water about 30 Ma (Donnelly 1988).

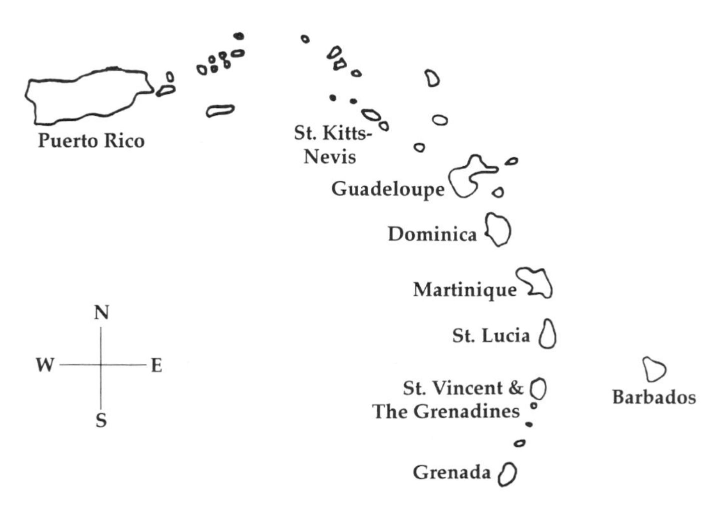

Fig. 8.2 A map of Puerto Rico and the Lesser Antilles.

In contrast to the ancient and chaotic formation of the Greater Antilles, the Lesser Antilles were built up by a gradual accumulation of volcanic material starting 20 Ma to the present. Although built in part on fragments of the Greater Antilles arc, the Lesser Antilles were never subjected to elaborate plate tectonic movements and essentially developed where they are situated today. Water levels between islands of the Lesser Antilles are high and always have been, indicating that land connections never existed between these islands. In addition, no land bridges are thought to have linked South America with the Lesser Antilles. Therefore, similar to the Hawaiian archipelago, all colonization of the Lesser Antilles must have occurred by over-water dispersal (Donnelly 1988).

Fossil representatives of Caribbean Drosophilidae preserved in Dominican amber indicate that ancestral groups have existed in the Greater Antilles since the early Miocene, 23 Ma (Grimaldi 1987), prior to the formation of the Lesser Antilles. The endemic Caribbean Drosophilidae consists of a total of 58 Antillean species spread among nine different genera, of which only one, *Mayagueza*, is endemic to the region (Grimaldi 1988). All other species belong to groups that have representatives outside the immediate Caribbean islands. Only two Caribbean *Drosophila* species groups are represented by more than a handful of species endemic to the Antilles, the *Drosophila repleta* group and the *Drosophila cardini* group (Grimaldi 1988). Because the *repleta* group has been described in detail recently (Grimaldi 1988), I will confine my attention to the patterns of speciation in the *cardini* group which consists of 16 species, eight of which are confined to the Greater and Lesser Antilles and eight of which have ranges in continental tropical America (Heed and Krishnamurthy 1959; Heed 1962).

The *cardini* group has been divided into two subgroups, the *Drosophila dunni* subgroup which has species endemic to Puerto Rico, Jamaica, and the islands of the Lesser Antilles, and the *Drosophila cardini* subgroup which has species which generally do not inhabit the islands (Table 8.1; Heed 1962). These *Drosophila* are particularly interesting because the island species of the *dunni* subgroup show a more or less regular cline in abdominal pigmentation, a rare phenomenon in *Drosophila* that suggests the operation of natural selection (Fig. 8.3; Heed and Krishnamurthy 1959). In addition to the interesting biogeographical distribution of species within this group, hybridization studies show the island species to represent the entire range of possible genetic relationships from complete fertility in the F_2 generation to complete reproductive isolation (both pre- and post-mating) (Heed and Krishnamurthy 1959; Heed 1962; Futch 1962). Relationships among the different species have also been analysed through examination of male genitalia (Heed 1962), inversion polytene chromosome analysis and interspecific hybridizations (Heed and Krishnamurthy 1959; Heed and Russell 1971), and more recently in my laboratory, analysis of DNA sequence from the mitochondrial genome.

Figures 8.4 and 8.5 show how the different methods of analysis compare for assigning relationships between the different species in the group. Although the different data sets favour slightly different relationships, there are certain groupings that are strongly supported by all the data. The greatest discrepancy involves the inclusion of *D. belladunni* (the Jamaica species) in the *dunni* subgroup. This relationship was not supported by the mitochondrial tree, although the molecular data do pair this species

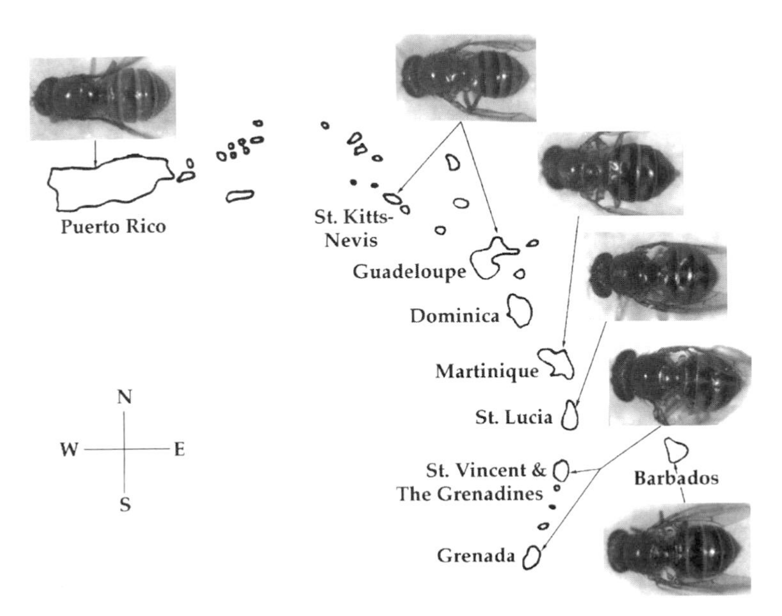

Fig. 8.3 A map of Puerto Rico and the Lesser Antilles showing female abdominal pigmentation patterns of the *dunni* subgroup. The species corresponding to the different islands are as follows: *D. dunni* (Puerto Rico), *D. arawakana* (St Kitts and Guadeloupe), *D. caribiana* (Martinique), *D. antillea* (St Lucia), *D. similis* (Grenada and St Vincent), and *D. nigrodunni* (Barbados).

with *D. acutilabella* and reveal them to be the sister taxa to all the other species in the *dunni* subgroup. In this sense, these two species form an insular bridge between the continental species of the *cardini* subgroup and the insular species of the *dunni* subgroup. This supports the idea that all the insular species are derived from a common ancestor and have not resulted from independent colonizations by different species on the mainland. Although physically close to the mainland, the Caribbean islands are sufficiently isolated to reduce gene flow significantly and allow for differentiation to occur between the different island and mainland species. The data also show that colonization by the insular forms has not followed a simple stepping-stone model from one end of the island chain to the other. Instead, species spread from Jamaica (or some other related source) to the two ends of the islands chain (Puerto Rico and St Thomas at one end and Grenada and St Vincent at the other). This was then followed by colonization of Martinique from Grenada/St Vincent with subsequent radiation to the islands of Guadeloupe, St Lucia, and Barbados.

In general, the different traits that distinguish the species do not always couple tightly with the pattern of colonization. The pattern of island hopping in the *dunni* subgroup described above contrasts with the regular cline in abdominal pigmentation. Although colonizations have obviously played an important role in speciation in this group since sister taxa occur on different islands rather than within islands, natural selection has also worked on these species as shown by the changes in the abdominal pigmentation.

Table 8.1 Species of the *Drosophila cardini* group and their distributions

Neotropical mainland	West Indies
1. Western hemisphere tropics	1. Florida, Cuba, Jamaica, Hispaniola
D. cardini	*D. acutilabella*
2. Mexico to Brazil	2. Jamaica
D. cardinoides	*D. belladunni*
3. Mexico to Trinidad	3. Puerto Rico–St Thomas
D. parthenogenetica	*D. dunni dunni*
D. neomorpha	*D. dunni thomasensis*
D. bedicheki	
4. South America	4. St Kitts–Guadeloupe
D. polymorpha	*D. arawakana arawakana*
D. neocardini	*D. arawakana kittensis*
5. Andes of Bolivia and Peru	5. Martinique
D. procardinoides	*D. caribiana*
	6. St Lucia
	D. antillea
	7. Barbados
	D. nigrodunni
	8. St Vincent–Grenada
	D. similis similis
	D. similis grenadensis

Light and dark species tend to be more closely related to each other than the dark species are to each other or the light species are to each other. Therefore, pigmentation pattern is not historically constrained and selection on this trait may be separable from the initial effects of the founder event. At this point we are not able to determine whether the colonization event itself may be able to facilitate changes in abdominal pigmentation; however, a closer examination of the evolution of genes responsible for the differences in pigmentation may lead to an answer in the future.

Reproductive isolation follows the pattern of colonization more linearly, yet there are still inconsistencies that warrant closer inspection. For example, *D. caribiana,* which inhabits Martinique, shows greater post-mating isolation with all the species in the *dunni* subgroup than is expected through chromosome analysis (Heed and Russell 1971) and mitochondrial sequence analysis (Hollocher *et al.* unpublished data), indicating that post-mating isolation may have evolved faster in this particular species. In addition, the two species that show the closest affinities in terms of mitochondrial sequence data, *D. nigrodunni* and *D. antillea*, show very different levels of reproductive isolation when mated with the other species in the *dunni* subgroup (Heed 1962; Hollocher *et al.* unpublished results), with *D. antillea* being incompatible with most other species whereas *D. nigrodunni* forms fertile hybrid females with most (Heed and Krishnamurthy 1959; Heed 1962).

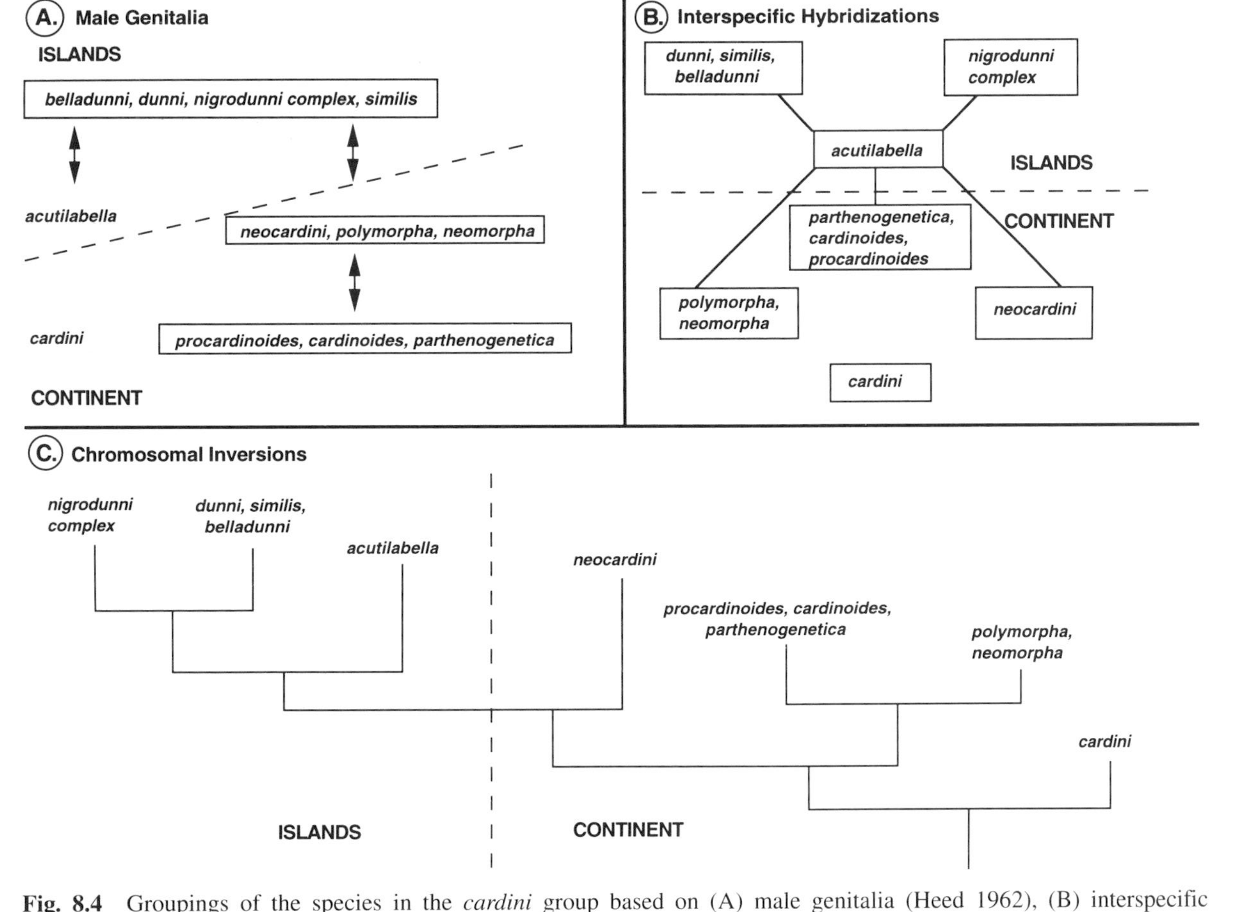

Fig. 8.4 Groupings of the species in the *cardini* group based on (A) male genitalia (Heed 1962), (B) interspecific hybridizations (Heed and Krishnamurthy 1959; Heed and Russell 1971), and (C) cytology (Heed and Russell 1971)

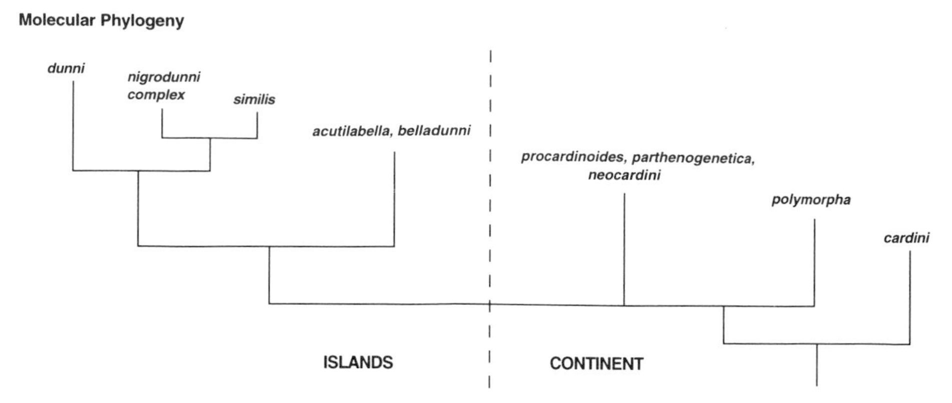

Fig. 8.5 Groupings of the species in the *cardini* group, based on mitochondrial sequence analysis (Hollocher *et al.* unpublished data).

Processes of speciation

Carson and Templeton (1984) make a point that not all species are equally prone to founder effect speciation. Even within *Drosophila*, different groups are not equally susceptible to speciation via founder events due to differences in population structure, system of mating, and even their chromosomal constitution. Therefore, it is important to determine whether the insular species in the *D. cardini* group meet the criteria that suggest they would be susceptible to genetic change via a founder event. Cosmopolitan species or generalists are not expected to be able to respond readily to a founder event (Carson and Templeton 1984). Species that are particularly good at invading new habitats usually have a 'general purpose genotype' characterized by a balanced system of heterosis. Because of the balancing selection involved in maintaining these systems, they can be easily carried through the founder event without much change and therefore would be most likely not to experience a genetic shift. The species of the *dunni* subgroup are not considered generalists. Although specific details of the breeding sites of the different species are not known, the flies have not been found in areas which commonly attract other cosmopolitan species. They are confined to middle elevation, native, and secondary growth forests (Heed and Krishnamurthy 1959; Hollocher *et al.* personal observation).

Carson and Templeton also hypothesized that founder effect speciation is more effective if crossover suppressors are eliminated or fixed so that recombination is free to establish new gene complexes (Carson 1975; Templeton 1980, 1982; Carson and Templeton 1984). As positive evidence that free recombination may be important, they cite how inversion polymorphisms are not carried over during speciation in the Hawaiian *Drosophila*. Examination of the species in the *dunni* subgroup show them to have the same low level of inversion polymorphism relative to the continental species as seen in the Hawaiian *Drosophila* (Heed and Russell 1971). Therefore if founder effect speciation is important, it would not be inhibited by blocked recombination in the insular species of the *D. cardini* group. It is possible that founder effect speciation may be operating within the *dunni* subgroup; however, its role in this group cannot be properly evaluated without analysis of the underlying genetic architecture of the traits that distinguish the different species.

In two important respects, the *dunni* subgroup does show distinct differences from the Hawaiian *Drosophila*, differences that may affect how the two species groups have responded genetically to colonization. First, the species in the Antilles are not as isolated from each other as are the Hawaiian species. Therefore, some speciation in the *dunni* subgroup may not have involved complete isolation from the source population during the entire process of differentiation. Of course, recurring gene flow would counteract the genetic effects resulting from the founding event. However, because the species of the Lesser Antilles are single-island endemics, colonization is rare relative to the formation of new species on each island. If the genetic reorganization triggered by a founder event is rapid enough, then subsequent gene flow may have little impact on the system. Other species of the *cardini* group in general do not occur on the Lesser Antilles, except for *D. cardini*, which is unable to crossbreed with any of the other species in the group and therefore would not be able to contribute to gene flow. Because species from adjacent islands are sometimes able to produce fertile offspring in the laboratory, there exists the formal possibility that gene flow between more distant islands could result through island intermediaries. However, laboratory crosses performed to test how effective this mechanism of gene flow would be showed that there was strong selection against the formation of genetic combinations from more distant islands. Only a very small subset of the possible genetic combinations could be produced through a series of crosses, indicating that there are strong incompatibilities between these species at several levels. This makes it unlikely that continuous gene flow had occurred during their formation (Heed 1962).

A second important distinction between the Hawaiian *Drosophila* and the Caribbean *Drosophila* is how responsive the mating behaviour has been to speciation. Although differences in male genitalia have occurred in both groups, the elaboration and diversification of courtship in the Hawaiian picture-winged *Drosophila* is not paralleled in the Caribbean *Drosophila*. Founder effect speciation can be very effective when coupled with sexual selection (Kaneshiro 1980, 1983, 1989), and once elaborate mating behaviour first evolved in the Hawaiian *Drosophila*, that system was more susceptible to founder effects (Ringo 1977; Templeton 1979*b*; Giddings and Templeton 1983). It is necessary to determine how tightly founder effect speciation needs to be coupled both to complete allopatry and to sexual selection to determine how generally it can be applied to other systems outside of the Hawaiian islands.

In addition to the genetic considerations outlined above, the Hawaiian and Caribbean systems differ drastically with respect to their community ecology which could also impact speciation. The extreme isolation of the Hawaiian islands means that many members of mainland communities that serve as sources of colonists simply do not make it out to the Hawaiian islands. In addition, the Hawaiian islands were formed sequentially in space and time; therefore, colonization from old to new islands most likely represented the filling of unoccupied niches by whatever flora and fauna happened to be available from nearby islands and more rarely from the mainland. Coupling these two effects, the community dynamics of the islands is very different from the mainland and the *Drosophila* fauna that results is very species rich, but poor. The Caribbean islands, on the other hand, were formed simultaneously an much closer to the corresponding mainland. Taking these two effects into account, can see that colonizations in the Caribbean are more likely to involve displacement of species which already occupy the available niches. Because of their proximity to the mainland, the Caribbean islands contain a better representation of the entire mainland community, meaning that opportunities for colonists to respond to open niches are limited in the Caribbean relative to more remote island systems. The *Drosophila* fauna that resulted in the Caribbean fits this general pattern by being very genera rich, but species poor. The ecological differences between the two island systems should not be construed as a negation of the importance of founder effect speciation, but serve to emphasize how speciation is a complex process that depends not only on the opportunity for genetic changes to occur, but also the ecological context surrounding those changes.

8.4 Conclusions

Inferring process from pattern is and always will be challenging. It is clear from the above discussion that a final decision regarding the importance of founder effect speciation in natural populations cannot yet be made. The formulation of the model relied principally on characteristics of the Hawaiian *Drosophila*. How general the process is outside this system has not been adequately explored to decide which elements must co-occur for founder effect speciation to be plausible. However, there are important concepts contained in the model which do not only apply to founder effects, and need to addressed in all aspects of speciation. Reducing the controversy to the relative importance of drift versus selection undervalues the genetic issues of speciation that the founder effect model of speciation was originally formulated to address.

It is becoming increasingly obvious that the genetic architecture of traits involved in speciation is a central issue which begs for more empirical work. The genetics of species divergence in general is still a little-understood area of research, although some headway has been made recently with respect to the genetic basis of reproductive isolation in *Drosophila* (Coyne 1992; Wu and Palopoli 1994; Coyne *et al.* 1994; Wu *et al.* 1995). Not only does genetic architecture play a role in determining the effectiveness of different evolutionary processes, but it also appears that different traits considered important in speciation, such as post-mating reproductive isolation, sexual isolation, and morphological differences, all show varying susceptibility to different evolutionary processes as well, because of the nature of the traits themselves. Post-mating reproductive isolation is never directly selected during speciation, but rather evolves as a pleiotropic consequence of divergence being caused by some other evolutionary process. Therefore it consistently shows an entirely different genetic patterning than pre-mating isolation, which is more likely to be directly selected during speciation, even when both these traits occur in the same species pair (Wu and Palopoli 1994; Coyne *et al.* 1994). Similarly, hybrid sterility has been shown to have evolved quite differently from hybrid inviability (Wu and Davis 1993; Orr 1993; Hollocher and Wu 1996; True *et al.* 1996).

It is now clear that hybrid sterility and inviability are fundamentally different physiologically and evolutionarily even though they are often lumped together under the rubric of post-mating reproductive isolation. Therefore, what are truly needed are more systematic analyses of the genetic architecture of several different traits simultaneously for different related species, in both the Hawaiian and the Caribbean species groups, to help resolve the genetic issues tackled by founder effect speciation models, issues which need to be confronted for all aspects of speciation.

8.5 Summary

Radiation of *Drosophila* along the Hawaiian archipelago has resulted in an astounding array of diversity. The speciation in this group corresponds well to the geological history of the region and colonization events appear to have been a major contributing factor. Although much less impressive in terms of diversity, *Drosophila* have also radiated throughout the Caribbean islands. In contrast to the pattern exhibited in the Hawaiian islands, major changes that distinguish the species in the Caribbean are not always coupled to colonization events. The patterns of speciation for these two island groups are compared and contrasted in light of founder effect speciation models.

Acknowledgements

I would like to thank Peter Simon, Anna Maria Hibbs, and David Kutzler for their technical assistance in working with the Caribbean *Drosophila*. I would also like to thank Alan R. Templeton and all the graduate students and postdoctoral students at Washington University and the University of Chicago for useful discussion concerning founder effect speciation during the time I was at both of these universities. I thank Bill Heed for our discussions of the *Drosophila cardini* group.

References

Barton, N. H. (1989). Founder effect speciation. In *Speciation and its consequences*, (ed. D. Otte and J. A. Endler), pp. 229–56. Sinauer, Sunderland, MA.

Barton, N. H. and Charlesworth, B. (1984). Genetic revolutions, founder effects, and speciation. *Annual Review of Ecology and Systematics*, **15**, 133–64.

Bryant, E. H. and Meffert, L. M. (1988). Effect of an experimental bottleneck on morphological integration in the housefly. *Evolution*, **42**, 698–707.

Bryant, E. H. and Meffert, L. M. (1990). Multivariate phenotypic differentiation among bottleneck lines of the housefly. *Evolution*, **44**, 660–8.

Bryant, E. H. and Meffert, L. M. (1996). Morphometric differentiation in serially bottlenecked populations of the housefly. *Evolution*, **50**, 935–40.

Carson, H. L. (1975). The genetics of speciation at the diploid level. *American Naturalist*, **109**, 83–92.

Carson, H. L. (1983). Chromosomal sequences and interisland colonizations in Hawaiian *Drosophila*. *Genetics*, **103**, 465–82.

Carson, H. L. (1987). Tracing ancestry with chromosomal sequences. *Trends in Ecology and Evolution*, **2**, 203–7.

Carson, H. L. and Clague, D. A. (1995). Geology and biogeography of the Hawaiian islands. In *Hawaiian biogeography: evolution on a hot-spot archipelago*, (ed. W. L. Wagner and V. A. Funk), pp. 14–29. Smithsonian Institution Press, Washington, D. C.

Carson, H. L. and Templeton, A. R. (1984). Genetic revolutions in relation to speciation phenomena: The founding of new populations. *Annual Review of Ecology and Systematics*, **15**, 97–131.

Cheverud, J. M. and Routman, E. J. (1996). Epistasis as a source of increased additive genetic variance at population bottlenecks. *Evolution*, **50**, 1042–51.

Coyne, J. A. (1992). Genetics and speciation. *Nature*, **355**, 511–15.

Coyne, J. A., Crittenden, A. P., and Mah, K. (1994). Genetics of a pheromonal difference contributing to reproductive isolation in *Drosophila*. *Science*, **265**, 1461–4.

DeSalle, R. (1992). The origin and possible time of divergence of the Hawaiian Drosophilidae: evidence from DNA sequences. *Molecular Biology and Evolution*, **9**, 905–16.

DeSalle, R. (1995). Molecular approaches to biogeographic analysis of Hawaiian Drosophilidae. In *Hawaiian biogeography: evolution on a hot-spot archipelago*, (ed. W. L. Wagner and V. A. Funk), pp. 72–89. Smithsonian Institution Press, Washington, D. C.

DeSalle, R. and Giddings, L. V. (1986). Discordance of mitochondrial and nuclear DNA phylogenies of Hawaiian *Drosophila*. *Proceedings of the National Academy of Sciences USA*, **83**, 6902–6.

Donnelly, T. W. (1988). Geologic constraints on Caribbean biogeography. In *Zoogeography of Caribbean insects*, (ed. J. K. Liebherr), pp. 15–37. Cornell University Press, Ithaca, NY.

Futch, D. G. (1962). Hybridization tests within the *cardini* species group of the genus *Drosophila*. *Univiversity of Texas Publication*, No. 6205, 539–54.

Gavrilets, S. and Hastings, A. (1996). Founder effect speciation: a theoretical reassessment. *American Naturalist*, **147**, 466–91.

Giddings, L. V. and Templeton, A. R. (1983). Behavioral phylogenies and the direction of evolution. *Science*, **220**, 372–7.

Goodnight, C. J. (1987). On the effect of founder events on epistatic genetic variance. *Evolution*, **41**, 80–91.

Goodnight, C. J. (1988). Epistasis and the effect of founder events on the additive genetic variance. *Evolution*, **42**, 441–54.

Grimaldi, D. A. (1987). Amber fossil Drosophilidae (Diptera), with particular reference to the Hispaniolan taxa. *American Museum Novitates*, No. 2880, 1–23.

Grimaldi, D. A. (1988). Relicts in the Drosophilidae (Diptera). In *Zoogeography of Caribbean insects*, (ed. J. K. Liebherr), pp. 183–213. Cornell University Press, Ithaca, NY.

Heed, W. B. (1962). Genetic characteristics of island populations. *University of Texas Publication*, No. 6205, 173–206.

Heed, W. B. and Krishnamurthy, N.B. (1959). Genetic studies on the *cardini* group of *Drosophila* in the West Indies. *University of Texas Publication*, No. 5914, 155–79.

Heed, W. B. and Russell, J. S. (1971). Phylogeny and population structure in island and continental species of the *cardini* group of *Drosophila* studied by inversion analysis. *University of Texas Publication*, No. 7103, 91–130.

Hollocher, H. and Wu, C.-I (1996). The genetics of reproductive isolation in the *Drosophila simulans* clade: X *vs.* autosomal effects and male *vs.* female effects. *Genetics*, **143**, 1243–55.

Hollocher, H., Templeton, A. R., DeSalle, R., and Johnston, J. S. (1992). The molecular through ecological genetics of *abnormal abdomen*. IV. Components of genetic variation in a natural population of *Drosophila mercatorum*. *Genetics*, **130**, 355–66.

Kaneshiro, K. Y. (1976). Ethological isolation and phylogeny in the *planitibia* subgroup of Hawaiian *Drosophila*. *Evolution*, **30**, 740–5.

Kaneshiro, K. Y. (1980). Sexual isolation, speciation and the direction of evolution. *Evolution*, **34**, 437–44.
Kaneshiro, K. Y. (1983). Sexual selection, and direction of evolution in the biosystematics of Hawaiian Drosophilidae. *Annual Review of Entomology*, **28**, 161–78.
Kaneshiro, K. Y. (1989). The dynamics of sexual selection and founder effects in species formation. In *Genetics, speciation, and the founder principle*, (ed. L. V. Giddings, K. Y. Kaneshiro, and W. W. Anderson), pp. 279–96. Oxford University Press.
Kaneshiro, K. Y. (1993). Habitat-related variation and evolution by sexual selection. In *Evolution of insect pests*, (ed. K. C. Kim and B. A. McPheron), pp. 89–101. John Wiley & Sons, New York.
Kaneshiro, K. Y., Gillespie, R. G., and Carson, H. L. (1995). Chromosomes and male genitalia of Hawaiian *Drosophila*: Tools for interpreting phylogeny and geography. In *Hawaiian biogeography: evolution on a hot-spot archipelago*, (ed. W. L. Wagner and V. A. Funk), pp. 57–71. Smithsonian Institution Press, Washington, D. C.
Mayr, E. (1963). *Animal species and evolution*. Harvard University Press, Cambridge, MA.
McDougal, I. (1979). Age of shield-building volcanism of Kauai and linear migration of volcanism on the Hawaiian Island chain. *Earth and Planetary Science Letters*, **46**, 31–42.
Meffert, L. M. and Bryant, E. H. (1991). Mating propensity and courtship behavior in serially bottlenecked lines of housefly. *Evolution*, **45**, 292–306.
Orr, H. A. (1993). Haldane's rule has multiple genetic causes. *Nature*, **361**, 532–3.
Powell, J. R. (1989). The effects of founder-flush cycles on ethological isolation in laboratory populations of *Drosophila*. In *Genetics, speciation, and the founder principle*, (ed. L. V. Giddings, K. Y. Kaneshiro, and W. W. Anderson), pp. 239–51. Oxford University Press.
Provine, W. B. (1989). Founder effects and genetic revolutions in microevolution and speciation: an historical perspective. In *Genetics, speciation, and the founder principle*, (ed. L. V. Giddings, K. Y. Kaneshiro, and W. W. Anderson), pp. 43–76. Oxford University Press.
Rice, W. R. and Hostert, E. E. (1993). Laboratory experiments on speciation: what have we learned in 40 years? *Evolution*, **47**, 1637–53.
Ringo, J. M. (1977). Why 300 species of Hawaiian *Drosophila*? The sexual selection hypothesis. *Evolution*, **31**, 694–6.
Ringo, J., Wood, D., Rockwell, R., and Dowse, H. (1985). An experiment testing two hypotheses of speciation. *American Naturalist*, **126**, 642–61.
Templeton, A. R. (1977). Analysis of head shape differences between two interfertile species of Hawaiian *Drosophila*. *Evolution*, **31**, 630–42.
Templeton, A. R. (1979*a*). The unit of selection in *Drosophila mercatorum*. II. Genetic revolutions and the origin of coadapted genomes in parthenogenetic strains. *Genetics*, **92**, 1283–93.
Templeton, A. R. (1979*b*). Once again, why 300 species of Hawaiian *Drosophila*? *Evolution*, **33**, 513–17.
Templeton, A. R. (1980). The theory of speciation via the founder principle. *Genetics*, **91**, 1011–38.
Templeton, A. R. (1981). Mechanisms of speciation—a population genetic approach. *Annual Review of Ecology and Systematics*, **12**, 23–48.
Templeton, A. R. (1982). Genetic architectures of speciation. In *Mechanisms of speciation*, (ed. C. Barigozzi), pp. 105–121. Liss, New York.
Templeton, A. R. (1989). Founder effects and the evolution of reproductive isolation. In *Genetics, speciation, and the founder principle*, (ed. L. V. Giddings, K. Y. Kaneshiro, and W. W. Anderson), pp. 329–44. Oxford University Press.
Templeton, A. R. (1996). Experimental evidence for the genetic-transilience model of speciation. *Evolution*, **50**, 909–15.
Templeton, A. R. and Rankin, M. A. (1978). Genetic revolutions and control of insect populations. In *The screwworm problem*, (ed. R. H. Richardson), pp. 83–112. University of Texas Press, Austin, TX.

Templeton, A. R., Davis, S. K., and Read, B. (1987). Genetic variability in a captive herd of Speke's gazelle (*Gazella spekei*). *Zoo Biology*, **6**, 305–13.

Thomas, R. H. and Hunt, J. A. (1991). The molecular evolution of the alcohol dehydrogenase locus and the phylogeny of Hawaiian *Drosophila*. *Molecular Biology and Evolution*, **8**, 687–702.

Throckmorton, L. (1975). The phylogeny, ecology and geography of *Drosophila*. In *Handbook of genetics*, Vol. 3, (ed. R. C. King), pp. 421–69. Plenum, New York.

True, J. R., Weir, B. S., and Laurie, C. C. (1996). A genome-wide survey of hybrid incompatibility factors by the introgression of marked segments of *Drosophila mauritiana* chromosomes into *Drosophila simulans*. *Genetics*, **142**, 819–37.

Val, R. C. (1977). Genetic analysis of the morphological differences between two interfertile species of Hawaiian *Drosophila*. *Evolution*, **31**, 611–29.

Wagner, A., Wagner G. P., and Similion, P. (1994). Epistasis can facilitate the evolution of reproductive isolation by peak shifts: a two-locus two-allele model. *Genetics*, **138**, 533–45.

Wu, C.-I and Davis, A. W. (1993). Evolution of postmating reproductive isolation: The composite nature of Haldane's rule. *American Naturalist*, **142**, 187–212.

Wu, C.-I and Palolopi, M. F. (1995). Genetics of postmating reproductive isolation in animals. *Annual Review of Genetics*, **28**, 283–308.

Wu, C.-I, Hollocher, H., Begun, D. J., Aquadro, C. F., Xu, Y., and Wu, M.-L. (1995). Sexual isolation in *Drosophila melanogaster*: A possible case of incipient speciation. *Proceedings of the National Academy of Sciences USA*, **92**, 2519–23.

9

Speciation and hybridization of birds on islands

Peter R. Grant and B. Rosemary Grant

9.1 Introduction

Speciation is a process of lineage transformation (anagenesis) or splitting (cladogenesis). Lineage splitting, through either division or budding, culminates in the formation of two or more non-interbreeding populations or species from an original one. We observe the results, species, and attempt to infer the process from their genetic, morphological, ecological, and behavioural characteristics. At best only parts of the process are observed, which is why there is no universal agreement on how new species are formed.

The most widely favoured explanation is encapsulated in the allopatric model of speciation. Studies of island bird taxa have made important contributions to the development of the model, beginning with observations made by Darwin and Wallace, and extending into this century with the work of Perkins (1903), Rensch (1933), Stresemann (1936), Mayr (1940, 1942, 1963), Lack (1947), and Amadon (1950). In this chapter we illustrate the main features of the allopatric model by applying it to the Darwin's finches of the Galápagos Islands. This application forms the basis of a discussion of four general issues or problems raised by the model, which are not restricted to finches, or even to birds.

The first problem is whether reproductive isolation evolves entirely in allopatry, or whether it is initiated in allopatry and is completed in sympatry. The second is related to the first: are differences that evolve in allopatry reinforced by a regime of divergent selection on the taxa in sympatry, and if so does the reinforcement arise from ecological pressures or from a degree of reproductive (genetic) incompatibility? Third, does interbreeding (i.e. hybridization) at the secondary contact phase and subsequently simply eliminate the differences that arose in isolation, or does it play a creative role in facilitating further divergence? Fourth, does genetic restriction and reorganization embodied in the founder principle make a significant contribution to the divergence that leads to the formation of a new species?

Darwin's finches are the most suitable group of island birds for addressing these issues, so we pay special but not exclusive attention to them. Other islands or groups of islands, such as the West Indies, New Guinea, Hawaiian and western Pacific islands, are rich areas of avian speciation, but they have suffered more losses from human influences than have the Galápagos (Steadman 1995). None of the species of Darwin's finches are known to have been driven extinct. The Hawaiian archipelago, by contrast, lost almost half its species of honeycreeper-finches before the first western people arrived, and several more since then (James and Olson 1991; Steadman 1995). Therefore whatever can be learned about the evolution of Darwin's finches on these islands serves as a model for understanding speciation of birds on islands in general, and perhaps elsewhere. They serve less well as a model of long-term evolution on islands because they do not display some of the patterns, such as gigantism and flightlessness, manifested in other taxa or on older, larger, and more isolated islands (e.g. moas, rails, and the dodo: McNab 1994*a*, *b*; Roff 1994).

First, a few words to clarify use of the term allopatric speciation. Mayr (1982) has used the special term peripatric for speciation that starts with the colonization of a new area by a few individuals (founders; Mayr 1954). This would apply to Darwin's finches on the Galápagos and the vast majority of speciation events on islands. However, since peripatric speciation has become equated with founder speciation, rightly or wrongly, we use the more general term allopatric speciation in order to treat separately the issue of whether speciation occurs as a result of founder effects (Section 9.7).

9.2 Speciation

The essence of the speciation problem is to account for the generation of two species from one. In the dendrogram in Fig. 9.1, for example, there are 10 divisions or splits of

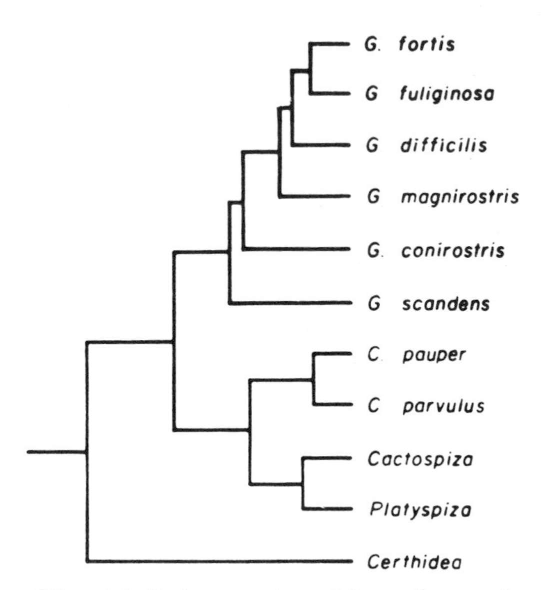

Fig. 9.1 Phenogram of Darwin's finches, constructed from allozyme data (based on Yang and Patton 1981). A recent re-analysis of the data by cladistic methods shows that statistical support for the three major groups is moderately strong, but is weak for all other features of the reconstruction (Stern and Grant 1996).

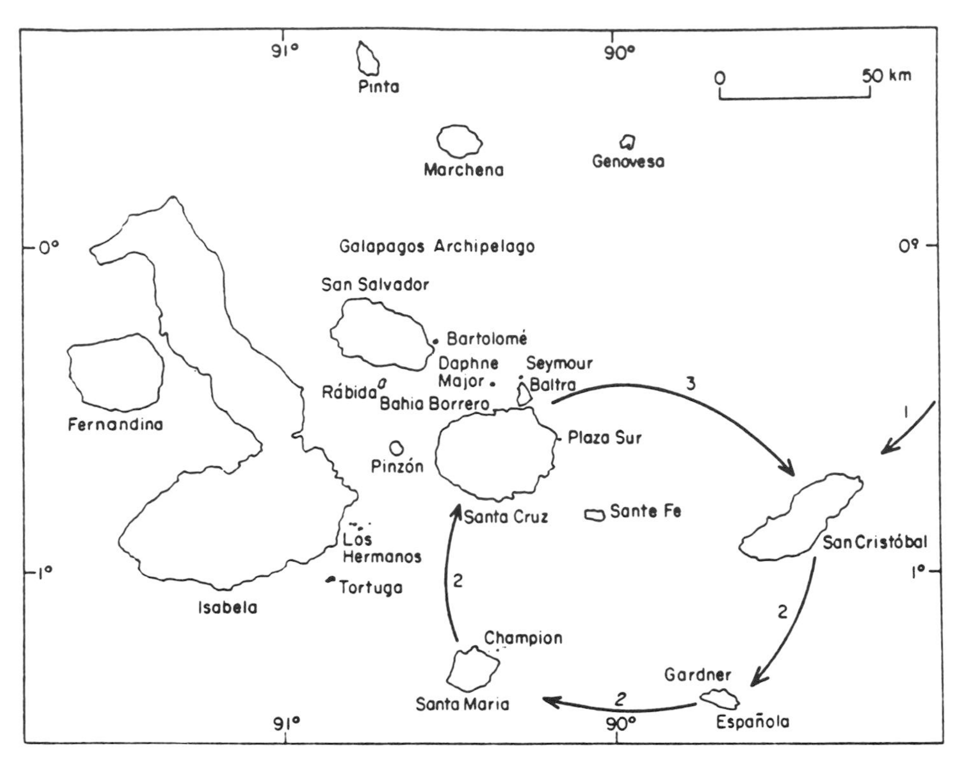

Fig. 9.2 The speciation cycle applied to Darwin's finches. Initially it comprises three steps: colonization of the archipelago (1), establishment of allopatric populations (2) followed by establishment of sympatry (3). Choice of the islands is arbitrary. Populations that are sympatric with little or no interbreeding are different species. Additional species are formed by evolution in a repetition of steps 2 and 3. From Grant (1981).

a lineage, giving rise to 11 species (missing from the figure are three other species: Lack 1947; Grant 1986). The dendrogram is not a literal representation of the actual evolution and multiplication of species, but it starkly portrays the phenomenon of lineage splitting to be explained. If an explanation is found for any one branching point then the diversification of the entire group can be explained by combining the principles involved in the speciation process, which should be common to all, with the historical and ecological details which will differ from one lineage split to another. The main task is to identify those principles. Figure 9.2 was prepared in order to capture the essence and to explore the possible variations (Grant 1981).

In step 1, the archipelago is colonized from continental South or Central America. A breeding population becomes established, and its size increases. In step 2 some individuals disperse to another island and establish a new breeding population. Some evolutionary change is expected to take place through selection and drift in the new environment. Step 2 may be repeated several times. A single breeding population has become converted into two or more allopatric, partly differentiated, populations. Step 3 is the contact through dispersal of members of two populations; the secondary sympatric phase. The choice of islands to illustrate these steps in Fig. 9.2 is arbitrary.

Step 3 is crucial to the speciation process. If evolutionary divergence in allopatry has been minor the members of the two populations are likely to treat each other as potential mates, interbreeding will ensue, without fitness loss, and panmixia will result. No speciation will have occurred. On the other hand evolutionary divergence might have proceeded to the point where members of the two groups do not recognize each other as potential mates, and do not even attempt to interbreed. Or they might interbreed but gain no fitness as a result of inviability or perhaps sterility of the offspring. If this happened then the process of speciation will have been completed, in allopatry. This was Stresemann's view. Lack (1947), following Dobzhansky (1937), pointed out a third possibility. Speciation may have been initiated in allopatry and completed in sympatry as a result of natural selection against interbreeding. This would happen, he suggested, if fitness loss was marked but not so profound as complete inviability or infertility of the offspring produced by interbreeding. He reasoned that divergence in sympatry would be produced by selection reinforcing the initial differences, culminating in a cessation of interbreeding. This has been referred to as the partial allopatric variant of the standard model (Grant 1986). The reinforcement process has been recently modelled and shown to be capable of completing the evolution of pre-mating isolation (Liou and Price 1994).

9.3 Reinforcement

The reinforcement hypothesis makes two testable predictions. First, if we could find a situation similar to that in step 2 just prior to step 3 we could artificially combine partly differentiated allopatric populations with the expectation of observing mixed courtship and breeding. Second, if we could find a situation similar to that modelled in the early stages of step 3 we should observe interbreeding (hybridization) occurring, with relatively poor performance of the offspring (hybrids).

The first prediction was tested, not with live birds, which would be unethical, but with specimens of dead birds (females) prepared in a courting posture. The test was performed in two stages. First, the sympatric and morphologically similar species small ground finch (*Geospiza fuliginosa*) and sharp-beaked ground finch (*G. difficilis*) on the island of Pinta were each tested with a pair of female models, one conspecific the other heterospecific. Males of each of the responding species made a clear discrimination between the models, directing far more attention, including attempts at copulation, to the conspecific than to the heterospecific model (Ratcliffe and Grant 1983*a*). Second, allopatric populations of the two species were tested in the same way, simulating the immigration of a closely related species. In agreement with the prediction of reproductive 'confusion', neither *G. difficilis* on Genovesa or *G. fuliginosa* on Plaza discriminated between the local conspecific model and the invader (Ratcliffe and Grant 1983*b*). Lack's reinforcement hypothesis is supported by these results (Grant 1986) and others (Ratcliffe and Grant 1983*a*, *b*).

The second prediction has been tested more recently by the accumulation of breeding data from a long-term study on the island of Daphne Major. Hybridization occurs at a low frequency between the medium ground finch (*G. fortis*) and two others; the cactus finch (*G. scandens*), a resident species (Fig. 9.3), and the small ground finch (*G. fuliginosa*), an uncommon immigrant. The natural hybridization of *G. fortis* and *G. fuliginosa* is as close to step 3 of the speciation model as we have been able to observe anywhere in the archipelago. Contrary to expectation from the reinforcement hypothesis, hybrids are both viable and fertile to a degree similar to that of the offspring of conspecific matings (Grant and Grant 1992), at least under the conditions prevailing after an exceptional amount of rain fell in 1983 (Grant and Grant 1993). Hybridization leads to introgression (Grant and Grant 1992, 1994, 1997*a*, 1998). Supporting data have been obtained on the island of Genovesa, where the large cactus finch (*G. conirostris*) interbreeds rarely with the sharp-beaked ground finch (*G. difficilis*) and the large ground finch (*G. magnirostris*) (Grant and Grant 1989). Thus all six species of Darwin's ground finches (genus *Geospiza*) are known to hybridize with at least one other congeneric species. Some intergeneric crosses are known among the tree finches and warbler finch as well (Grant 1986).

Two conclusions follow from these tests. First, speciation does not run its full course in allopatry; at least in some cases, and perhaps all, the potential for interbreeding at the secondary contact exists. Second, the partial allopatric variant of the model is appropriate for the investigated species but not always for the reason identified by Lack. Absence of a loss of fitness through interbreeding on Daphne means that the conditions for natural selection against interbreeding usually do not exist (see also Grant and Grant 1997*b, c*). The hypothesis of reinforcement is not supported. On both Daphne and Genovesa there have been long periods when hybrids have survived well, and backcrossed to the parental species. They have survived less at other times of generally low survival, for ecological reasons (Grant and Grant 1989, 1993).

Lack changed his views about reinforcement (Boag and Grant 1984), and concluded that ecological interactions at secondary contact are likely to have been primarily important in determining whether fusion or fission and coexistence would occur (Lack 1947). Our data are consistent with this alternative view. Thus divergence may have occurred under natural selection in step 3 of past speciation cycles, but it did so because the morphologically most similar individuals suffered greater effects from competition for food. Supporting evidence for this interpretation includes: (1) enhancement of beak size differences between species in sympatry, (2) a positive correlation between beak size differences and dietary differences that is predicted from the distribution of seed sizes on each island (Schluter and Grant 1984), and (3) the direct documentation of natural selection on genetically variable beak traits caused by a change in food supply (Grant 1986; Gibbs and Grant 1987; Grant and Grant 1995*a*).

9.4 Evidence from other islands

Continental islands often support populations of birds that are morphologically different from mainland populations though clearly related to them. Populations on small satellite islands differ from those on nearby large oceanic islands in a similar way. They are treated as subspecies, occasionally species, and represent the early stage of

Fig. 9.3 Hybridizing Darwin's finches on Daphne Major Island, Galápagos; *Gesopiza scandens* above, *G. fortis* below, and an F_1 hybrid between them

speciation cycles (step 2). Divergence from presumed relatives is not random but shows a strong size trend: island birds tend to have large beaks and long tarsi (Murphy 1938; Grant 1965*a*). The absence of ecological competitors has often been claimed as a key factor explaining the size trends and the associated broader feeding niches (Grant 1965*a*).

Step 3 of the speciation cycle on continental and satellite islands occurs when a second colonization takes place by members of the same ancestral species. From several studies around the world the double invasion phenomenon has been inferred from the observation that pairs, rarely triplets, of island species resemble each other and only one mainland species (Mayr 1942). A feature of these island species is a pronounced ecological difference, often associated with a pronounced morphological difference (Grant 1966, 1968, 1972). Whether adaptation to the island environment by the first species preceded the arrival of the second and permitted its establishment, or whether competitive interactions were partly responsible for the divergence, is a recurring problem that is usually difficult to resolve (for example see Lack and Southern 1949; Grant 1979, 1980; Baker *et al.* 1990*a*; Dennison and Baker 1991; Carrascal *et al.* 1992). In addition, introgressive hybridization may contribute to the evolutionary dynamics in the early stages of coexistence. An example of hybridization between double invaders is given in Section 9.6.

It is possible to reconstruct step 3 because in most cases colonization routes are easy to identify: mainland to island. Some have been documented directly (Mayr 1942; Degnan 1993*a*), others have been reconstructed from patterns of distribution and morphology (Mayr 1942, 1969). They are not easy to identify in archipelagos, even those in which the islands are linearly arrayed as in the Hawaiian archipelago. Nevertheless Lack (1947) used the double (multiple) invasion argument to explain the evolution of a large difference in beak size between coexisting tree finches (*Camarhynchus*) on the island of Floreana (Charles), and Bock (1970) used it to explain large differences between coexisting honeycreeper finch (*Loxops*) species in Hawaii.

Sequences of colonization and differentiation can be reconstructed with molecular data in order to test hypotheses of directional evolution (Grant 1980). Mitochondrial DNA data (Tarr and Fleischer 1993, 1995) are consistent with Bock's interpretation of the sequence of colonization events in the *Loxops* species group. Similar reconstructions have been attempted for the colonization of the Azores, Canaries, and Madeira (Baker *et al.* 1990*a*; Dennison and Baker 1991), and for the Caribbean (Klein and Brown 1994; Seutin *et al.* 1995). They have revealed cryptic evidence of multiple colonizations of the same island (Klein and Brown 1994; see also Edwards 1993). Assays such as these will be valuable in testing the controversial idea of a taxon cycle of evolution, which is an evolutionary trend from a single, broadly adapted, generalist species that colonizes an archipelago to several, locally adapted, specialist, and restricted species. Eventually the specialist species become replaced by a new generalist species, and the cycle begins again (Greenslade 1968; Ricklefs and Cox 1972; Pregill and Olson 1981). The approach has not been tried yet with Darwin's finches (the data are in hand), either as a test of the multiple invasion hypothesis of Lack (1947) or the taxon cycle idea applied to the finches by Cox (1990).

9.5 Hybridization

Hybridization, the interbreeding of two species, could play a creative role in speciation in two ways: directly, simply by producing a new species, or indirectly, by providing genetic variation to each of the interbreeding species and thereby facilitating further evolutionary change in each.

The first possibility can be examined in the context of the allopatric model (Fig. 9.2) by noting a fourth possible outcome at step 3 of the model; the mixing of two, well differentiated, populations to form a new, hybrid, species, differing from all other allopatric populations. In other words fusion and speciation would occur instead of fission and speciation. If this happened it would be a manifestation of the creative role of hybridization in speciation. It is well known in plants, and usually involves polyploidy. Is it likely to happen in birds?

The possibility is very difficult to assess, given the allopatric status of the new species. Lack (1940) considered the highly variable large ground finch on the island of Darwin (Culpepper) to have been formed in this way, by the mixing of *G. magnirostris* and *G. conirostris* to form *G. darwini*. He initially interpreted the medium ground finch on Daphne to have been formed similarly by a mixing of *G. fortis* and *G. fuliginosa*, as Vagvolgyi and Vagvolgyi (1990) have done recently (Grant 1993). Mayr (1942, p. 270), while recognizing that several species of different taxa show signs of being influenced by hybridization, could find no avian example of a clear hybrid origin of a new species. Recently *G. darwini* has been treated as a differentiated form of *G. magnirostris* (Grant 1986). We agree with Mayr that hybridization, by itself, is unlikely to produce a new species. There is no evidence of polyploidy in Darwin's finches (Jo 1983).

On the other hand modern studies have confirmed the suspected but previously undemonstrated hybridization of Darwin's finches. They have also shown that hybrids may be favoured by selection at times, and some may be disfavoured at others. Thus a population may be formed by the interbreeding of residents of one with immigrants of another, but is unlikely to remain a passive product of the interbreeding. Rather its characteristics will be moulded by natural selection to the local environment, with perhaps continuing input of genes from another species through hybridization. This leads to the intriguing second possibility that those genes could facilitate further evolutionary change by selection, perhaps along a new trajectory and culminating in the formation of a new species. New additive genetic variance introduced by hybridization (hybridizational variance) may be orders of magnitude greater than mutational variance (Grant and Grant 1994). Furthermore allometries can be altered by hybridization and backcrossing (Grant and Grant 1994). Thus hybridization could contribute to the speciation process, without being solely responsible for it, by enhancing genetic variation and relaxing genetic constraints on particular directions of evolutionary change.

9.6 Implications of hybridization

Recent studies of introgressive hybridization have done more than point to a creative role in evolution while invalidating the reinforcement hypothesis applied to Darwin's finches; they have cast doubt on previous attempts to use genetic data to reconstruct phylogenies (Avise 1989; Degnan 1993*b*). For the same reason there must be doubt about some reconstructions of colonization events that are performed with the aid of molecular data (Section 9.4).

An exchange of genes through introgressive hybridization makes two taxa more similar than they would be if they remained allopatric. This appears to have occurred in Darwin's finches. For example, Yang and Patton (1981) found that *Geospiza fuliginosa* and *G. fortis* were more similar to each other in allozymes at a locality on Santiago Island and at another on Santa Cruz Island than either was to a conspecific elsewhere. Thus dating the point at which two taxa such as these diverged from a common ancestral lineage is potentially flawed. If all taxa are affected by hybridization to the same extent then the dating of all branching points is wrong to the same (unknown) extent. This seems less likely than a greater opportunity for some taxa to share genes through horizontal transfer than others, perhaps as a result of ecological factors (Grant 1986) and especially early in their divergence. A bias will therefore enter the comparison of taxa and their branching points. If this has happened with Darwin's finches the order of branching points in the phylogeny may have been incorrectly determined (Stern and Grant 1996), and the accumulation of finch species may have occurred earlier than is shown in Fig. 9.1.

The scale of the problem in general for birds can be appreciated from two facts. First, almost one in ten species of birds is known to have hybridized at least once (Grant and Grant 1992). Admittedly introgression is less common and widespread than hybridization itself, because some hybridization occurs without gene exchange. Second, the potential to hybridize remains for an extraordinarily long time after speciation in birds. Prager and Wilson (1975) calculated the length of the hybridization period to be 22 million years on average, based on an analysis of albumin and transferrin immunological distances between 36 pairs of hybridizing bird taxa (more than half were different genera). Their maximum value was an estimated 50 million years. This is truly remarkable, and put in perspective by noting that the most successful avian order, the Passeriformes (passerines), comprising more than half of the nine to ten thousand species of birds present today, has been in existence for little more than that (54 million years; Boles 1995). A similarly long period of potential hybridization characterizes amphibia (Prager and Wilson 1975) and turtles (Bowen and Avise 1995), though not mammals (Prager and Wilson 1975).

An inference of the Prager and Wilson (1975) results is that pre-zygotic isolation arises much faster than post-zygotic isolation, at least in birds and some other vertebrates, and that Darwin's finches have not had enough time to evolve the physiological mechanisms responsible for post-zygotic isolation. Indeed no post-zygotic isolation among Darwin's finches is known. The scope for reinforcement is therefore restricted to those cases where little divergence has taken place over the long period prior to the evolution of post-zygotic incompatibilities. The *evolution* of pre-zygotic isolation may also take a long time, following a period in which it occurs for non-genetic reasons, for example as a result of species-discriminatory mate choices that are formed through imprinting in early life on culturally inherited parental traits such as song (Grant and Grant 1996*a*, 1997*a*, 1998).

Hybridization of island birds is not restricted to the Galápagos. The third species of *Zosterops* (white-eyes) to colonize Norfolk Island hybridized with its predecessor (Gill 1970). Hybridization without fitness loss has been reported in *Nesospiza* buntings on Inaccessible Island in the Tristan da Cunha group (Ryan *et al.* 1994). Some degree of post-zygotic isolation has evolved in *Ficedula albicollis* and *F. hypoleuca*, two species of flycatchers which hybridize on the Swedish island of Gotland. Apparently this has not led to reinforcement of pre-existing differences in plumage, although the song sung by male *F. albicollis* may have diverged in sympatry (Wallin 1986; Alatalo *et al.* 1990). In accordance with Haldane's rule, female hybrids are sterile (Tegelström and Gelter 1990; Gelter *et al.* 1992). Mitochondrial DNA differences suggest the species shared a common ancestor 5 million years ago (Tegelström and Gelter 1990). This puts an upper bound on the time to evolve partial sterility, at least in these two species. The whole of the radiation of Darwin's finches (~3 million years) occurred within this length of time. Dates for the radiation of the Hawaiian honeycreeper-finches (probably at least 50 species) have been estimated, with appropriate caveats, at ~7–8 million years by Johnson *et al.* (1989) and ~3.5 million years by Tarr and Fleischer (1995).

9.7 Founder effects and peripatric speciation

Mayr (1992, and earlier) has written that speciation is likely to occur rapidly on islands through (1) a loss of genetic variation in the founding of a new population by just a few individuals, (2) the setting up of new epistatic balances in the early stages of population establishment when inbreeding is likely to be common and further genetic loss will occur through selection and random drift, and (3) selection on genes against altered genetic backgrounds.

This model of speciation, termed peripatric or founder effect speciation, was developed from the observation that peripheral (often insular) populations of a polytypic species tended to be the most distinct, and from the conviction that non-additive genetic factors were more important in evolutionary divergence than had been generally acknowledged (Mayr 1954, 1992).

The theory of peripatric speciation (allopatric speciation with founder effects; Mayr 1982) is very difficult to test because the difference between it and the allopatric model (without founder effects) we have articulated is one of difficult-to-measure genetic architecture, epistasis and loss of recessive alleles being important in peripatric speciation and additive genetic variation being important in our version of allopatric speciation. In a sense the theory has been tested repeatedly, experimentally and unwittingly. Alien species have been introduced to islands around the world for over a century, usually in very small numbers (e.g. Moulton and Pimm 1987). A few instances of rapid and substantial phenotypic change, indicative of incipient speciation, would be evidence in favour of founder effect speciation. None has been reported. Only minor phenotypic changes have been noted (see Baker and Moeed 1987; Baker *et al.* 1990*a, b*), although a systematic study of evolutionary change in the introduced populations has not yet been undertaken. The only direct study of a founder event from origin onwards provided evidence of non-random colonization, inbreeding, and drift, but nothing that could be interpreted as reflecting a genetic reorganization (Grant and Grant 1995*b*).

The theory of genetic reorganization (founder effects) has been criticized on theoretical grounds (Lande 1980; Barton and Charlesworth 1984), and although it has its supporters it will remain conjectural until the genetic changes involved in speciation can actually be determined (Provine 1989). The issues and alternatives to Mayr's scheme are discussed in Chapters 6–8.

Frequent opportunities arise for the founding of new populations by a few individuals as a result of sea-level fluctuations and the creation of numerous small islands (e.g. Fig. 9.4). But is the theory of founder effects necessary to account for speciation of birds on islands? It was developed because in Mayr's experience (1) major evolutionary changes occur on islands that are not biotically or physically distinctive, yet (2) the majority of peripheral isolates show no evidence of major evolutionary change despite living in novel environments, and likewise (3) pronounced ecological shifts within widespread species are often not accompanied by phenotypic shifts (Mayr 1992). Thus evolutionary change does not always covary with ecological change. The lack of covariation undermines the claim that speciation can be explained solely in terms of selection pressures from the environment (exogenous factors). By default endogenous (genetic) factors take on special importance.

The reasoning may be flawed in three ways. First, covariation may have been broken in the fairly recent past by the large human-induced changes in habitats and distributions of birds on both islands and continents (Steadman 1995). Second, the lack of covariation may have been overemphasized (see also Coyne 1994). Observations that led Mayr to develop his theory were made in the era before ecological variables were measured. Evolutionary change may be interpretable in terms of current environmental factors when measurements of salient ecological and behavioural variables replace non-quantitative observations (e.g. Conant 1988; Grant and Grant 1995*a*). Finally, covariation would not be expected if sexual selection had contributed to divergence in allopatry, as seems to have been the case in many avifaunas (Grant and Grant 1997*b*). An example may be the highly colourful and patterned *Tanysiptera* kingfishers on New Guinea and its satellite islands, which figured prominently in Mayr's (1954) explanation of ideas about founder effects. For birds in general some plumage traits are under simple genetic control and exhibit dominance and recessiveness, although most are polygenic, and epistasis appears to be rare (Grant and Grant 1997*b*).

In conclusion, while not rejecting founder effect speciation outright we see no reason why the standard allopatric model we have described for islands cannot explain observed patterns of island differentiation and speciation, and why it is necessary to invoke genetic reorganizations produced in founder events. Similarly, sympatric (Grant 1986; Grant and Grant 1989) and parapatric (Ryan *et al.* 1994) alternatives to the allopatric model appear to have at most limited applicability to the speciation of birds on islands.

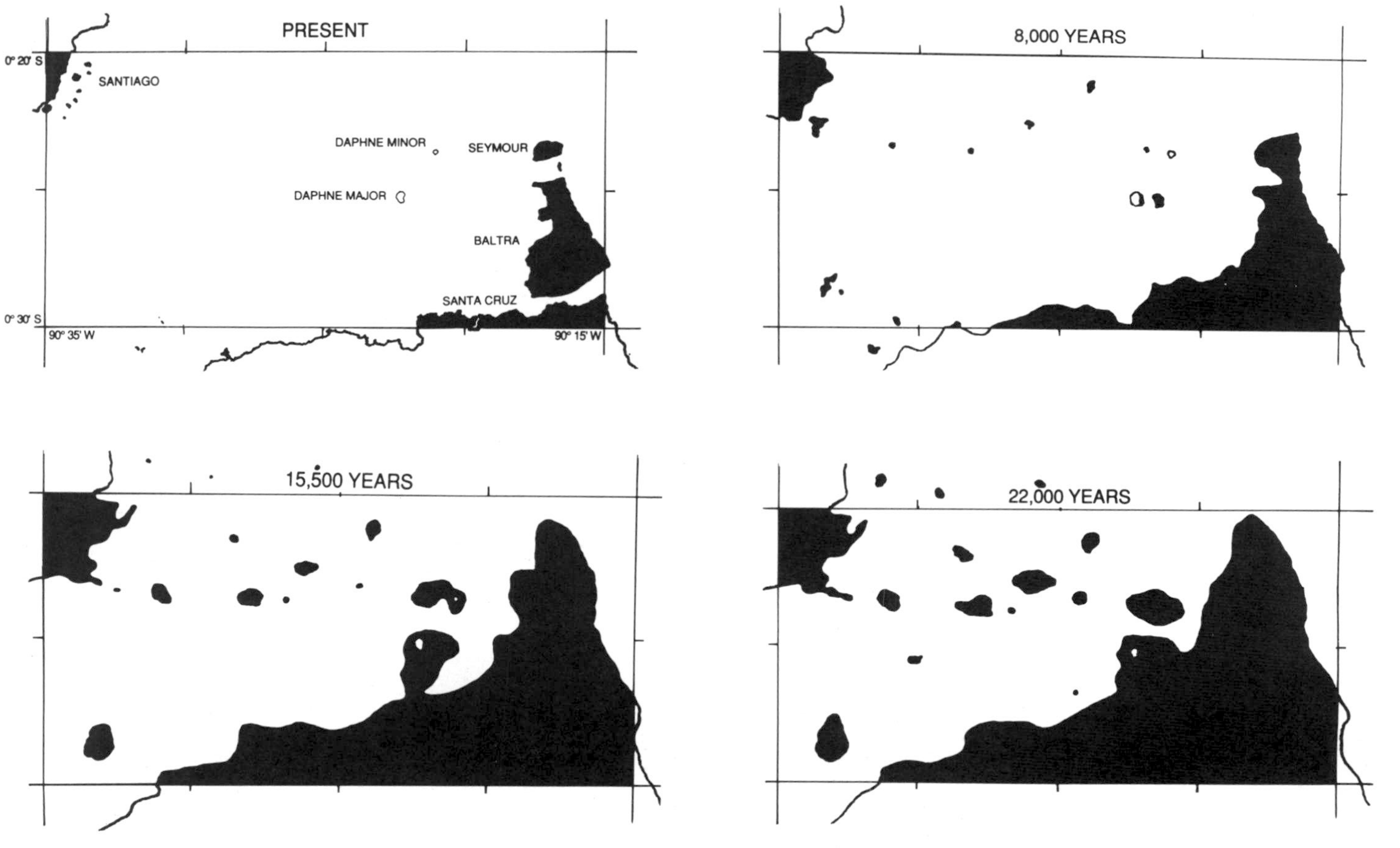

Fig. 9.4 Geological history of the central islands of the Galápagos archipelago over the last 22 000 years, based on estimates of sea-level changes from coral cores (Bard *et al.* 1990).

9.8 Sexual selection

Darwin's finches are unusual among island birds in one respect; most closely related species do not differ in plumage. Our explanation of the allopatric model of speciation (Section 9.3) was silent on a possible role of plumage variation and sexual selection.

Island birds tend to have dull plumage, and sexual dichromatism, if present, tends to be reduced (Mayr 1942; Grant 1965*b*; Herremans 1990). The absence of congeneric species of similar appearance has been identified as a possible factor in the evolution of these plumage trends (Grant 1965*b*). Recent attention has been given to the role of parasites in the evolution of plumage traits under sexual selection (Andersson 1994). The probable absence or scarcity of parasite species on islands, especially remote ones, removes or reduces one particular component of sexual selection. Therefore loss of brightness in step 2 of many speciation cycles may have been accomplished by genetic drift in the absence of parasite-driven sexual selection. The directional effects of drift may be explained by mutations at gene loci affecting plumage being predominantly biased in the direction of simplification of biosynthetic pathways, drifting to fixation.

The Hawaiian honeycreeper-finches provide some of the best examples of step 3, and these run counter to the trend by displaying a variety of bright plumage patterns (Amadon 1950; Freed *et al.* 1987). The patterns suggest that, like their mainland cardueline relatives, these birds were subjected to strong sexual selection pressures during the adaptive radiation, and possibly during speciation processes. Causes of these pressures are obscure. For example it is doubtful if parasites were instrumental in contributing to these pressures in view of the vulnerability of the honeycreeper-finches to protozoan and bacterial or viral parasites introduced from the mainland (van Riper *et al.* 1986). Female preference for novel and exaggerated plumage traits of courting males may have given directionality to the evolutionary trends in this archipelago (e.g. see Andersson 1994). Establishment of sympatry in the secondary contact phase of speciation (Fig. 9.2) depended upon ecological divergence, as reflected in beak sizes (Amadon 1950; Bock 1970), so the plumage and beak variation indicates that sexual and natural selection jointly contributed to the radiation. The role of song would be worth investigating in these species because in Darwin's finches song learning and sexual imprinting appear to have been important in the speciation process (Grant and Grant 1996*a*, 1997*a*, *c*, 1998).

9.9 The timescale of evolution

The thirteen species may have been formed in a burst of speciation early in the history of the group on Galápagos, at a more or less constant rate, or more rapidly in recent times. In the absence of fossils from all but the last few thousand years (Steadman 1986), the only way to determine the pattern of speciation is to use genetic data, or its proxy, allozymes. It appears from these (Fig. 9.1) that few of the extant species were formed in the first half of the diversification.

The starting point of the Darwin's finch radiation was a splitting of the ancestral stock into a lineage leading to the modern warbler finch (*Certhidea olivacea*) and all of the rest (Fig. 9.1). This was dated by Yang and Patton (1981) to little more than half a million years ago, using Nei's method of equating genetic distance to time on the basis of clock-like behaviour of the substitution of neutral alleles. Thus Darwin's finches are young, and their radiation was rapid. However, recent work with other groups of birds suggests that the calibration of the clock was in error in this calculation, and that the period over which Darwin's finches have diversified is much longer than originally thought. On the basis of the new information the first split occurred about 2.8 million years ago (Grant 1994), and a doubling of the number of species took place on average about once every 0.75–0.90 million years. These estimates may be a little low if hybridization has produced a bias (Section 9.6).

This new finding has an important implication. For as long as the radiation was thought to be confined to the last half million years the geological history of the islands could be ignored (Grant 1986; Cox 1990). With this greater period, however, it cannot be ignored. At the time of arrival of the first finches, approximately three million years ago, there were only five islands in existence (Fig. 9.5); San Cristóbal (possibly split into two or more at that time), Española and three others that are now submerged (see Christie *et al.* 1992; Grant and Grant 1996*b*). Colonization may have been influenced by climate, as a global cooling period began about three million years ago (deMenocal 1995).

Figure 9.6 maps the finch species diversification onto the pattern of island diversification. The number of islands has apparently always exceeded the number of species. One can interpret the relationship to mean that the increase in number of islands facilitated an increase in number of species through speciation. This gives a new twist to the old idea of the role of ecological opportunity in governing speciation and adaptive radiations (Lack 1947). For example, the slow initial diversification (Fig. 9.1) can be explained by the small number of islands. The scheme in Fig. 9.2 is therefore not strictly correct, and should be modified, although fortunately two of the islands arbitrarily chosen to illustrate the initial cycle of speciation happen to be ones in existence when the ancestral species arrived. The archipelago was not a fully formed ecological theatre, passively waiting for the arrival of the finch evolutionary play. The two developed together. The same applies to Hawaiian evolution (see Tarr and Fleischer 1995).

9.10 Conclusions

The allopatric model of speciation fits the facts of Darwin's finch evolution on the Galápagos islands, and the evolution of other species of birds on islands elsewhere. Evolutionary change is initiated in allopatry and continues in sympatry, the driving force is directional natural selection arising from ecological factors, principally food factors, and a low level of introgressive hybridization without appreciable fitness loss occurs during and after the initial secondary contact for an unknown, but probably long,

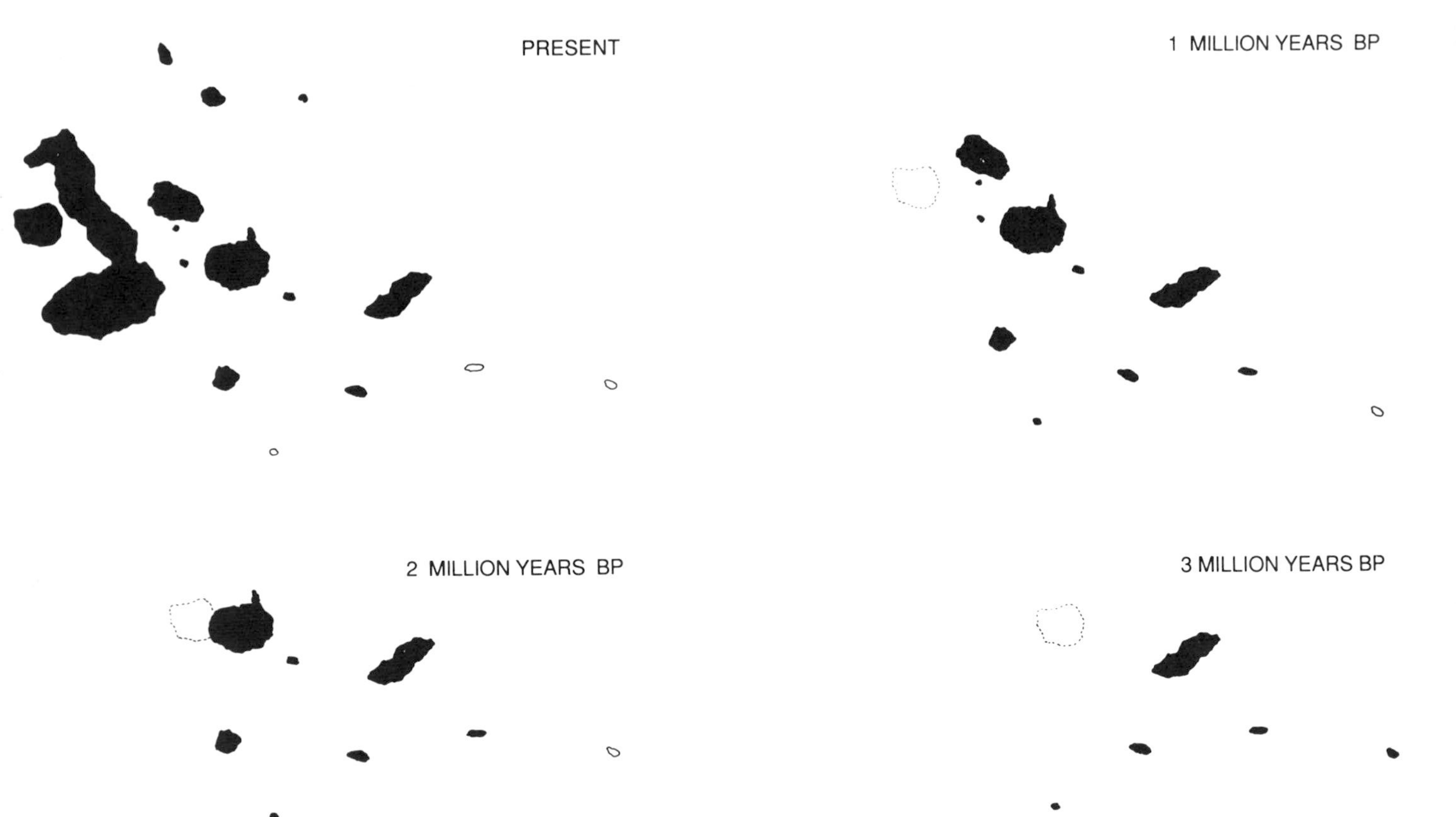

Fig. 9.5 Geological history of the Galápagos archipelago over the last three million years, since the time they were colonized by ancestral Darwin's finches. The reconstruction is based on data in Christie *et al.* (1992) (see also Grant and Grant 1996*b*), and assumes south-easterly plate movement of constant direction and speed over a fixed hotspot beneath modern Fernandina (position shown by a broken line). Submerged seamounts are shown in outline.

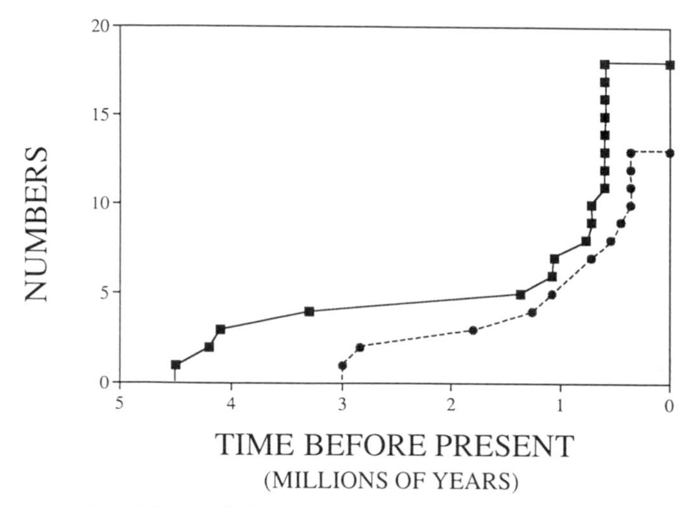

Fig. 9.6 Accumulation of Darwin's finch species (circles) in relation to the increase in number of Galápagos islands (squares); see Grant and Grant (1996*b*) for details. Both sets of ages are subject to errors that we have ignored. Similarly the possibility of unknown extinctions, while acknowledged, is ignored.

time. Thus species, perhaps for millions of years, are not completely reproductively isolated from each other, and isolating mechanisms do not set up impregnable barriers to gene exchange thereby preserving the genetic integrity of the species.

Our proposal makes no allowance for several other factors or processes held to be important in the speciation of other groups (e.g. see Endler 1989). There is no explicit statement on the evolution of post-zygotic barriers to gene exchange through inviability or sterility factors, no provision for coadapted gene complexes and epistatic gene interactions being perturbed by inbreeding, and no provision for genetic reorganization occurring through founder-flush demographic cycles. We have invoked no founder effects or loss of genetic variation in colonization events. All could play a role in the speciation of birds on some islands. None appear to be necessary.

9.11 Summary

The process of speciation in birds can be inferred from the pattern of diversification on islands, especially in archipelagos. The basic model is one of initial differentiation of allopatric populations, with further differentiation taking place at the time that sympatry is established. Differences that evolve in allopatry are reinforced by a regime of divergent selection on the taxa in sympatry arising from ecological pressures and not from reproductive (genetic) incompatibility. A low level of interbreeding (hybridization) at the secondary contact phase and subsequently may occur with little or no fitness loss. Introgressive hybridization has the potential of playing a creative role in evolution, facilitating further divergence by enhancing genetic variation and relaxing genetic constraints on new directions of evolutionary change under natural selection. Hybridization potential may last for many millions of years after two taxa diverge, implying that post-zygotic isolation evolves slowly. The main alternative model of speciation in island birds is the founder effects model. It emphasizes major changes in non-additive genetic factors taking place in the founding of a new population by a small number of individuals. There is no direct evidence that would make it preferable to the standard allopatric model for island speciation without founder effects.

Acknowledgements

This research was supported by grants from the National Science Foundation (USA). We thank Dave Christie for advice and Kornelia Rassmann for discussion on the geological history of the Galápagos, and Ernst Mayr for critical comments on the manuscript.

References

Alatalo, R. V., Eriksson, D., Gustafsson, L., and Lundberg, A. (1990). Hybridization between Pied and Collared Flycatchers—sexual selection and speciation theory. *Journal of Evolutionary Biology*, **3**, 375–89.

Amadon, D. (1950). The Hawaiian honeycreepers (Aves: Drepaniidae). *Bulletin of the American Museum of Natural History*, **95**, 151–262.

Andersson, M. (1994). *Sexual selection*. Princeton University Press.

Avise, J. C. (1989). Gene trees and organismal trees: a phylogenetic approach. *Evolution*, **43**, 1192–208.

Baker, A. J. and Moeed, A. (1987). Rapid genetic differentiation and founder effect in colonizing populations of Common Mynas (*Acridotheres tristis*). *Evolution*, **41**, 525–38.

Baker, A. J., Dennison, M. D., Lynch A., and LeGrand G. (1990*a*). Genetic divergence in peripherally isolated populations of chaffinches in the Atlantic islands. *Evolution*, **44**, 981–99.

Baker, A. J., Peck, M. K., and Goldsmith, M. A. (1990*b*). Genetic and morphometric differentiation in introduced populations of Common Chaffinch (*Fringilla coelebs*) in New Zealand. *Condor*, **92**, 76–88.

Bard, E., Hamelin, B., Fairbanks, R. G., and Zindler, A. (1990). Calibration of the ^{14}C timescale over the last 30,000 years using mass spectrometric U-Th ages from Barbados corals. *Nature*, **345**, 405–10.

Barton, N. H. and Charlesworth, B. (1984). Genetic revolutions, founder effects, and speciation. *Annual Review of Ecology and Systematics*, **15**, 133–64.

Boag, P. T. and Grant, P. R. (1984). The classical case of character release: Darwin's finches (*Geospiza*) on Isla Daphne Major, Galápagos. *Biological Journal of the Linnean Society*, **22**, 243–87.

Bock, W. (1970). Microevolutionary sequences as a fundamental concept in macroevolutionary models. *Evolution*, **24**, 704–22.

Boles, W. E. (1995). The world's oldest songbird. *Nature*, **374**, 21–2.

Bowen, B. W. and Avise, J. C. (1995). Conservation genetics of marine turtles. In *Conservation genetics: case histories from nature*, (ed. J. C. Avise and J. L. Hamrick), pp. 190–237. Chapman and Hall, New York.

Carrascal, L. M., Tellería, J. L., and Valido, A. (1992). Habitat distribution of Canary chaffinches among islands: competitive exclusion or species-specific habitat preferences? *Journal of Biogeography*, **19**, 383–90.

Christie, D. M., Duncan, R. A., McBirney, A. R., Richards, M. A., White, W. M., Harpp, K. S., and Fox, C. G. (1992). Drowned islands downstream from the Galapagos hotspot imply extended speciation times. *Nature*, **355**, 246–8.

Conant, S. (1988). Geographic variation in the Laysan finch (*Telyspiza cantans*). *Evolutionary Ecology*, **2**, 270–82.

Cox, G. (1990). Centres of speciation and ecological differentiation in the Galapagos land bird fauna. *Evolutionary Ecology*, **4**, 130–42.

Coyne, J. A. (1994). Ernst Mayr and the evolution of species. *Evolution*, **48**, 19–30.

Degnan, S. (1993*a*). Genetic variability and population differentiation inferred from DNA fingerprinting in silvereyes (Aves: Zosteropidae). *Evolution*, **47**, 1105–17.

Degnan, S. (1993*b*). The perils of single gene trees—mitochondrial versus single-copy nuclear DNA variation in white-eyes (Aves: Zosteropidae). *Molecular Ecology*, **2**, 219–25.

Dennison, M. and Baker, A. J. (1991). Morphometric variability in continental and Atlantic island populations of chaffinches (*Fringilla coelebs*). *Evolution*, **45**, 29–39.

deMenocal, P. B. (1995). Plio-Pleistocene African climate. *Science*, **270**, 53–9.

Dobzhansky, T. (1937). *Genetics and the origin of species.* Columbia University Press, New York.

Edwards, S. V. (1993). Mitochondrial gene genealogy and gene flow among island and mainland populations of a sedentary songbird, the Gray-Crowned Babbler (*Pomatostomus temporalis*). *Evolution*, **47**, 1118–37.

Endler, J. A. (1989). Conceptual and other problems in speciation. In *Speciation and its consequences*, (ed. D. Otte and J.A. Endler), pp. 625–48. Sinauer, Sunderland, MA.

Freed, L., Conant, S., and Fleischer, R. C. (1987). Evolutionary ecology and radiation of Hawaiian passerine birds. *Trends in Ecology and Evolution*, **2**, 196–203.

Gelter, H. P., Tegelström, H., and Gustafsson, L. (1992). Evidence from hatching success and DNA fingerprinting for the fertility of hybrid Pied × Collared Flycatchers *Ficedula hypoleuca × albicollis*. *Ibis*, **134**, 62–8.

Gibbs, H. L. and Grant, P. R. (1987). Oscillating selection on Darwin's Finches. *Nature*, **327**, 511–13.

Gill, F. B. (1970). Hybridization in Norfolk Island White-eyes (*Zosterops*). *Condor*, **72**, 481–2.

Grant, B. R. and Grant, P. R. (1989). *Evolutionary dynamics of a natural population: the large cactus finch of the Galápagos.* University of Chicago Press.

Grant, B. R. and Grant, P. R. (1993). Evolution of Darwin's finches caused by a rare climatic event. *Proceedings of the Royal Society of London* B, **251**, 111–17.

Grant, B. R. and Grant, P. R. (1996*a*). Cultural inheritance of song and its role in the evolution of Darwin's Finches. *Evolution*, **50**, 2471–87.

Grant, B. R. and Grant, P. R. (1998). Hybridization and speciation in Darwin's Finches: the role of sexual imprinting on a culturally transmitted trait. In *Endless forms: species and speciation*, (ed. D. J. Howard and S. H. Berlocher). Oxford University Press.

Grant, P. R. (1965*a*). The adaptive significance of some island size trends in island birds. *Evolution*, **19**, 355–67.

Grant, P. R. (1965*b*). Plumage and the evolution of birds on islands. *Systematic Zoology*, **14**, 47–52.

Grant, P. R. (1966). Ecological compatibility of bird species on islands. *American Naturalist*, **100**, 451–62.
Grant, P. R. (1968). Bill size, body size, and the ecological adaptations of bird species to competitive situations on islands. *Systematic Zoology*, **17**, 319–33.
Grant, P. R. (1972). Bill dimensions of the three species of *Zosterops* on Norfolk Island. *Systematic Zoology*, **21**, 289–91.
Grant, P. R. (1979). Evolution of the chaffinch, *Fringilla coelebs*, on the Atlantic Islands. *Biological Journal of the Linnean Society*, **11**, 301–32.
Grant, P. R. (1980). Colonization of the Atlantic islands by chaffinches (*Fringilla* spp.). *Bonner zoologische Beiträge*, **31**, 311–17.
Grant, P. R. (1981). Speciation and the adaptive radiation of Darwin's Finches. *American Scientist*, **69**, 653–63.
Grant, P. R. (1986). *Ecology and evolution of Darwin's finches*. Princeton University Press.
Grant, P. R. (1993). Hybridization of Darwin's finches on Isla Daphne Major, Galápagos. *Philosophical Transactions of the Royal Society of London* B, **351**, 127–39.
Grant, P. R. (1994). Population variation and hybridization: comparison of finches from two archipelagos. *Evolutionary Ecology*, **8**, 598–617.
Grant, P. R. and Grant, B. R. (1992). Hybridization of bird species. *Science*, **256**, 193–7.
Grant, P. R. and Grant, B. R. (1994). Phenotypic and genetic effects of hybridization in Darwin's Finches. *Evolution*, **48**, 297–316.
Grant, P. R. and Grant, B. R. (1995*a*). Predicting microevolutionary responses to directional selection on heritable variation. *Evolution*, **49**, 241–51.
Grant, P. R. and Grant, B. R. (1995*b*). The founding of a new population of Darwin's Finches. *Evolution*, **49**, 229–40.
Grant, P. R. and Grant, B. R. (1996*b*). Speciation and hybridization in island birds. *Philosophical Transactions of the Royal Society of London* B, **351**, 765–72.
Grant, P. R. and Grant, B. R. (1997*a*). Hybridization, sexual imprinting and mate choice. *American Naturalist*, **149**, 1–28.
Grant, P. R. and Grant, B. R. (1997*b*). Genetics and the origin of bird species. *Proceedings of the National Academy of Sciences USA*, **94**, 7768–75.
Grant, P. R. and Grant, B. R. (1997*c*). Mating patterns of Darwin's finches determined by song and morphology. *Biological Journal of the Linnean Society*, **60**, 317–43.
Greenslade, P. J. M. (1968). Island patterns in the Solomon islands bird fauna. *Evolution*, **22**, 751–61.
Herremans, M. (1990). Trends in the evolution of insular land birds, exemplified by the Comoro, Seychelle and Mascarene Islands. In *Vertebrates in the tropics*, (ed. G. Peters and R. Hutterer), pp. 249–60. Alexander Koenig Zoological Research Institute and Zoological Museum, Bonn.
James, H. F. and Olson, S. L. (1991). Descriptions of thirty-two new species of birds from the Hawaiian islands: part II. Passeriformes. *Ornithological Monographs* No. 46. American Ornithologists' Union, Washington, D.C.
Jo, N. (1983). Karyotypic analysis of Darwin's Finches. In *Patterns of evolution in Galápagos organisms*, (ed. R. I. Bowman, M. Berson, and A. E. Leviton), pp. 201–17. American Association for the Advancement of Science, Pacific Division, San Francisco.
Johnson, N. K., Marten, J. A., and Ralph, C. J. (1989). Genetic evidence for the origin and relationships of Hawaiian honeycreepers (Aves: Fringillidae). *Condor*, **91**, 379–96.
Klein, N. K. and Brown, W. M. (1994). Intraspecific molecular phylogeny in the Yellow Warbler (*Dendroica petechia*), and its implications for avian biogeography in the West Indies. *Evolution*, **48**, 1914–32.
Lack, D. (1940). Evolution of the Galápagos finches. *Nature*, **146**, 324–7.
Lack, D. (1947). *Darwin's finches*. Cambridge University Press.
Lack, D. and Southern, H. N. (1949). Birds on Tenerife. *Ibis*, **91**, 607–26.

Lande, R. (1980). Genetic variation and phenotypic evolution during allopatric speciation. *American Naturalist*, **116**, 463–79.
Liou, L. W. and Price, T. D. (1994). Speciation by reinforcement of premating isolation. *Evolution*, **48**, 1451–9.
Mayr, E. (1940). Speciation phenomena in birds. *American Naturalist*, **74**, 249–78.
Mayr, E. (1942). *Systematics and the origin of species*. Columbia University Press, New York.
Mayr, E. (1954). Change of genetic environment and evolution. In *Evolution as a process*, (ed. J. Huxley, A. C. Hardy, and E. B. Ford), pp. 157–80. Allen and Unwin, London.
Mayr, E. (1963). *Animal species and evolution*. Belknap Press, Cambridge, MA.
Mayr, E. (1969). Bird speciation in the tropics. *Biological Journal of the Linnean Society*, **1**, 1–17.
Mayr, E. (1982). Processes of speciation in animals. In *Mechanisms of speciation*, (ed. C. Barigozzi), pp. 1–19. A. R. Liss Inc., New York.
Mayr, E. (1992). Controversies in retrospect. *Oxford Surveys in Evolutionary Biology*, **8**, 1–34.
McNab, B. K. (1994*a*). Energy conservation and the evolution of flightlessness in birds. *American Naturalist*, **144**, 628–42.
McNab, B. K. (1994*b*). Resource use and the survival of land and freshwater vertebrates on oceanic islands. *American Naturalist*, **144**, 643–60.
Moulton, M. P. and Pimm, S. L. (1987). A morphological assortment in introduced Hawaiian passerines. *Evolutionary Ecology*, **1**, 113–24.
Murphy, R. C. (1938). The need for insular exploration as illustrated by birds. *Science*, **88**, 533–9.
Perkins, R. C. L. (1903). Vertebrata. In *Fauna Hawaiiensis*, (ed. D. Sharpe), pp. 365–466. Cambridge University Press.
Prager, E. R. and Wilson, A. C. (1975). Slow evolutionary loss of the potential for interspecific hybridization in birds: a manifestation of slow regulatory evolution. *Proceedings of the National Academy of Sciences USA*, **72**, 200–4.
Pregill, G. K. and Olson, S. L. (1981). Zoogeography of West Indian vertebrates in relation to Pleistocene climatic cycles. *Annual Review of Ecology and Systematics*, **12**, 75–98.
Provine, W. B. (1989). Founder effects and genetic revolutions in microevolution and speciation: an historical perspective. In *Genetics, speciation, and the founder principle*, (ed. L. V. Giddings, K. Y. Kaneshiro, and W. W. Anderson), pp. 43–76. Oxford University Press.
Ratcliffe, L. M. and Grant, P. R. (1983*a*). Species recognition in Darwin's Finches (*Geospiza*, Gould). I. Discrimination by morphological cues. *Animal Behaviour*, **31**, 1139–53.
Ratcliffe, L. M. and Grant, P. R. (1983*b*). Species recognition in Darwin's Finches (*Geospiza*, Gould). II. Geographic variation in mate preference. *Animal Behaviour*, **31**, 1154–65.
Rensch, B. (1933). Zoologische Systematik und Artbildungsproblem. *Zoologische Anzeiger*, Supplement, **14**, 180–222.
Ricklefs, R. E. and Cox, G. W. (1972). Taxon cycles in the West Indian avifauna. *American Naturalist*, **106**, 195–219.
Roff, D. A. (1994). The evolution of flightlessness: is history important? *Evolutionary Ecology*, **8**, 639–57.
Ryan, P. G., Moloney, C. L., and Hudon, J. (1994). Color variation and hybridization among *Nesospiza* buntings on Inaccessible Island, Tristan da Cunha. *Auk*, **111**, 314–27.
Schluter, D. and Grant, P. R. (1984). Determinants of morphological patterns in communities of Darwin's Finches. *American Naturalist*, **123**, 175–96.
Seutin, G., Klein, N., Ricklefs, R. E., and Bermingham, E. (1995). Historical biogeography of the Bananaquit (*Coereba flaveola*) in the Caribbean region: a mitochondrial DNA assessment. *Evolution*, **48**, 1041–61.
Steadman, D. W. (1986). Holocene vertebrate fossils from Isla Floreana, Galápagos. *Smithsonian Contributions to Zoology*, No. 413.
Steadman, D. W. (1995). Prehistoric extinctions of Pacific island birds: biodiversity meets zooarchaeology. *Science*, **267**, 1123–31.

Stern, D. L. and Grant, P. R. (1996). A phylogenetic reanalysis of allozyme variation among populations of Galápagos Finches. *Zoological Journal of the Linnean Society*, **118**, 119–34.

Stresemann, E. (1936). Zur Frage der Artbildung in der Gattung *Geospiza*. *Orgaan der Club Van Nederlandische Vogelkunde*, **9**, 13–21.

Tarr, C. L. and Fleischer, R. C. (1993). Mitochondrial-DNA variation and evolutionary relationships in the Amakihi complex. *Auk*, **110**, 825–31.

Tarr, C. L. and Fleischer, R. C. (1995). Evolutionary relationships of the Hawaiian honeycreepers (Aves: Drepanidinae). In *Hawaiian biogeography: evolution on a hot-spot archipelago*, (ed. W. L. Wagner and V. A. Funk), pp. 147–59. Smithsonian Institution Press, Washington, D. C.

Tegelström, H. and Gelter, H. P. (1990). Haldane's rule and sex biased gene flow between two hybridizing flycatcher species (*Ficedula albicollis* and *F. hypoleuca*, Aves: Muscicapidae). *Evolution*, **44**, 2012–21.

Vagvolgyi, J. and Vagvolgyi, M. W. (1990). Hybridization and evolution in Darwin's Finches of the Galapagos Islands. *Accademia Nazionali dei Lincei Atti dei Convegni Lincei*, **85**, 749–72.

van Riper III, C., van Riper S. G., Goff, M. L., and Laird, M. (1986). The epizootiology and ecological significance of malaria in Hawaiian birds. *Ecological Monographs*, **56**, 327–44.

Wallin, L. (1986). Divergent character displacement in the song of two allospecies: the Pied Flycatcher *Ficedula hypoleuca* and the Collared Flycatcher *Ficedula albicollis*. *Ibis*, **128**, 251–9.

Yang, S.-Y. and Patton, J. L. (1981). Genic variability and differentiation in Galápagos finches. *Auk*, **98**, 230–42.

10

Ecological speciation in postglacial fishes

Dolph Schluter

10.1 Theories of ecological and non-ecological speciation

The ecological processes that drive speciation are poorly understood. The topic has received only sporadic attention over the past few decades, confined largely to a fraction of literature devoted to the possibility of speciation in sympatry (Rice and Hostert 1993; Bush 1994). Progress in identifying environmental mechanisms has been slow regardless of the putative geographic mode of speciation. By 'speciation' I mean the evolution of reproductive isolation. 'Ecological processes' are the interactions between individual organisms and their environment that cause natural selection and that drive population establishment, growth, and decline.

This gap in our understanding is most glaring in the context of adaptive radiation in novel environments, such as remote archipelagos and newly formed lakes. Rates of species accumulation are often accelerated in such circumstances, and speciation is typically accompanied by substantial shifts in phenotype (morphology, physiology, and resource use). Phenotypic differentiation in adaptive radiation is probably the outcome of strong natural selection (Lack 1947; Simpson 1953; Grant 1986; Schluter and Grant 1984; Schluter 1988, 1994), but what drives speciation?

Perhaps the oldest hypothesis for the origin of biological species (Mayr 1942) is that reproductive isolation evolves ultimately from the same forces that cause phenotypic change: divergent selection stemming from the exploitation of alternative resource environments and from resource competition. For example, Dobzhansky (1951) believed that

> the genotype of a species is an integrated system adapted to the ecological niche in which a species lives. Gene recombination in the offspring of species hybrids may lead to discordant gene patterns.

I will use 'ecological speciation' to refer to the process whereby reproductive isolation evolves from divergent selection, to emphasize the vital contribution of resource environment. 'Non-ecological' modes of speciation are those in which reproductive isolation evolves by genetic drift, founder events, or fixation of alternative advantageous alleles in separate populations experiencing the same natural selection pressures (cf. Barton 1989; see also Chapter 7). Despite reasonable arguments and experimental models for ecological speciation (Barton 1989; Rice and Hostert 1993; Coyne 1994), I am only too aware of studies that successfully link the evolution of reproductive isolation in nature to niche-based divergent natural selection: Galápagos finches (Grant 1986; see also Chapter 9) and monkey flowers (Macnair and Christie 1983).

In this chapter I show how recent studies of fish in postglacial lakes are beginning to provide these links. I begin with a summary of island-like properties of postglacial lakes from the perspective of the fish inhabiting them. I then review examples of rapid fish speciation in postglacial lakes and their common features. Finally, I summarize tests of ecological speciation in these fishes. Throughout, I focus on the broad issue of ecological speciation rather than on the subsidiary questions of whether it took place in allopatry or sympatry, and whether premating reproductive isolation evolved solely as an incidental by-product of divergent selection or additionally required reinforcement.

10.2 Fish speciation in postglacial lakes

Recent lakes as islands

Most of the vast numbers of lakes scattered over the northern parts of North America and Eurasia were formed after the immense sheets of ice that covered the region retreated about 15 000 years ago (e.g. see maps in Harrison (1982) and Pielou (1992)). The lakes were subsequently colonized by fish from glacial refugia, a process that was impeded by the limited number and duration of passage routes. Marine access to coastal lakes and rivers was limited to salt-tolerant species, and changes in sea level meant that many lakes were accessible for only a short period (McPhail and Lindsey 1986). Species intolerant of salt water dispersed inland via infrequent changes in the drainage patterns of lakes and rivers during deglaciation (McPhail and Lindsey 1986). Lakes and watersheds of previously glaciated areas are therefore like islands in important ways: movement between them was sporadic, their faunas are depauperate and heterogeneous, and separate drainages constitute evolutionarily independent units.

Fish species pairs in recently glaciated lakes share four remarkable attributes that together implicate ecological selection in the speciation process: rapid evolution of reproductive isolation, persistence (and perhaps origin) in the face of gene flow, a high degree of niche differentiation, and high intrinsic viability and fertility of hybrids. I summarize each of these elements in turn.

Rapid evolution of reproductive isolation

Elevated rates of speciation are indicated by the variety of cases in which two very closely related species coexist in lakes less than 15 000 years old. The examples in Table 10.1 are the subset of cases for which genetic distances (Nei's D or equivalent) are small but significant ($0 < D < 0.06$), independent evidence of assortative mating exists (indicated by a paucity of morphological hybrids, separate breeding times or localities, and/or mating observations), and evidence suggests that morphological differences are genetic rather than environmentally induced. Not yet added to the table

Table 10.1 Examples of young sympatric species pairs in lakes and rivers of recently glaciated areas

Nominal species	Region	Trophic characteristics	Genetic difference
Three-spined stickleback[a] *Gasterosteus aculeatus*	British Columbia	limnetic (planktivore) benthic (benthivore)	0.02/—/0.02–0.10
Lake whitefish[b] *Coregonus clupeaformis*	E. Canada, Maine	dwarf (planktivore) normal (benthivore)	0.01/0.5/—
Lake whitefish[c] *C. clupeaformis*	Yukon, Alaska low gill rakers (benthivore)	high gill rakers (planktivore)	0.01–0.02/—/0.02–0.30
Sockeye salmon[d] *Oncorhynchus nerka*	W. Canada, Alaska	sockeye (anadromous) kokanee (freshwater resident)	0.02/—/—
Atlantic salmon[e] *Salmo salar*	Newfoundland	anadromous freshwater resident	0.06/—/—
Brown trout[f] *Salmo trutta*	Ireland	sonaghen (planktivore) gillaroo (benthivore)	0.04/—/0.08
Brown trout[g] *S. trutta*	Sweden	planktivore benthivore	0.03/—/—
Arctic charr[h] *Salvelinus alpinus*	Scotland	planktivore benthivore	0.02/—/—
Arctic charr[i] *S. alpinus*	Iceland	planktivore and piscivore small and large benthivore	0.001/—/ 0.01
Rainbow smelt[j] *Osmerus mordax*	E. Canada, Maine	dwarf (planktivore) normal (benthivore, piscivore)	—/—/0.01–0.10

Genetic differences are in the format *x/y/z* where *x* is Nei's distance *D* based on electrophoresis, *y* is mtDNA sequence divergence estimated using restriction enzymes, and *z* is %mtDNA 'nucleotide divergence' (Nei and Miller 1990), which combines differences in both nucleotide sequence and haplotype frequency.

[a] Enos and Paxton Lakes. McPhail (1992), Schluter and McPhail (1992), Taylor *et al.* (1996).

[b] Fenderson (1964), Kirkpatrick and Selander (1979), Bernatchez and Dodson (1990).

[c] Bodaly (1979), Bodaly *et al.* (1992), Bernatchez *et al.* (1996).

[d] Ricker (1940), Foote *et al.* (1989), Taylor *et al.* (1996).

[e] Verspoor and Cole (1989), Claytor and Verspoor (1991).

[f] Lough Melvin. Ferguson and Taggart (1991), Hynes *et al.* (1996). A third piscivorous species (ferox), somewhat more distantly related to the other two (Nei's D=0.08) is also present.

[g] Lake Bunnersjoarna. Ryman *et al.* (1979), N. Ryman (personal communication 1995).

[h] Loch Rannoch. Gardner *et al.* (1988), Walker *et al.* (1988), Hartley *et al.* (1992).

[i] Lake Thingvallavatn. Magnusson and Ferguson (1987), Danzmann *et al.* (1991), Malmquist *et al.* (1992), Skulason *et al.* (1992), Snorrason *et al.* (1994). Four trophic forms are present, and their relationships and species status are still unclear. However, preliminary indications are that two genetically distinct lineages are present, each of which has two developmental morphs, as listed above (Skulason *et al.* 1992).

[j] Taylor and Bentzen (1993).

are cases in which evidence on genetic differences is still uncertain or lacking; these include omul in Lake Baikal (*Coregonus autumnalis*; Smirnov *et al.* 1992), cisco in Quebec (*Coregonus artedii*; Henault and Fortin 1989), European whitefish (*C. lavaretus*) and vendace (*C. albula*) in Scandanavia (Svärdson 1979), kokanee salmon in Kamchatka (*Oncorhynchus nerka kennerlyi*; Kurkenov 1977), and pygmy whitefish in Alaska (*Prosopium coulteri*; McCart 1970).

These examples include a diversity of fish taxa, although salmonids (salmon, trout, whitefishes) predominate. In all instances both species retain the same Latin binomial, underscoring the confusion surrounding their taxonomic status. I am unaware of an equivalently large set of examples from fish in unglaciated regions of the temperate zone. Low levels of interspecific genetic divergence are occasionally seen there, such as in the desert pupfishes of the American south-west (Soltz and Hirshfield 1981), but the most closely related species are not as often sympatric.

Gene flow and species origins

Phylogenetic studies suggest that rates of allopatric, sympatric, and parapatric speciation are simultaneously elevated in post-Pleistocene lakes. Lake whitefishes of eastern Canada and Maine offer the clearest case of species pairs that resulted from separate invasions by previously allopatric ancestors. Sympatric 'dwarf' and 'normal' whitefishes in one Maine lake are fixed for alternative mtDNA haplotype groups whose main geographical distributions lie to the south-west and north-east, respectively, in association with different late-Pleistocene refugia (Bernatchez and Dodson 1990). The situation is more confusing in nearby lakes. Both haplotype groups occur in one of two species in a second lake, suggesting mtDNA gene flow after secondary contact. Both mtDNA groups are also present in the sole species present in a third lake, indicating that it is a collapsed species pair.

Sockeye salmon represent the most likely case of sympatric or parapatric speciation. Anadromous sockeye salmon spend the first year or two of their life in large lakes before migrating to sea where they attain a large size. Kokanee salmon reside permanently in lakes and grow to a much smaller size. The two forms overlap in breeding season and spawning localities yet are genetically distinct (Foote *et al.* 1989; Taylor *et al.* 1996; Wood and Foote 1996). Phylogenies based on allozymes, minisatellite DNA, and mtDNA all show that kokanee are polyphyletic and arose independently several times from anadromous sockeye within different river systems (Fig. 10.1). This conclusion is bolstered by the spontaneous appearance of a small-bodied, freshwater resident salmon in river systems outside the natural range of sockeye following artificial transplants of only the anadromous form (Taylor *et al.* 1996).

Molecular surveys in most other cases do not clearly identify the geographic modes of speciation. Genetic data for several are consistent with sympatric speciation but can also be explained by double colonization whose trace has been obscured by subsequent gene flow (see Chapter 11 for other examples in which gene flow confounds phylogeny). Lake whitefish of Yukon and Alaska (Table 10.1) highlight the difficulty of distinguishing these processes (Fig. 10.2). Two mtDNA haplotype groups are present, one of which (II) is found exclusively in or near remnants of former Beringia whereas the other

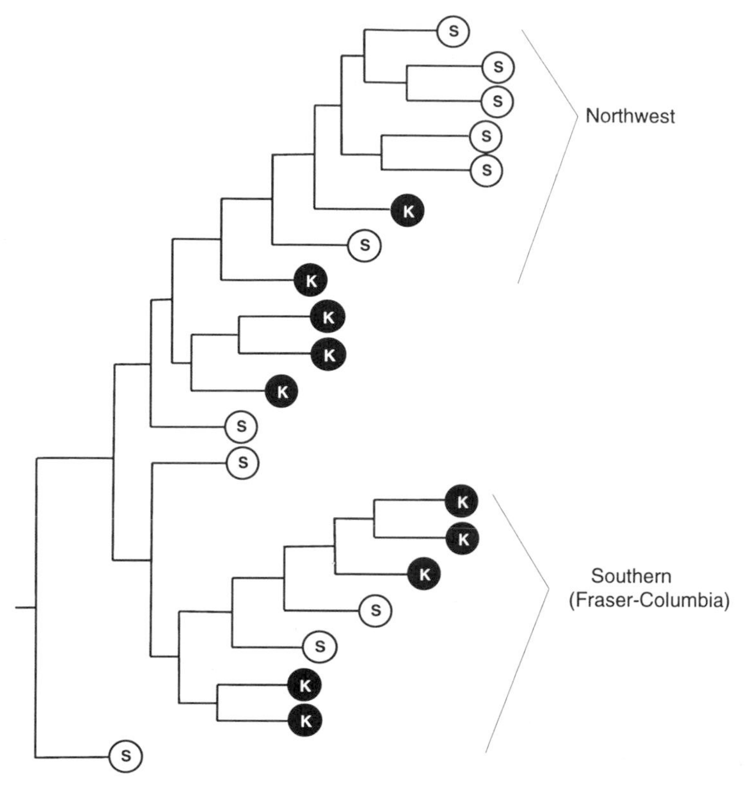

Fig. 10.1 Phylogeny of sockeye (S) and kokanee (K) populations in the Pacific North-west (*Oncorhynchus nerka*). Two major clades include populations associated with Pleistocene glacial refuges in Beringia ('Northern') and the Columbia River basin ('Southern'). Sockeye and kokanee are represented in both clades and appear to be polyphyletic. Based on variability at two microsatellite loci and one RFLP restriction site. After Taylor *et al.* (1996).

(III) is chiefly Eurasian (Bernatchez *et al.* 1996). In Little Teslin Lake in the Yukon River system the limnetic whitefish species is mainly group II whereas the benthic is fixed for group III, supporting an allopatric origin of species with scant gene flow after secondary contact. But in Squanga Lake, a few kilometres upriver, a single group III haplotype predominates in both species; in Dezadeash Lake in the Alsek River system (which was temporarily joined to the Yukon during deglaciation), the limnetic and benthic species are fixed for (different) group III haplotypes (Fig. 10.2). Gene flow has evidently occurred between forms within lakes, but additional data are needed to determine whether it occurred prior to sympatric speciation within each lake or following double invasion. Allozyme frequencies support the former interpretation—they are virtually identical between sympatric species in all three lakes—but this too could have resulted from gene flow after secondary contact (Bernatchez *et al.* 1996).

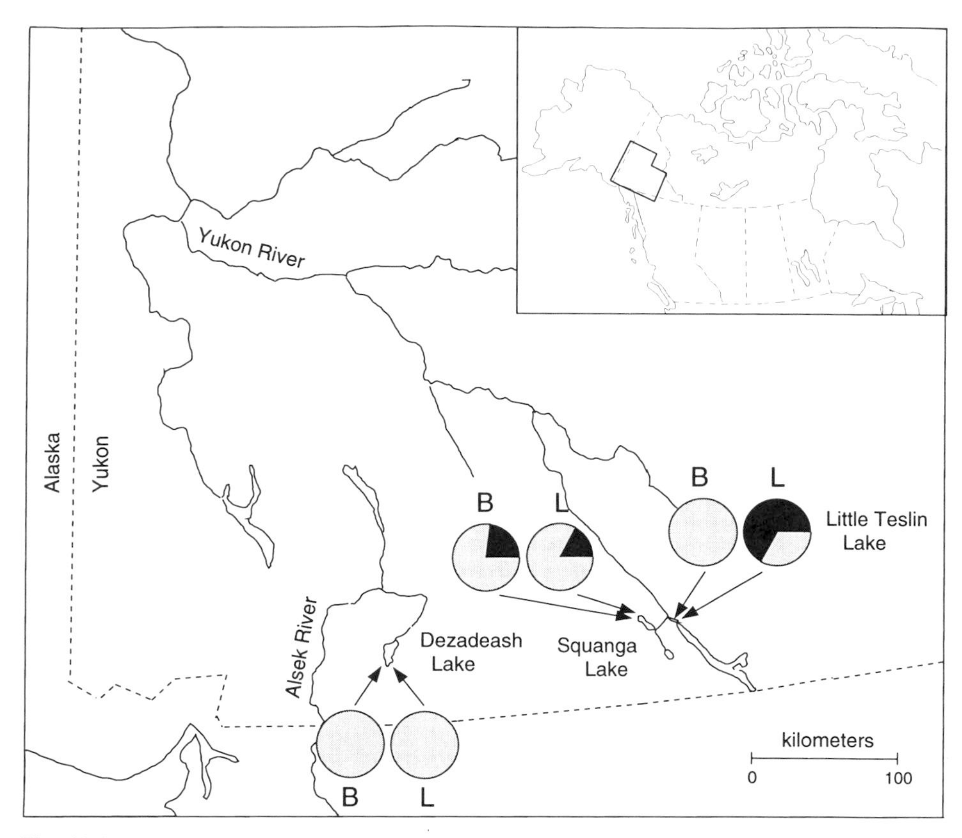

Fig. 10.2 Map of the Yukon and Alsek River systems, and lakes containing limnetic (L) and benthic (B) species of lake whitefish (*Coregonus clupeaformis*). Pies indicate fraction of individuals belonging to mtDNA haplotype groups II (filled) and III (light shading). After Bernatchez *et al.* (1996).

The origin of limnetic–benthic pairs of threespined sticklebacks in British Columbia (Table 10.1) is currently in doubt. A phylogeny based on mtDNA restriction fragments suggests that each pair originated independently by sympatric speciation (Fig. 10.3). However, the result contradicts earlier evidence that the limnetic species in each two-species lake is more similar to the modern marine species of stickleback in allozyme frequencies and in salinity tolerance of embryos than is the second (McPhail 1992; Kassen *et al.* 1996), a pattern more consistent with double invasion from the sea. Origins of sonaghen and gillaroo (Table 10.1) is also uncertain. Different mtDNA haplotypes predominate in the two species, and most are unique to Lough Melvin (Hynes *et al.* 1996). Those in sonaghen tend to be the sister haplotypes to those in gillaroo, which seems most consistent with sympatric divergence. However, Hynes *et al.* (1996) suspect that the overall level of genetic difference between the two forms is too great to have evolved in the past 13 000 years—the age of Lough Melvin.

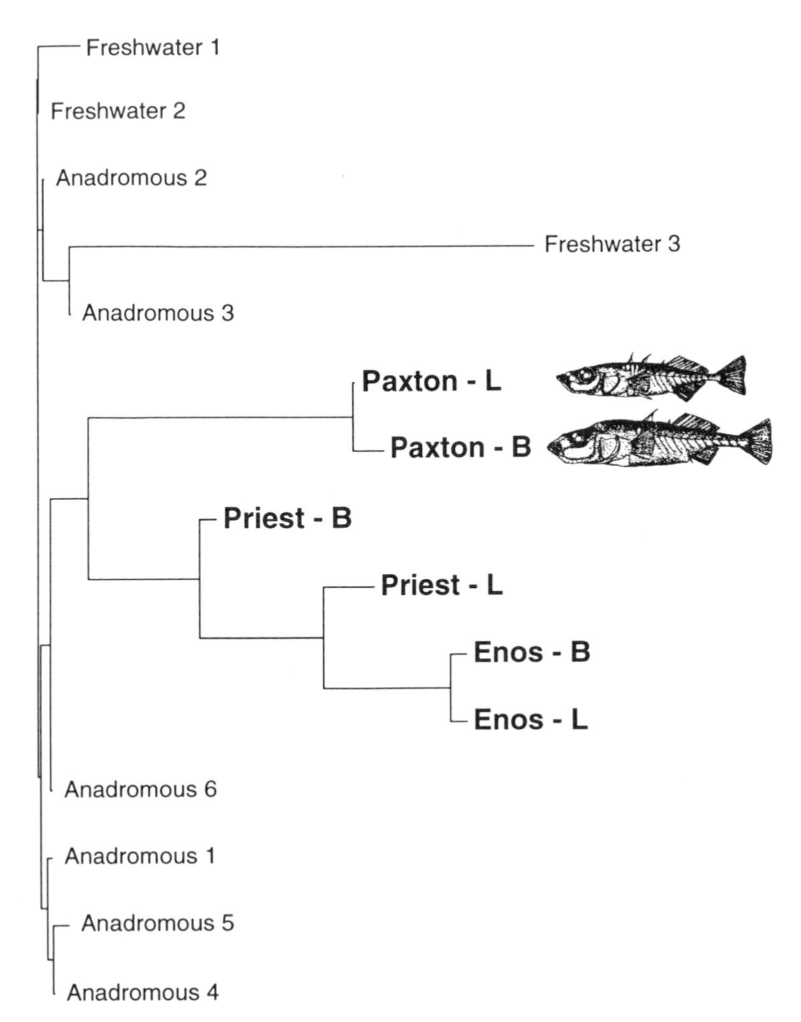

Fig. 10.3 Phylogeny of stickleback populations (*Gasterosteus aculeatus*) in the Strait of Georgia region of British Columbia, Canada, based on mtDNA RFLPs. Species pairs of limnetics (L) and benthics (B) in Paxton, Priest, and Enos lakes are indicated in bold. Remaining populations are six anadromous, two resident in freshwater streams (Freshwater 1, 2), and a solitary lake species (Freshwater 3). After Taylor *et al.* (1997).

Two conclusions may be drawn from the genetic data, despite uncertainty over origins. First, most sympatric species pairs have persisted in the face of gene flow. Whether gene flow diminished gradually in sympatry from initially high levels (sympatric speciation) or was only moderate to begin with (double invasion) remains to be determined. In either case selection against intermediate phenotypes is implicated—otherwise premating isolation would have decayed through time. Second, even if most sympatric species pairs were formed out of previously allopatric populations, the small genetic differences between alternative mtDNA lineages in most cases (<0.5% sequence divergence; Bernatchez and Dodson 1990; Bernatchez *et al.* 1996; Taylor *et al.* 1996) ensures that the period of allopatry could not have been prolonged.

Ecological and morphological differentiation

A large degree of ecological differentiation is a consistent feature of sympatric species pairs. Most remarkable are the repeated instances of divergence along similar ecological gradients (Schluter and McPhail 1993; Robinson and Wilson 1994). In the majority of cases one of the species is pelagic and specializes on zooplankton whereas the other is more benthic, consuming larger invertebrates obtained from sediments or on plants in more littoral habitats (Table 10.1). A consistent set of morphological differences is associated with this habitat split: planktivores are smaller and more slender than benthivores and tend to have narrower mouths and longer, more numerous gill rakers. (Gill rakers are protuberances along the gill arches that seive ingested prey or direct water currents and particle movement within the buccal cavity (Sanderson *et al.* 1991).) Such differences are arguably important only because they lower incidence of competitive exclusion. However, ecological and morphological differentiation may reflect divergent selection pressures between resource environments, the basis of ecological speciation.

Evidence for divergent selection between limnetic and benthic habitats is threefold. First, morphological divergence has rendered the species differently capable of exploiting these two habitats (Schluter 1993; Malmquist 1992; see also Robinson *et al.* 1996). In open water (presented in large aquaria with natural prey densities) limnetic sticklebacks captured plankton much more efficiently than benthics, an advantage that was reversed in the littoral zone (Fig. 10.4). Second, growth rates measured in a transplant experiment in the wild mirrored results on foraging efficiency: limnetic sticklebacks grew at double the rate of benthics in open water; benthics held an equivalent advantage in the littoral zone (Schluter 1995). Third, observations and experiments suggest that ecological and morphological divergence in sympatry is greater now than in the past and was driven by competition for food. Stickleback species occurring alone in small lakes ('solitary') are morphologically intermediate between limnetics and benthics and exploit both habitats (Schluter and McPhail 1992). In a pond experiment natural selection favoured the more benthic-like individuals within a solitary species following introduction of a planktivore (Schluter 1994).

These examples refer only to foraging habitats; differences in breeding habitats also exist but are often less. For example, both sticklebacks breed in close proximity in the littoral zone, albeit in different microhabitats (Hatfield 1995). Anadromous salmon return from the ocean to breed in many of the streams in which kokanee also spawn (Ricker 1940). This dissociation between feeding and breeding contrasts with many specialized insects for which divergence onto a new host yields habitat-based assortative mating as a by-product (Bush 1994). In postglacial fishes the link between divergent selection on foraging niche and the evolution of habitat-based premating isolation seems less straightforward. Species pairs usually differ in time and location of spawning, and possibly this evolved independently of foraging habitats and for different

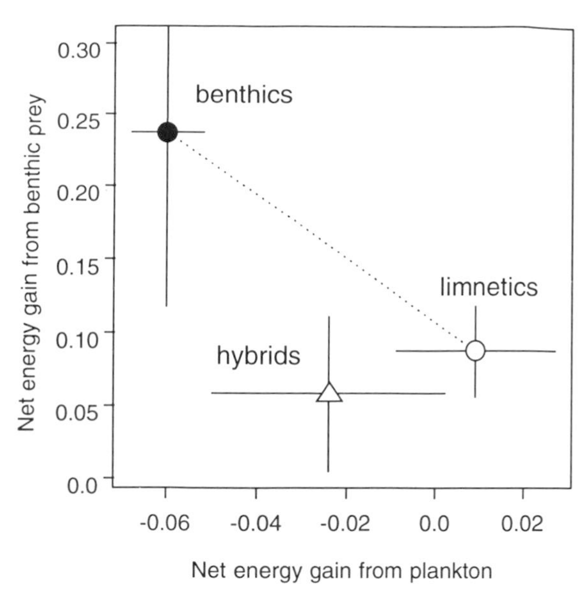

Fig. 10.4 Estimated net energy intake, in J/min, of threespined stickebacks foraging in the two major habitats of Paxton Lake. Estimates were obtained by recording prey capture rates of individual sticklebacks foraging in blocks of open water or littoral zone sediment transplanted from the lake to large aquaria. Points are medians ±1 SE. After Schluter (1993). Hybrids are 'F_{10}' hybrids—individuals sampled from a new population established using limnetic × benthic F_1 hybrids ten years previously.

selective reasons. Or, divergence in feeding niche may have led to separation in breeding habitat and time by one of three mechanisms: genetic hitch-hiking, altered ensuing selection pressures on life history, and reinforcement of premating isolation. Finally, it is conceivable that divergent selection on breeding habitat and time in some instances facilitated subsequent divergence in foraging niche. The existence of genetically different stocks of salmon that do not overlap in breeding time but exhibit little or no ecological differentiation (Tallman and Healey 1994) lend credence to the last possibility.

High viability and fertility of hybrids

Finally, study of a small but growing number of species pairs indicate that intrinsic viability and fertility of interspecific hybrids is high. Laboratory growth rates of F_1 and F_2 hybrids between limnetic and benthic sticklebacks are simply the average of that of their parents; growth of backcrosses is only slightly lower (Hatfield 1995). F_1 hybrids between sockeye salmon and kokanee are highly viable in a laboratory setting (Wood and Foote 1990). Hybrid breakdown arising from incompatability of species genomes is therefore not the basis of postmating isolation between these young species. Instead, it is likely that hybrid inferiority has an ecological basis, a possibility discussed in the next section.

10.3 Tests of ecological speciation

The above patterns, while suggestive, beg for more direct tests of ecological speciation. Unfortunately, progress in this direction has been limited. Here I summarize preliminary tests of three predictions.

Ecological basis of postmating isolation

The first expectation is that selection against hybrids should have an ecological basis. This prediction stems from the observation that postmating isolation in the laboratory is virtually lacking, yet hybrids have an intermediate phenotype that is expected to compromise their efficiency of resource exploitation. The prediction was first tested using wild limnetic, benthic, and hybrid sticklebacks in blocks of natural habitat transplanted to large aquaria (Fig. 10.4). Feeding efficiency of hybrids was poorer than benthics in the littoral zone and worse than limnetics in open water, supporting the expectation. This reduction in hybrid feeding efficiency is the result of a reduced success at capturing plankton compared to limnetics, and a reduced prey size in the littoral compared to benthics (Schluter 1993). These feeding differences translate into slower growth of hybrids, as shown by transplants of individuals to open water and littoral zone enclosures (Schluter 1995; Hatfield 1995). The same growth disadvantage was seen in field transplants using limnetic, benthic, and F_1 hybrids raised from eggs to adult in the laboratory (Hatfield 1995). Subsequent experiments on smaller size classes of sticklebacks suggest that these differences are weaker in small juveniles (Vamosi 1996).

Less information is available from other systems, as hybrid transplants to the wild have not been attempted. Hybrids between sockeye salmon and kokanee are intermediate in saltwater capability, and their larvae hatch earlier or later than larvae of the parent species, depending on the identity of the maternal and paternal parents (Wood and Foote 1990). Presumably, selection for different saltwater capabilities and synchronous hatching times between ocean-going and freshwater resident salmon acts against hybrids in the wild, but the crucial experiments have not yet been done.

Morphology-based assortative mating

Under the hypothesis of ecological speciation, premating isolation also evolves as a consequence of divergent selection between resource environments. But this may happen via many routes—mating preferences may diverge because natural selection directly favours it, because the preferences are correlated genetically with phenotypic traits under divergent selection, or for other reasons—which means that a general procedure won't suffice to test all cases. In postglacial fishes, two kinds of test have been attempted. The first examines whether mating preferences are dependent on the very phenotypic traits that have diverged between species. The second, summarized in the next section, asks whether the same premating isolating mechanism has evolved in parallel multiple times under similar environmental selection pressures.

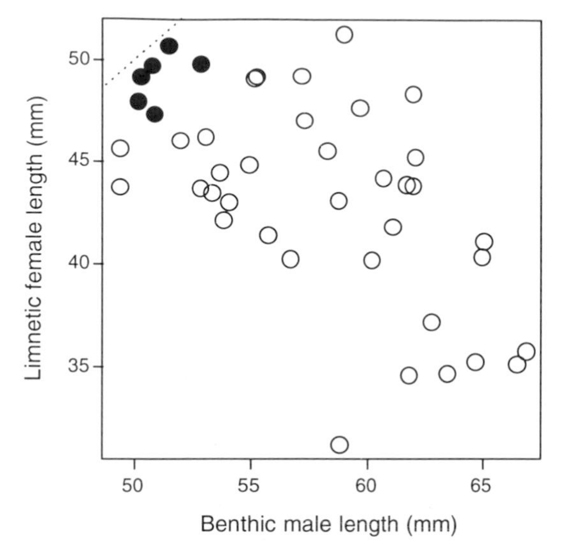

Fig. 10.5 Spawning probability between limnetic female and benthic male sticklebacks in relation to their body size. A filled symbol indicates that the pair spawned within 30 minutes; open symbols indicate no spawning. A trial involved introducing a single female to an aquarium containing a single nesting male. The dotted line indicates $Y = X$. Data from Paxton and Priest Lakes were combined. After Nagel and Schluter (1998).

Evidence that speciation resulted from divergent natural selection is gained when traits conferring adaptation to alternative environments are the basis of reproductive isolation (e.g. Macnair and Christie 1983). At least three studies suggest that morphological traits, especially body size, strongly influence the probability of mating. First, differences between stickleback species in body size are linked to differences in feeding efficiency in the two main habitats (Schluter and McPhail 1992; Schluter 1993, 1995). In laboratory mating experiments heterospecific matings took place only between the largest females of the smaller species (limnetics) and the smallest males of the larger species (benthics) (Fig. 10.5); and between the largest limnetic males and smallest benthic females (Schluter and Nagel 1995; Nagel and Schluter 1998). Second, assortative mating in sockeye salmon is based on the large differences in body size between the anadromous and freshwater resident forms (Foote and Larkin 1988).

Finally, male F_1 hybrid sticklebacks are less preferred as mates by limnetic and benthic females, possibly because they are morphologically intermediate (Hatfield and Schluter 1996; Vamosi 1996). Such phenotype-based mate choice could generate sexual selection against hybrids, further diminishing their fitness (Hatfield and Schluter 1996; Vamosi 1996).

Parallel speciation

Among the most remarkable evidence to be gained in favour of an ecological basis to speciation is the repeated evolution of the same mechanism of premating isolation independently in different lakes, under similar ecological conditions ('parallel speciation'; Schluter and Nagel 1995). Such data support the hypothesis of natural selection, for the same reason that other forms of parallel evolution are evidence for adaptation: genetic drift is unlikely to produce repetitive shifts in the same direction under a specific environmental setting.

As an example scenario for parallel speciation, consider an ancestral species that independently gives rise to multiple new populations at the periphery of its range in two different types of environment. Parallel speciation by natural selection occurs when reproductive isolation evolves between descendant forms adapting to different environments, but not between descendants adapting to similar environments.

Indirect evidence that parallel speciation has occurred is that mate choice is often based on body size (see above), and that divergence in size has occurred independently on multiple occasions (Schluter and Nagel 1995). For example kokanee salmon, which are small in size, have evolved repeatedly from large-bodied anadromous sockeye. If size is a major determinant of assortative mating (Foote and Larkin 1988) then separate, independently derived kokanee populations should not be reproductively isolated from one another. The crucial test of assortative mating between kokanee populations remains to be done.

A second apparent case of parallel speciation occurs in the sticklebacks, where limnetics and benthics have apparently evolved independently multiple times (Schluter and McPhail 1992; Taylor *et al.* 1997), and where assortative mating is also size-based (Schluter and Nagel 1995). Nagel (1994) showed that assortative mating between limnetics from different lakes (Paxton and Priest Lakes, Texada Island, British Columbia) was weak, and also that there was no detectable assortative mating between benthics from these same two lakes nor between them and benthics from Enos Lake, Vancouver Island, British Columbia. In other words, fish of similar body size and shape that exploit similar ecological environments have similar mate preferences. These results are still tentative evidence for parallel speciation because the phylogenetic relationships among these species is still uncertain (Taylor *et al.* 1997).

10.4 Discussion

The idea was developed in the first half of this century that divergent natural selection may cause reproductive isolation to evolve between populations exploiting different resource environments. Yet evidence from nature remains scarce in some measure because few have tested it. This lack of progress is partly the result of a historical preoccupation with other issues, especially the geographic mode of speciation (sympatric versus allopatric) and the contributions of 'genetic' mechanisms other than selection, i.e. drift and founder events. Establishment of the principle that speciation is chiefly allopatric might even have discouraged the study of natural selection in speciation. Random drift and other non-ecological mechanisms inevitably cause sterility and inviability in crosses between populations geographically isolated for a sufficiently long time; ecological processes are therefore not needed to explain the fact of species. However, divergent selection can cause reproductive isolation much sooner than drift, and may allow stable sympatry after a period of geographic isolation too brief for drift to effect much change.

The present chapter summarizes the contributions that studies of fish in postglacial lakes have made to the question of speciation by divergent selection. Postglacial lakes contain a wealth of examples of coexisting species exhibiting marked levels of ecological and morphological differentiation despite low levels of genetic divergence. Phenotypic differences in sympatry are primarily the outcome of markedly different selection pressures in alternative lake habitats, and by competition for food. Gene flow, either prior to sympatric divergence or following secondary contact between previously separated lineages, was habitual but did not prevent speciation. Nor did gene flow cause species pairs, once created, to collapse to hybrid 'swarms' despite high fertility and viability of hybrids. Hybrid fitness is strongly environment specific, and appears to stem directly from the disadvantage of having an intermediate phenotype. Mate choice is based in part on morphology, and hence premating isolation evolved in concert with morphological differentiation. Premating isolation may have evolved in parallel in independent populations experiencing similar selection pressures.

Many of these data are preliminary and await detailed confirmation. As well, much more effort is necessary to clarify the mechanisms of selection and the precise basis of pre- and postmating isolation. Nevertheless the results strongly point to ecological selection pressures as the primary cause of rapid speciation in these fishes. At this point, it is difficult to conceive of alternative non-ecological modes of speciation that would account for such a consistent set of findings. The context of these results is the depauperate lake environment; hence these findings give insights into the causes of speciation especially during adaptive radiation in novel environments. Moreover, these results are not unique. Several of the patterns summarized here echo findings in the Galápagos ground finches, especially the role of resources and competition in divergence, the sensitivity of hybrid fitness to ecological conditions, and the depedence of mate choice on trophic morphology (see Chapter 9).

At the very least these results cry for a reappraisal of the role of ecology in speciation, and for a renewed effort to understand mechanisms of selection leading to the evolution of reproductive isolation. There has been a tendency in recent years for researchers to infer mechanisms of speciation in nature entirely from genetic measurements—particularly the genetic basis of reproductive isolation and the deduced strength of selection acting on hybrids. Such evidence is undeniably crucial, but will of itself lead to a poor understanding of the causes of speciation. A complementary natural history of the evolution of reproductive isolation is needed in addition. In the same way that field studies of selection on beak size and shape have revealed much about the causes of morphological diversity, so too might field studies of selection on traits conferring degrees of reproductive isolation reveal much about the ecological causes of speciation.

Indeed, such field studies might change the questions we ask in genetic studies of speciation. For example, laboratory measures of hybrid sterility and inviability (Dobzhansky's 'discordant gene patterns') are standard traits for dissecting the genetic basis of postmating isolation. Yet, such measures may evolve comparatively slowly, if the high fertility and viability of hybrid postglacial fishes is any guide; indeed they may often evolve after speciation is already complete. It makes sense to focus genetic studies instead on earlier stages of ecological speciation. For example, when viability of hybrids is determined by the efficiency with which they can exploit environmental resources, or when mating success of hybrids is determined by degree of resemblance to the parent species, the most interesting genes—those underlying postmating isolation—are those that cause hybrids to be phenotypically intermediate. Because of their tractability for ecological experiments and the simplicity of their environments, studies of postglacial fishes can contribute much to a joint endeavour.

10.5 Summary

A venerable view of speciation is that reproductive isolation ultimately evolves from contrasting selection pressures between populations exploiting different resource environments. Yet, this 'ecological' mode of species origins has rarely been tested in nature. Here I describe emerging evidence of divergent selection and rapid speciation in fishes of postglacial lakes. Examples of very closely related species pairs within such lakes are known from over a dozen independent fish lineages, representing elevated rates of speciation in novel environments. The fish species pairs display four remarkable attributes which together suggest ecological speciation: rapid evolution of assortative mating in sympatry and/or allopatry; persistence (and in some cases origin) in sympatry despite a history of gene flow; a high degree of niche differentiation (usually one species is planktivorous and the other is benthic) accompanied by differences in body size and shape; and high intrinsic viability and fertility of interspecific hybrids. More direct evidence for ecological speciation comes from preliminary confirmation of three predictions: selection against hybrids has an ecological basis (morphologically intermediate hybrid sticklebacks have a reduced ability to exploit the main resources on which the parent species are specialized); premating isolation is linked to the morphological traits that have diverged between species (hybridization in sticklebacks occurs only between the morphologically most similar individuals of the two parent species); premating isolation mechanisms evolve in parallel in similar environments (interspecific mate preferences appear to have diverged in parallel in stickleback species pairs that evolved independently). These findings argue that ecological selection pressures are responsible for the origin of many species in adaptive radiation, and perhaps more generally. Field studies of selection on traits determining reproductive isolation are sorely needed, and would complement (and perhaps transform) traditional genetic approaches to speciation.

Acknowledgements

Thanks to E. B. Taylor, L. Bernatchez, and S. Skulason for keeping me up to date on their work and allowing me to cite manuscripts in press. My work on sticklebacks is supported by research grants from the Natural Sciences and Engineering Research Council (Canada).

References

Barton, N. H. (1989). Founder effect speciation. In *Speciation and its consequences*, (ed. D. Otte and J. A. Endler), pp. 229–56. Sinauer, Sunderland, MA.

Bernatchez, L. and Dodson, J. J. (1990). Allopatric origin of sympatric populationsof lake whitefish (*Coregonus clupeaformis*) as revealed by mitochondrial-DNA restriction analysis. *Evolution*, **44**, 1263–71.

Bernatchez, L., Vuorinen, J. A., Bodaly, R. A., and Dodson, J. J. (1996). Genetic evidence for reproductive isolation and multiple origins of sympatric trophic ecotypes of whitefish (*Coregonus*). *Evolution*, **50**, 624–35.

Bodaly, R. A. (1979). Morphological and ecological divergence within the lake whitefish (*Coregonus clupeaformis*) species complex in Yukon Territory. *Journal of the Fisheries Research Board of Canada*, **36**, 1214–22.

Bodaly, R. A., Clayton, J. W., Lindsey, C. C., and Vuorinen, J. (1992). Evolution of lakewhitefish (*Coregonus clupeaformis*) in North America during the Pleistocene: genetic differentiation between sympatric populations. *Canadian Journal of Fisheries and Aquatic Sciences*, **49**, 769–79.

Bush, G. (1994). Sympatric speciation in animals—new wine in old bottles. *Trends in Ecology and Evolution*, **9**, 285–8.

Claytor, R. R. and Verspoor, E. (1991). Discordant phenotypic variation in sympatric resident and anadromous Atlantic salmon (*Salmo salar*) populations. *Canadian Journal of Zoology*, **69**, 2846–52.

Coyne, J. A. (1994). Ernst Mayr and the origin of species. *Evolution*, **48**, 19–30.

Danzmann, R. G., Ferguson, M. M., Skulason, S., Snorrason, S. S., and Noakes, D. L. G. (1991). Mitochondrial DNA diversity among four sympatric morphs of Arctic charr, *Salvelinus alpinus* L., from Thingvallavatn, Iceland. *Journal of Fish Biology*, **39**, 649–59.

Dobzhansky, T. (1951). *Genetics and the origin of species*, 3rd edn. Columbia University Press, New York.

Fenderson, O. C. (1964). Evidence of subpopulations of lake whitefish, *Coregonus clupeaformis*, involving a dwarfed form. *Transactions of the American Fisheries Society*, **93**, 77–94.

Ferguson, A. and Taggart, J. B. (1991). Genetic differentiation among the sympatric brown trout (*Salmo trutta*) populations of Lough Melvin, Ireland. *Biological Journal of the Linnean Society*, **43**, 221–37.

Foote, C. J. and Larkin, P. A. (1988). The role of male choice in the assortative mating of anadromous and non-anadromous sockeye salmon (*Oncorhynchus nerka*). *Behaviour*, **106**, 43–62.

Foote, C. J., Wood, C. C., and Withler, R. E. (1989). Biochemical genetic comparisons of sockeye salmon and kokanee, the anadromous and nonanadromous forms of *Oncorhynchus nerka*. *Canadian Journal of Fisheries and Aquatic Sciences*, **46**, 149–58.

Frost, W. E. (1965). Breeding habits of Windermere charr, *Salvelinus willughbii* (Gunther), and their bearing on speciation of these fish. *Proceedings of the Royal Society of London* B, **163**, 232–84.

Gardner, A. S., Walker, A. F., and Greer, R. B. (1988). Morphometric analysis of two ecologically distinct forms of Arctic charr, *Salvelinus alpinus* (L.), in Loch Rannoch, Scotland. *Journal of Fish Biology*, **32**, 901–10.

Grant, P. R. (1986). *Ecology and evolution of Darwin's finches*. Princeton University Press.

Harrison, C. (1982). *An atlas of the birds of the western palearctic*. Princeton University Press.

Hartley, S. E., McGowan, C., Greer, R. B., and Walker, A. F. (1992). The genetics of sympatric Arctic charr [*Salvelinus alpinus* (L.)] populations from Loch Rannoch, Scotland. *Journal of Fish Biology*, **41**, 1021–31.

Hatfield, T. (1995). Speciation in sympatric sticklebacks: hybridization, reproductive isolation and the maintenance of diversity. Unpubl. Ph.D thesis, University of British Columbia, Vancouver.

Hatfield, T. and Schluter, D. (1996). A test for sexual selection on hybrids of two sympatric sticklebacks. *Evolution*, **50**, 2429–34.

Henault, M. and Fortin, R. (1989). Comparison of meristic and morphometric characters among spring- and fall-spawning ecotypes of cisco (*Coregonus artedii*) in southern Quebec, Canada. *Canadian Journal of Fisheries and Aquatic Sciences*, **46**, 166–73.

Hynes, R. A., Ferguson A., and McCann, M. A. (1996). Variation in mitochndrial DNA and post-glacial colonization of north western Europe by brown trout. *Journal of Fish Biology*, **48**, 54–67.

Kassen, R., Schluter, D., and McPhail, J. D. (1996). Evolutionary history of threespine sticklebacks (*Gasterosteus spp.*) in British Columbia: insights from a physiological clock. *Canadian Journal of Zoology*, **73**, 2154–8.

Kirkpatrick, M. and Selander, R. (1979). Genetics of speciation in lake whitefishes in the Allegash Basin. *Evolution*, **33**, 478–85.

Kurenkov, S. I. (1977). Two reproductively isolated groups of kokanee salmon, *Oncorhynchus nerka kennerlyi*, from Lake Kronotskiy. *Journal of Ichthyology*, **17**, 526–34.

Lack, D. (1947). *Darwin's finches*. Cambridge University Press.

Macnair, M. R. and P. Christie. (1983). Reproductive isolation as a pleiotropic effect of copper tolerance in *Mimulus guttatus*? *Heredity*, **50**, 295–302.

Magnusson, K. P. and Ferguson, M. M. (1987). Genetic analysis of four sympatric morphs of Arctic charr, *Salvelinus alpinus*, from Thingvallavatn, Iceland. *Environmental Biology of Fishes*, **20**, 67–73.

Malmquist, H. J. (1992). Phenotype-specific feeding behaviour of two arctic charr *Salvelinus alpinus* morphs. *Oecologia*, **92**, 354–61.

Malmquist, H. J., Snorrason, S. S., Skulason, S., Jonsson, B., Sandlund, O. T., and Jonasson, P. M. (1992). Diet differentiation in polymorphic Arctic charr in Thingvallavatn, Iceland. *Journal of Animal Ecology*, **61**, 21–35.

Mayr, E. (1942). *Systematics and the origin of species*. Columbia University Press, New York.

McCart, P. (1970). Evidence for the existence of sibling species of pygmy whitefish (*Prosopium coulteri*) in three Alaskan lakes. In *Biology of coregonid fishes*, (ed. C. C. Lindsey and C. S. Wood), pp. 81–98. University of Manitoba Press, Winnipeg.

McPhail, J. D. (1992). Ecology and evolution of sympatric sticklebacks (*Gasterosteus*): evidence for a species pair in Paxton Lake, Texada Island, British Columbia. *Canadian Journal of Zoology*, **70**, 361–9.

McPhail, J. D. and Lindsey, C. C. (1986). Zoogeography of the freshwater fishes of Cascadia (the Columbia system and rivers north to the Stikine). In *The zoogeography of North American freshwater fishes*, (ed. C. H. Hocutt and E. O. Wiley), pp. 615–37. Wiley, Chichester.

Nagel, L. M. (1994). The parallel evolution of reproductive isolation in threespine sticklebacks. Unpubl. M.Sc thesis, University of British Columbia, Vancouver.

Nagel, L. M. and Schluter, D. (1998). Body size, natural selection and speciation in sticklebacks. *Evolution*, in press.

Nei, M. and Miller, J. C. (1990). A simple method for estimating average number of nucleotide substitutions within and between populations from restriction data. *Genetics*, **125**, 873–9.

Partington, J. D. and Mills, C. A. (1988). An electrophoretic and biometric study of Arctic charr, *Salvelinus alpinus* (L.), from ten British lakes. *Journal of Fish Biology*, **33**, 791–814.

Pielou, E. C. (1992). *After the ice age*. University of Chicago Press.

Rice, W. R. and Hostert, E. E. (1993). Laboratory experiments on speciation: what have we learned in 40 years? *Evolution*, **47**, 1637–53.

Ricker, W. E. (1940). On the origin of kokanee, a fresh-water type of sockeye salmon. *Transactions of the Royal Society of Canada*, **34**, 121–35.

Robinson, B. W. and Wilson, D. S. (1994). Character release and displacement in fishes: a neglected literature. *American Naturalist*, **144**, 596–627.

Robinson, B. W., Wilson, D. S., and Shea, G. O. (1996). Trade-offs of ecological specialization: An intraspecific comparison of pumpkinseed sunfish phenotypes. *Ecology*, **77**, 170–8.

Ryman, N., Allendorf, F. W., and Staahl, G. (1979). Reproductive isolation with little genetic divergence in sympatric populations of brown trout (*Salmo trutta*). *Genetics*, **92**, 247–62.

Sanderson, S. L., Cech, J. J., and Patterson, M. R. (1991). Fluid dynamics in suspension- feeding blackfish. *Science*, **251**, 1346–8.

Schluter, D. (1988). Character displacement and the adaptive divergence of finches on islands and continents. *American Naturalist*, **131**, 799–824.

Schluter, D. (1993). Adaptive radiation in sticklebacks: size, shape, and habitat use efficiency. *Ecology*, **74**, 699–709.

Schluter, D. (1994). Experimental evidence that competition promotes divergence in adaptive radiation. *Science*, **266**, 798–801.

Schluter, D. (1995). Adaptive radiation in sticklebacks: trade-offs in feeding performance and growth. *Ecology*, **76**, 82–90.

Schluter, D. and Grant, P. R. (1984). Determinants of morphological patterns in communities of Darwin's finches. *American Naturalist*, **123**, 175–96.

Schluter, D. and McPhail, J. D. (1992). Ecological character displacement and speciation in sticklebacks. *American Naturalist*, **140**, 85–108.

Schluter, D. and McPhail, J. D. (1993). Character displacement and replicate adaptive radiation. *Trends in Ecology and Evolution*, **8**, 197–200.

Schluter, D. and Nagel, L. M. (1995). Parallel speciation by natural selection. *American Naturalist*, **146**, 292–301.

Simpson, G. G. (1953). *The major features of evolution*. Columbia University Press, New York.

Skulason, S., Antonsson, T., Gudbergsson, G., Malmquist, H. J., and Snorrason, S. (1992). Variability in Icelandic charr. *Icelandic Agricultural Science*, **6**, 142–53.

Smirnov, V. V., Todd, T. N., and Luczynski, M. (eds.) (1992). Intraspecific structure of Baikal omul, *Coregonus autumnalis migratorius* (Georgi). *Polska Archiv Hydrobiologia*, **39**, 325–33.

Snorrason, S. S., Skulason, S., Jonsson, B., Malmquist, H. J., Jonasson, P. M., Sandlund, O. T., and Lindem, T. (1994). Trophic specialization in Arctic charr *Salvelinus alpinus* (Pisces; Salmonidae): Morphological divergence and ontogenetic niche shifts. *Biological Journal of the Linnean Society*, **52**, 1–18.

Soltz, D. L. and Hirshfield, M. F. (1981). Genetic differentiation of pupfishes (genus *Cyprinodon*) in the American Southwest. In *Fishes in North American deserts*, (ed. R. J. Naiman and D. L. Soltz), pp 291–333. Wiley, Chichester.

Svärdson, G. (1979). Speciation of Scandinavian *Coregonus*. *Report from the Institute of Freshwater Research, Drottningholm*, **57**, 1–95.

Tallman, R. F. and Healey, M. C. (1994). Homing, straying, and gene flow among seasonally separated populations of chum salmon (*Oncorhynchus keta). Canadian Journal of Fisheries and Aquatic Sciences*, **51**, 577–88.

Taylor, E. B. and Bentzen, P. (1993). Evidence for multiple origins and sympatric divergence of trophic ecotypes of smelt (*Osmerus*) in northeastern North America. *Evolution*, **47**, 813–32.

Taylor, E. B., Foote, C. J., and Wood, C. C. (1996). Molecular genetic evidence for parallel life history evolution within a Pacific salmon (sockeye salmon and kokanee, *Oncorhynchus nerka*). *Evolution*, **50**, 401–16.

Taylor, E. B., McPhail, J. D., and Schluter, D. (1997). History of ecological selection in sticklebacks: uniting experimental and phylogenetic approaches. In *Molecular evolution and adaptive radiation*, (eds. T. J. Givnish and K. J. Sytsma), pp. 511–34. Cambridge University Press.

Vamosi, S. (1996). Postmating isolation mechanisms between sympatric populations of three-spined sticklebacks. Unpubl. M.Sc. thesis, University of British Columbia, Vancouver.

Verspoor, E. and Cole, L. J. (1989). Genetically distinct sympatric populations of resident and anadromous Atlantic salmon, *Salmo salar*. *Canadian Journal of Zoology*, **67**, 1453–61.

Walker, A. F., Greer, R. B., and Gardner, A. S. (1988). Two ecologically distinct forms of Arctic charr *Salvelinus alpinus* (L.) in Loch Rannoch, Scotland. *Biological Conservation*, **43**, 43–61.

Wood, C. C. and Foote, C. J. (1990). Genetic differences in the early development and growth of sympatric sockeye salmon and kokanee (*Oncorhynchus nerka*) and their hybrids. *Canadian Journal of Fisheries and Aquatic Sciences*, **47**, 2250–60.

Wood, C. C. and Foote, C. J. (1996). Evidence for sympatric genetic divergence of anadromous and nonanadromous morphs of sockeye salmon (*Oncorhynchus nerka*). *Evolution*, **50**, 1265–79.

11

How 'molecular leakage' can mislead us about island speciation

Bryan Clarke, M. S. Johnson, and James Murray

11.1 Introduction

It is often supposed that to reconstruct the true phylogeny of organisms we should use characters or sequences that are, as nearly as possible, selectively neutral. Those under selection are thought to be unsatisfactory because different lineages can converge when they are exposed to similar environments. Here we report evidence that, in groups of sympatric species, neutral sequences may be seriously misleading about evolutionary relationships, while genes under selection may yet contain valid phylogenetic information.

The primary observations are on allelic frequencies in land snails of the genus *Partula* from the island of Moorea in French Polynesia. Since the pioneering studies of Garrett (1884) and Crampton (1932), the Partulae of Moorea and other Pacific islands have been exceptionally productive of information about the origin, variation, and differentiation of species. Studies on the evolution of the group have recently been reviewed (Cowie 1992; Johnson *et al.* 1993*a*).

The Society Islands, to which Moorea belongs, were formed successively as the Pacific plate moved in a north-westerly direction, at approximately 11 cm per year, over a 'hot-spot' in the mantle (Duncan and McDougall 1976). Consequently the islands at the north-west end of the chain (Bora Bora, about 3.3 Ma (million years) old, Tahaa, about 2.9 Ma old, and Raiatea and Huahine, about 2.5 Ma old) are older than those at the south-east (Moorea, about 1.5 Ma old, and Tahiti, about 1 Ma old). The snails seem to have populated the newer islands from the older ones, carried by birds or blown by typhoons. Until recent extinctions (Murray *et al.* 1988), each island typically harboured several endemic species. Moorea had seven. The Partulae of Moorea are now extinct on the island, but there are living representatives of most species in the laboratory, and in many zoos, as well as a large array of frozen and preserved populations. To avoid oscillating between the present and past tenses when referring to living material, collections, and populations, we will keep to the present tense.

A study of allozymic differences between the species in the Society archipelago showed that, with a few exceptions, the species from one island resemble each other more than they resemble any species from another island (Johnson *et al.* 1986*a*). It was therefore reasonable to conclude that most of the speciation occurred *in situ* on each island, through a series of successive radiations each starting with a single invasion, or with a 'burst' of invasions over a short period of evolutionary time. Subsequent invasions would then be less likely to succeed because they would encounter competitors already adapted to the local environment.

Our earlier analysis (Johnson *et al.* 1986*a*) was carried out using pooled samples from each species. The present chapter uses a larger set of data, from individual populations, and shows an unexpected cline in the degree of genetic difference between two sympatric taxa, *Partula suturalis* Pfeiffer and *P. taeniata* Mörch. The pattern of differences suggests that secondary invasions have occurred, and that the resemblances between species within islands may have been at least partly due to gene flow, rather than to a common origin *in situ*.. If this is correct, our results make an important general point that neutral genes can be unsatisfactory indicators of phylogenetic relationships among sympatric species.

11.2 The Moorean species of *Partula*

Anatomical and reproductive relationships between the Moorean species of *Partula* were studied by Murray and Clarke (1980), who recognized two major groups. The first, the *suturalis* group, contains four species (*P. suturalis, P. aurantia* Crampton *P. tohiveana* Crampton, and *P. mooreana* Hartman). The second, the *taeniata* group, contains two (*P. taeniata* and *P. exigua* Crampton). Within each of these groups there is clear evidence of natural hybridization, some of it sporadic, some of it more intense but restricted to particular populations. There are natural hybrids of *P. suturalis* with *P. aurantia*, *P. suturalis* with *P. tohiveana*, *P. aurantia* with *P. tohiveana*, and *P. taeniata* with *P. exigua*. The seventh species, *P. mirabilis* Crampton, forms a bridge between the two species groups, since it hybridizes naturally with *P. taeniata*, and in the laboratory with *P. aurantia* and *P. tohiveana.* Despite these possibilities for genetic exchange, most species in most places on Moorea are clearly distinct from each other. As many as four can coexist in one place without any loss of anatomical integrity (Murray and Clarke 1980). When they coexist, they are ecologically distinguishable (Murray *et al.* 1993)

The two commonest Moorean species are *Partula taeniata* and *P. suturalis* (Fig. 11.1) which both occur throughout the island. *P. taeniata* is found alone in some habitats at low altitudes, but otherwise the two are widely sympatric (Crampton 1932). They differ ecologically, *P. taeniata* favouring shrubs up to about 5 m in height, and *P. suturalis* favouring the trunks of the purau tree, *Hibiscus tiliaceus* L. (Murray *et al.* 1993). Despite these broad preferences, they are quite often found on the same host plant. There is no direct evidence of natural hybridization between them, and when they are grown together in the laboratory they do not cross (Murray and Clarke 1980). They are anatomically very different, *P. suturalis* being substantially larger. They even differ in the genetics of their shell colour and banding polymorphisms (Murray and Clarke 1976*a*, *b*).

Both species show north–south clines in the size and shape of the shell. The northern *P. taeniata* tend to be smaller and fatter than those from the south. On the other hand the southern *P. suturalis* tend to be smaller and fatter than those from the north (Crampton

Fig. 11.1 *Partula suturalis* (banded, left-handed) and *P. taeniata* (unbanded, right-handed) from the island of Moorea. Snails of opposite coiling types, as shown here, rarely hybridize, whereas snails of the same coiling type may hybridize more easily.

1932; Lundman 1947; Johnson *et al.* 1993*b*). These opposing clines argue against any simple explanation in terms of climatic or vegetational patterns, although the southern valleys tend to be wetter and lusher than the northern ones. The two clines are not exact mirror images of each other. The shapes of *P. taeniata* and *P. suturalis* are most alike in the north-eastern and mid-western valleys, and most different in the far southern and north-western ones. Shell sizes are most alike in the eastern and north-eastern valleys and most different in the centre and north-west. It is important for the present analysis that the north-western populations show particularly strong morphological differences between *P. taeniata* and *P. suturalis*.

There is another way in which northern and southern populations differ. The shells of *P. suturalis* from the north of the island are coiled to the left and those from the south are coiled to the right (Fig. 11.1 shows a left-handed *suturalis*). There is a narrow zone of transition, containing both dextral and sinistral shells, in between. The genetic change of coil from sinistral to dextral coincides with the sympatric presence of other species in the *suturalis* group that are themselves sinistral (*P. mooreana* and *P. tohiveana*). Snails of opposite coil are less likely to mate with each other, and the change represents a stage in the evolution of reproductive isolation (Clarke and Murray 1969; Johnson 1982; Johnson *et al.* 1990). *Partula taeniata* is dextral throughout the island, so we would expect the barriers to gene flow between *P. taeniata* and *P. suturalis* to be greater in the north than in the south.

11.3 Samples and analysis

Sixty-one samples, representing seven species of *Partula* (*P. taeniata*, *P. suturalis*, *P. mooreana*, *P. mirabilis*, *P. tohiveana*, *P. aurantia*, and *P. exigua*) were collected at 27

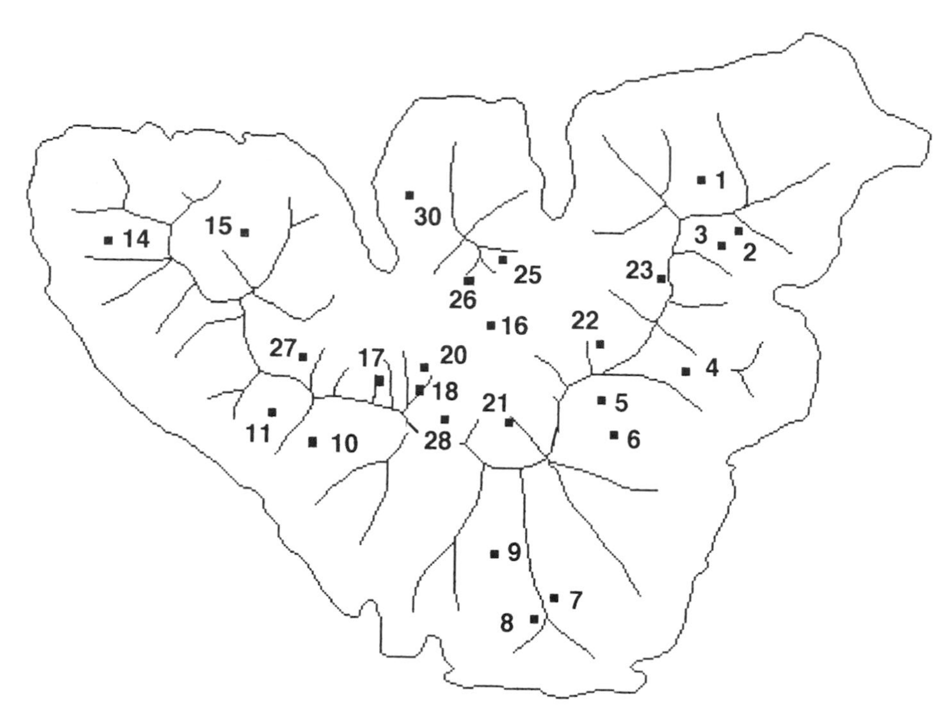

Fig. 11.2 A map of the island of Moorea, showing the locations at which samples of *Partula* were collected. The two samples whose locations are not shown (565 and 566) were collected near sample 3. Details about the species found in each sample are given in Table 11.1. The maximum diameter of the island is approximately 15 km.

localities on the island of Moorea (see Fig. 11.2). In addition nine samples, representing seven species (*P. otaheitana* (Bruguière), *P. jackieburchi* Kondo, *P. hyalina* Broderip, *P. filosa* Pfeiffer, *P. nodosa* Pfeiffer, *P. affinis* Pease, and *P. clara* Pease), were collected at three localities on the neighbouring island of Tahiti. Eight pooled samples representing each of three species from the island of Raiatea (*P. faba* Martin, *P. dentifera* Pfeiffer, and *P. hebe* Pfeiffer), three from the island of Huahine (*P. rosea* Broderip, *P. varia* Broderip, and *P. arguta* Pease), one from Bora Bora (*P. lutea* Lesson) and one from Rarotonga in the Cook Islands (*P. assimilis* Pease) were used to root the trees of Moorean and Tahitian species. All samples are listed in Table 11.1. Further details about localities and sampling methods are to be found in Johnson *et al.* (1986*a*, *b*, 1993*b*).

Tissues of individual snails from each sample were homogenized, electrophoresed, and stained using the techniques described by Johnson *et al.* (1977, 1986*b*). For the present analysis we took data on 19 variable enzyme loci, ignoring loci that were invariant across all 78 samples. The variable loci were *Alph, Est-1, Est-2, Got-1, Got-2, Idh-1, Idh-2, Mdh-1, Mdh-2, Mdh-3, Mpi, Np, Pep-2, Pep-4, Pep-6, 6pgd, Pgi, Pgm-1*, and *Pgm-2*. Further details about these loci are given by Johnson *et al.* (1977,

Table 11.1 Localities and species sampled

Island	Locality[a]	Species	Mean sample size (range)[b]
Moorea	1	*taeniata*	6
	1	*suturalis*	20
	2	*taeniata*	19
	3	*aurantia*	20
	3	*exigua*	4
	4	*taeniata*	23
	4	*suturalis*	11
	4	*tohiveana*	23
	5	*suturalis*	14.6 (13–19)
	5	*tohiveana*	11
	6	*taeniata*	19.9 (19–20)
	6	*suturalis*	20
	7	*taeniata*	9
	7	*suturalis*	20
	8	*taeniata*	21
	8	*suturalis*	20
	9	*taeniata*	21
	9	*suturalis*	19
	9	*mooreana*	14
	10	*taeniata*	20
	10	*suturalis*	19.9 (19–20)
	10	*mooreana*	22
	11	*taeniata*	20
	11	*suturalis*	22
	14	*taeniata*	19.7 (15–20)
	14	*suturalis*	20
	15	*taeniata*	19.7 (17–20)
	15	*suturalis*	22
	16	*taeniata*	17.3 (17–20)
	16	*mirabilis*	20
	17	*taeniata*	20
	17	*suturalis*	20
	18	*taeniata*	20
	18	*suturalis*	20
	18	*mirabilis*	15
	20	*taeniata*	20.3 (20–23)
	20	*suturalis*	19.9 (19–20)
	20	*mirabilis*	3
	21	*taeniata*	21.7 (11–23)
	21	*suturalis*	20
	21	*mirabilis*	8
	21	*tohiveana*	21
	22	*taeniata*	11.3 (11–14)
	22	*suturalis*	19.2 (19–20)

Table 11.1 *continued.*

Island	Locality[a]	Species	Mean sample size (range)[b]
	22	*tohiveana*	20
	23	*suturalis*	12
	23	*taeniata*	14.6 (13–20)
	23	*aurantia*	18
	25	*taeniata*	19.9 (19–20)
	25	*suturalis*	19.8 (17–20)
	26	*taeniata*	16.2 (11–20)
	26	*mirabilis*	20
	27	*taeniata*	19.9 (16–22)
	27	*suturalis*	22
	28	*taeniata*	21.9 (20–22)
	28	*suturalis*	71
	28	*mirabilis*	57
	28	*mooreana*	34
	30	*suturalis*	14.9 (3–16)
	565	*exigua*	20
	566	*exigua*	10
Tahiti	577	*otaheitana*	12
	578	*otaheitana*	19
	742	*jackieburchi*	24
	742	*affinis*	3
	577	*hyalina*	16
	742	*hyalina*	3
	577	*filosa*	21
	578	*nodosa*	22
	578	*clara*	16
Raiatea		*faba*	48
		dentifera	64
		hebe	115
Huahine		*rosea*	54
		varia	113
		arguta	30
Bora Bora		*lutea*	40
Rarotonga		*assimilis*	20

[a] For Moorean localities, see Fig. 11.1; for other islands, see text.
[b]The sample sizes are averaged over the numbers used for each of the 19 enzymes, and the figures in brackets represent the extremes. Where there are no bracketed numbers, the samples for all enzymes were identical in size. Note the small size of sample 54 with four individuals (eight chromosomes), and of samples 59, 65, and 67 with three individuals (six chromosomes) each.

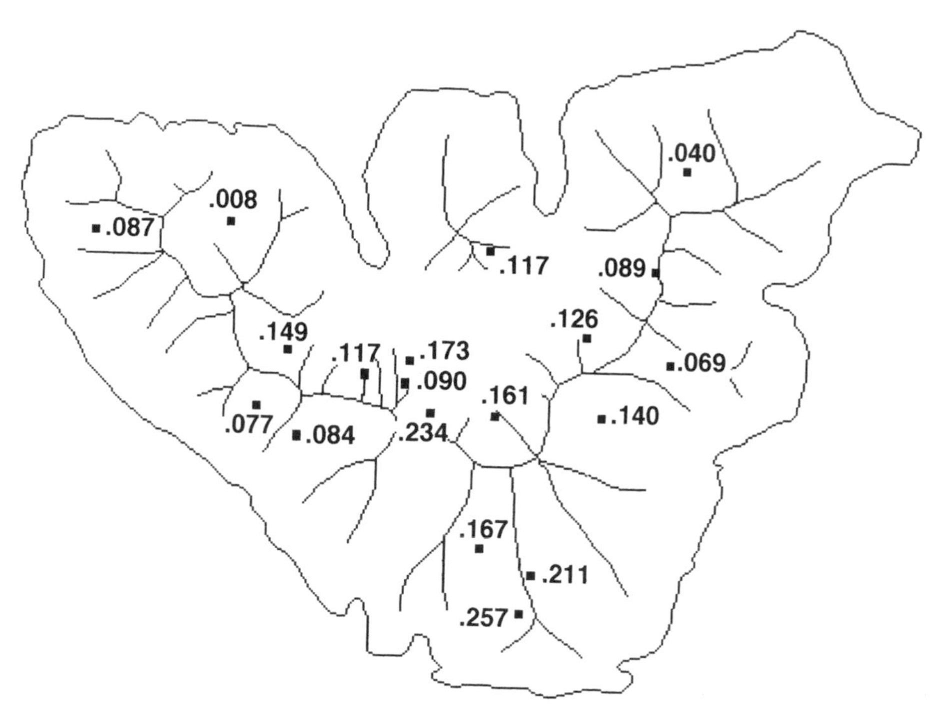

Fig. 11.3 A map of Moorea showing, for each locality at which *Partula taeniata* and *P. suturalis* were found sympatrically, Nei's coefficient of genetic distance between them. There is an obvious, if irregular, gradient from north to south.

1986*a*).The gene frequencies at the 19 loci allowed us to compare all samples with all others, using Nei's 'unbiased' coefficients of genetic distance (Nei 1978). The matrix of distances between populations is given by Clarke *et al.* (1996). It was used to generate UPGMA, Fitch–Margoliash, and neighbour-joining trees with the PHYLIP suite of programs (Felsenstein 1993).

11.4 Patterns of variation

Figure 11.3 is a map of the genetic distances between sympatric populations of *P. taeniata* and *P. suturalis*. It is remarkable in three ways. In the first place it shows a clear, if irregular, cline from larger differences in the south to smaller differences in the north. Second, by the standards of *Partula* the differences in the south are very large, as great as some of those between species on Moorea and Raiatea (about 270 km apart), or even on Raiatea and Saipan (about 7500 km apart). Third, by the same standards the differences in the north are very small, those in localities 1 and 15 being less than the median value of distances *within* Moorean species (0.041). We are therefore faced with trying to answer two, possibly separate, questions. Why are the differences in the south so large? Why are the differences in the north so small?

The south

The large differences between *P. taeniata* and *P. suturalis* in the south are due to both species having diverged from their northern conspecifics, as well as from each other, at many loci (Johnson *et al.* 1986*b*, 1993*a*). *P. taeniata* seems to have diverged more than *P. suturalis*, and indeed the southern *P. taeniata* populations differ as much from other *P. taeniata* as they do from *P. suturalis*. The maximum genetic distance within *P. taeniata* is 0.267, between localities 8 and 22. The maximum distance within *P. suturalis* is 0.148, between localities 8 and 30. It should be emphasized at this point that the southern *P. taeniata* are indeed good *P. taeniata*; over a few kilometres they grade, apparently without interruption, into the northern ones. They occupy the ecological niche of *taeniata*. The southern and northern *taeniata* interbreed freely in the laboratory. The same principles apply to the southern *P. suturalis*.. They too are good members of their species, despite containing high frequencies of an enzyme allele ($Pgm\text{-}2^{0.87}$) not found elsewhere on Moorea (Johnson *et al.* 1986*b*). Since there are no obvious ecological, physical, or geological reasons why the southern snails should be so different from other members of their own species, and from each other, we must seek clues about their history.

It is possible that the southern populations have been changed by an invasion from another island. In that case, because both species are aberrant, there must have been two sets of invaders, one that hybridized with *P. taeniata* and one that hybridized with *P. suturalis*. It is not unlikely that an event such as a typhoon could move several species simultaneously.

Because of its proximity, the most probable source of invaders is the island of Tahiti, about 20 km away. The next nearest island inhabited by *Partula* is Huahine, more than 200 km distant. There is a second reason to favour Tahiti as a source. It was probably populated from Moorea, which is half a million years older. Before Tahiti became inhabited, the Moorean ancestors of *P. taeniata* and *P. suturalis* might well have begun to differentiate. Thus the Tahitian descendants of proto-suturalis and proto-taeniata, when they returned to Moorea, would each be expected to be more easily compatible with its closest Moorean relatives than with members of the other species. This would explain why there was not a general mêlée of hybridization producing convergence between the taxa.

We tested the hypothesis of an invasion by reconstructing evolutionary trees of populations in Moorea and Tahiti, using the matrix of Nei's distances. All our trees agree, regardless of the method of their construction or the taxa used for rooting them, in showing that the extreme southern *P. taeniata* are more closely related to Tahitian species than to any other taxon in Moorea (see Fig. 11.4). They also agree in placing the southern *P. suturalis* firmly with the other Moorean *P. suturalis*. Thus the hypothesis of an invasion is supported by the data from *P. taeniata*, but not by the data from *P. suturalis*. The conflict between the two sets of data is resolved if we suppose that *P. suturalis*, the larger and more mobile species, has a higher level of gene flow with its northern conspecifics, obscuring the traces of a Tahitian ancestry. This supposition is consistent with an earlier study, in which a Wagner tree, based on pooled samples, showed the southern *suturalis*, rather than the southern *taeniata*, closest to the Tahitian

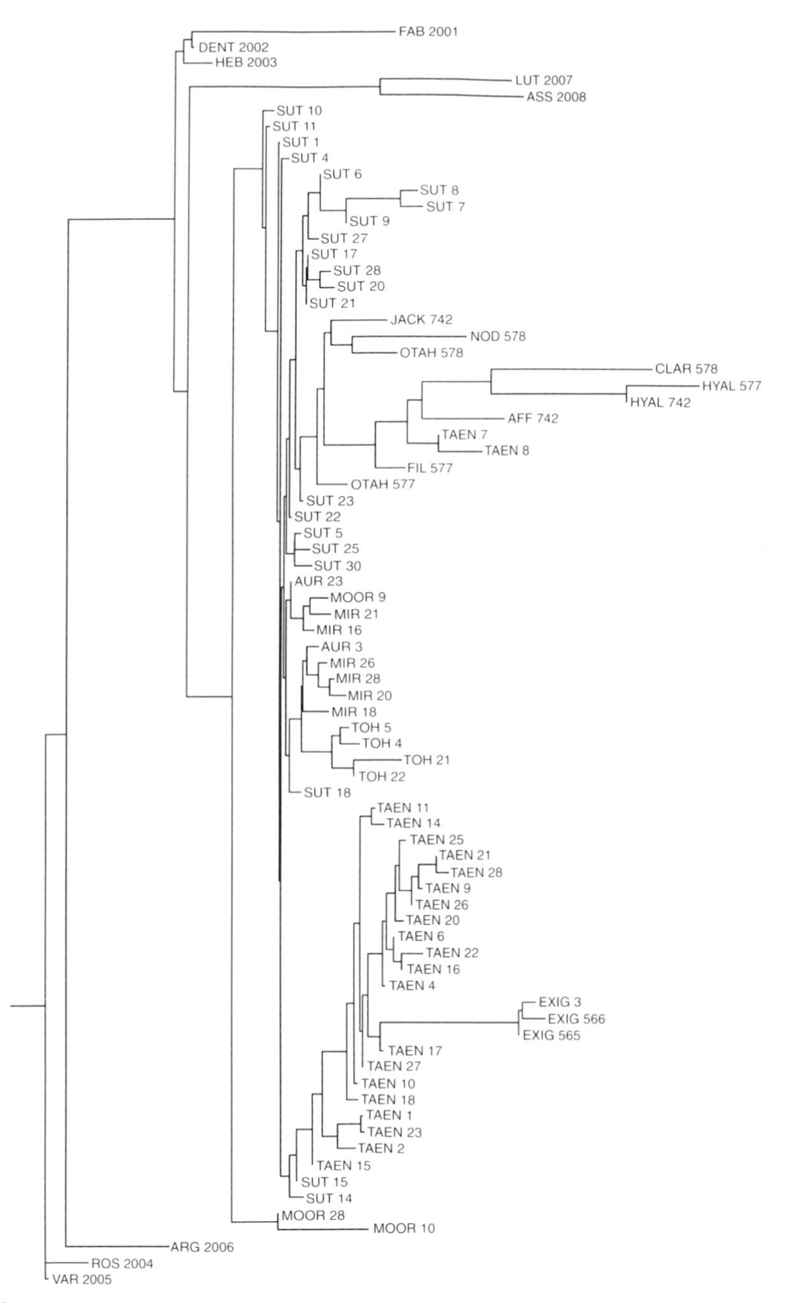

Fig. 11.4 A neighbour-joining tree of all the samples. The tree for Tahiti and Moorea is rooted by the inclusion of species from the older islands of Raiatea (species abbreviated as fab, dent, and heb), Huahine (those abbreviated as ros, var, and arg), Bora Bora (that abbreviated as lut) and Rarotonga (that abbreviated as ass). Note that populations of *P. taeniata* from the extreme south of Moorea (taen 7 and taen 8) group with the species from Tahiti (abbreviated as otah, jack, aff, hyal, fil, and nod). Equivalent populations of *P. suturalis* from the extreme south (sut 7 and sut 8), however, group with their Moorean conspecifics. The association of southern *taeniata* with the Tahitian species has been found consistently in all the trees studied, including those produced by UPGMA and the Fitch–Margoliash method. Bootstrapping the neighbour-joining tree by loci shows the association in 65 out of 100 trials.

taxa (Johnson *et al.* 1986*a*). It is also consistent with observations on restriction sites in the mitochondrial DNA (Murray *et al.* 1991). The southern *P. suturalis* show mitochondrial haplotypes that are as near to a Tahitian species (*P. nodosa*) as to any other Moorean taxon. The southern *P. taeniata* show a single haplotype that is common on both islands. Immigration from Tahiti seems to be the only explanation that accounts for all the facts.

The north

The extremely short genetic distance (0.008) between *P. taeniata* and *P. suturalis* at locality 15 may be a quirk of sampling, but it resembles the values at other northern localities in being more appropriate to differences within species than to differences between them. The most obvious explanation for such low values is that the species have recently diverged (as has been argued for the African cichlids discussed in Chapter 14). However, there are difficulties with this explanation in the present case. As pointed out above, the large differences in the south have apparently been due to invasions from Tahitian populations that originated *after* the two Moorean species had diverged. Our reconstructed trees suggest the same order of events. While an accelerated rate of divergence might have happened because of special selective factors in the Tahitian environment, this seems unlikely and there is no independent evidence of it. On Moorea the anatomical and ecological differences between *P. suturalis* and *P. taeniata* are very striking. The snails belong to different species groups, and even differ in the genetics of their shell characters. Paradoxically, their anatomical divergence seems to be greatest where they are most alike at the molecular level.

The lack of concordance between morphological and molecular characters extends to other pairs of species. *P. taeniata* and *P. exigua* are very similar in morphology, hybridize in nature, and freely cross in the laboratory (Murray and Clarke 1968; Johnson *et al.* 1977), yet the genetic distance between them in sympatry (about 0.200) is much greater than that between northern *P. taeniata* and *P. suturalis*. *P. mirabilis* is morphologically intermediate between *P. taeniata* and *P. suturalis*. It hybridizes naturally with *P. taeniata*, but genetically it seems always to be nearer *P. suturalis* (at six localities where populations of *P. mirabilis* and *P. taeniata* are sympatric, the genetic distances between them range from 0.110 to 0.185; at four of these localities we also have distances between *P. mirabilis* and *P. suturalis*, which range from 0.029 to 0.093).

11.5 Resolving the paradox of discordant variation

How can we resolve these paradoxes? One possible solution emerges naturally from the fact that we are observing speciation in progress. When newly arisen species are sympatric, or come into sympatry, their degree of reproductive isolation may be enough to ensure their distinctness, but not yet perfectly complete. Even a small amount of gene flow can have a very large effect over time. Let us consider the simplest possible case, a single locus with two neutral alleles. If two large populations of equal size differ in the frequencies of the alleles, and if they hybridize at a rate h (defined as the number of successful hybrid matings as a proportion of all matings, calculated in numbers of fertile offspring), the difference in allelic frequencies between the populations will decline at a rate of $2h$ per generation. Thus, if h is small, after $1/h$ generations the difference between the populations will have fallen to about 13.5%. of its starting value, and after $2/h$ generations to about 2%.

This simple calculation shows that a level of hybridization undetectable in the field or laboratory can have profound evolutionary consequences. As an extreme example, if the proportion of successful hybrid matings is 1 in 100 000, the difference between two large populations will decline to 14% of its starting value in 100 000 generations and to 2% in 200 000. The latter, for *Partula,* is about 400 000 years, or less than one-third of its possible span of life on Moorea.

Convergence will be opposed by random drift, but with little effect if the population sizes of the two species are larger than twice the reciprocal of the hybridization rate. The relevant population sizes are neither those of local populations nor those of the species as a whole, but intermediates that depend on the amounts of intraspecific gene flow. These values can only be estimated by detailed computer simulations, but for *Partula* they must well exceed a thousand, which is in the order of the local population sizes (Murray and Clarke 1984). Thus rates of hybridization substantially less than one in a thousand would be effective in bringing about homogenization, and would take only a few thousand years to do so.

It is now easy to see how discrepancies can arise between anatomical and molecular characters. Suppose that a new species originates on another island and then invades Moorea. Initially it will show both anatomical and molecular differences from the local species. If its reproductive isolation is less than perfect, however, any differences in the frequencies of genes that are selectively neutral, or unequivocally advantageous to both species, will gradually or quickly be lost. The longer any two species exist together, the more alike they will become, except only in those characters that are important to the ecological, reproductive, or behavioural distinctions between them. Such characters are expected progressively to diverge, but the genes determining them will almost certaily be a small proportion of the genome. During a long co-existence the two species will probably evolve to reduce hybridization, either through changes directly related to mating, like the reversal of coil in *P. suturalis*, or through ecological displacements. Thus as they grow more alike at the molecular level, they will become progressively less alike in their anatomy, behaviour, and ecology, and progressively less able to mate with each other. The relative speed at which complete isolation evolves will no doubt vary from case to case, but the process can evidently lead to a negative relationship between the extent of hybridization and the overall similarity of gene frequencies, and equally to a negative relationship between anatomical and molecular divergence.

This scenario of 'molecular leakage' fits *Partula* very well. It accommodates the known facts of interspecific hybridization. It not only explains the lack of concordance between morphology, molecules, and reproductive isolation, but also the exceptionally high levels of allozymic heterozygosity (Johnson *et al.* 1986*b*), which are to be expected after homogenization, and the widespread sharing of mitochondrial haplotypes between species on different islands (Murray *et al.* 1991). Evolutionary trees based upon neutral alleles may be misleading when they suggest that speciation on each island has occurred *in situ*.

11.6 General conclusions

If trees based upon neutral (or indeed generally advantageous) genetic differences in *Partula* can seriously mislead us about its evolutionary history, we have to ask whether this reflects a general problem. Of course, leakage is restricted to taxa in sympatry. Furthermore it might be argued that the matter is only serious when the taxa have diverged very recently, because otherwise they would have existed lived long enough for the evolution of complete isolation, and for their subsequent divergence in all characters, both anatomical and molecular. The available evidence suggests, albeit weakly, that this is not the case. Prager and Wilson (Prager *et al.* 1974; Prager and Wilson 1975), using 'immunological distances' between albumins, estimated that the mean time to the common ancestors of bird species known to be capable of hybridizing was about 21 million years. The figure in frogs was similar, but in mammals the estimated time was two million years. For our purposes these estimates have several pitfalls. 'Immunological clocks' may be set wrongly because of uncertainties in the fossil record, and they may not run to time. The viability of hybrids is not by itself enough to produce introgression, which requires them also to be fertile. Such considerations suggest that Prager and Wilson's times may be too long. On the other hand, taxa with very low rates of hybridization (10^{-4} or less) would not have been included in the survey, and from this viewpoint the times may be too short. In any case we know that the three or four million years available for the evolution of *Partula* in the Society Islands have not exhausted the possibilities of introgression. On Moorea there is the potential for gene flow between any species and any other, through bridging populations and taxa (Johnson *et al.* 1993*a*). A similar point is made with respect to the finches of the Galápagos, and the honeycreeper finches of the Hawaiian Islands, in Chapter 9. Riesberg and Soltis (1991), reviewing the evidence for 'reticulate evolution' in plants, have argued that the introgression of nuclear genes and cytoplasmic organelles may bias phylogenetic reconstruction at all taxonomic levels, and a similar case has been made by Dowling and De Marais (1993) for cyprinid fishes. Avise (1994) reviews other examples of introgression. Perhaps we need to re-examine cases of unusual genetic homogeneity between sympatric species of animals (such as the African and other cichlid fishes; see, for example, Meyer *et al.* (1990) and Schliewen *et al.* (1994)), and ask if they too are not due, at least in part, to 'leakage'.

How can we escape being misled? It will not be enough to look at a larger number of genes or other stretches of DNA, because if some loci have been rendered homogeneous the greatest part of the genome will probably have suffered the same fate. If we examine the sequences of individual alleles, some will reflect one ancestry and some the other, but once allelic frequencies have been made roughly equal we will not know which ancestry belongs to which species, and may erroneously conclude that the polymorphism is older than the speciation. Only when events have been relatively recent, like the invasion of southern Moorea from Tahiti, will we be able to detect them.

For events in the longer term, possible sources of information are the genes that evolved in allopatry to determine the adaptive distinctions between the species, and genes closely linked to them. The neutral parts of such sequences should preserve true records of their phylogenetic history. We may need to distinguish genes evolved in allopatry from those producing later adaptations, such as reinforcement or ecological displacement, by mutation from introgressed alleles. However, it may be good enough to separate out the selected loci, and to show that most of them are consistent in their phylogenies.

The only other hope is that the general progress of neutral and generally advantageous alleles towards uniformity might be retarded by linkage to loci with alleles selected for distinctness. However, we are not aware of any quantitative models that tell us how much of the genome can be preserved from introgressive convergence by linkage, or for how long. Intuition, and the general shortage of linkage disequilibrium, suggests that the amount may be small and the time short. If so, we are left with the challenge to detect and sequence 'adaptive' genes, so that we can reveal the true history of speciation.

11.7 Summary

Two species of land snails, *Partula taeniata* and *Partula suturalis*, occur sympatrically on the island of Moorea in French Polynesia. The genetic distance between them varies clinally from north to south. Their extreme difference in the south is attributed to an invasion from the neighbouring island of Tahiti. Their genetic closeness in the north, despite large morphological and ecological differences, is attributed to 'molecular leakage', convergence of the neutral and advantageous genes in the two species through occasional hybridization. Rates of hybridization as low as 1 in 100 000 can render two species nearly homogeneous in their gene frequencies over periods of time that are short on an evolutionary scale, and therefore mislead us about the phylogenetic history of the taxa concerned. In such circumstances the only valid phylogenetic information may be contained in genes that are kept distinct by natural selection.

Acknowledgements

We are grateful to John Brookfield, Sara Goodacre, Paul Sharp, Diogo Thomaz, and Mark Beaumont for very helpful discussions, and to the BBSRC for financial support.

References

Avise, J.C. (1994). *Molecular markers: natural history and evolution*. Chapman and Hall, New York.

Clarke, B. and Murray, J. (1969). Ecological genetics and speciation in land snails of the genus *Partula*. *Biological Journal of the Linnean Society*, **1**, 31–42.

Cowie, R. H. (1992). Evolution and extinction of Partulidae, endemic Pacific island land snails. *Philosophical Transactions of the Royal Society of London* B, **335**, 167–91.

Crampton, H. E. (1932). Studies on the variation, distribution, and evolution of the genus *Partula*. The species inhabiting Moorea. *Publications of the Carnegie Institution of Washington*, **410**, 1–335.

Clarke, B., Johnson, M. S., and Murray J. (1996). Clines in the genetic distance between two species of island land snails: how 'molecular leakage' can mislead us about speciation. *Philosophical Transactions of the Royal Society of London* B, **351**, 773–84.

Dowling, T. E. and De Marais, B. D. (1993). Evolutionary significance of introgressive hybridization in cyprinid fishes. *Nature*, **362**, 444–6.

Duncan, R. A. and McDougall, I. (1976). Linear volcanism in French Polynesia. *Journal of Volcanology and Geothermal Research*, **1**, 197–227.

Felsenstein, J. (1993). *PHYLIP (Phylogeny Inference Package)*, version 3.5c. Department of Genetics, University of Washington, Seattle. (Distributed by the author.)

Garrett, A. (1884). The terrestrial mollusca inhabiting the Society Islands. *Journal of the Academy of Natural Sciences, Philadelphia (Series 2)*, **9**, 17–114.

Johnson, M. S. (1982). Polymorphism for direction of coil in *Partula suturalis*: behavioural isolation and positive frequency-dependent selection. *Heredity*, **49**, 145–51.

Johnson, M. S., Clarke, B., and Murray, J. (1977). Genetic variation and reproductive isolation in *Partula*. *Evolution*, **31**, 116–26.

Johnson, M. S., Murray, J., and Clarke, B. (1986*a*). An electrophoretic analysis of phylogeny and evolutionary rates in the genus *Partula* from the Society Islands. *Proceedings of the Royal Society of London* B, **227**, 161–77.

Johnson, M. S., Murray, J., and Clarke, B. (1986*b*). Allozymic similarities among species of *Partula* on Moorea. *Heredity*, **56**, 319–27.

Johnson, M. S., Clarke, B., and Murray, J. (1990). The coil polymorphism in *Partula suturalis* does not favour sympatric speciation. *Evolution*, **44**, 459–64.

Johnson, M. S., Murray, J., and Clarke, B. (1993*a*). The ecological genetics and adaptive radiation of *Partula* on Moorea. *Oxford Surveys in Evolutionary Biology*, **9**, 167–238.

Johnson, M. S., Murray, J., and Clarke, B. (1993*b*). Evolutionary relationships and extreme genital variation in a closely-related group of *Partula*. *Malacalogia*, **35**, 43–61.

Lundman, B. (1947). Maps of the racial geography of some Partulae of the Society Islands based upon the material published by H. E. Crampton. *Zoolisk Bidrag frân Uppsala*, **25**, 517–33.

Meyer, A., Kocher, T. D., Basasibwaki, P., and Wilson, A. C. (1990). Monophyletic origin of Lake Victoria cichlid fishes suggested by mitochondrial DNA sequences. *Nature*, **347**, 550–3.

Murray, J. and Clarke, B. (1968). Partial reproductive isolation in the genus *Partula* (Gastropoda) on Moorea. *Evolution*, **22**, 684–98.

Murray, J. and Clarke, B. (1976*a*). Supergenes in polymorphic land snails. I. *Partula taeniata*. *Heredity*, **37**, 253–69.

Murray, J. and Clarke, B. (1976*b*). Supergenes in polymorphic land snails. II. *Partula suturalis*. *Heredity*, **37**, 271–82.

Murray, J. and Clarke, B. (1980). The genus *Partula* on Moorea: speciation in progress. *Proceedings of the Royal Society of London* B, **211**, 83–117.

Murray, J. and Clarke, B. (1984). Movement and gene flow in *Partula taeniata*. *Malacologia*, **25**, 343–8.

Murray, J., Murray, E., Johnson, M. S., and Clarke, B. (1988). The extinction of *Partula* on Moorea. *Pacific Science*, **42**, 150–3.

Murray, J., Stine, O. C., and Johnson, M. S. (1991). The evolution of mitochondrial DNA in *Partula*. *Heredity*, **66**, 93–104.

Murray, J., Clarke, B., and Johnson, M. S. (1993). Adaptive radiation and community structure of *Partula* on Moorea. *Proceedings of the Royal Society of London* B, **254**, 205–11.

Nei, M. (1978). Estimation of average heterozygosity and genetic distance from a small number of individuals. *Genetics*, **89**, 583–90.

Prager, E. M. and Wilson, A. C. (1975). Slow evolutionary loss of the potential for interspecific hybridization in birds: a manifestation of slow regulatory evolution. *Proceedings of the National Academy of Sciences USA*, **72**, 200–4.

Prager, E. M., Brush, A. H., Nolan, R. A., Nakanishi, M., and Wilson, A. C. (1974). Slow evolution of transferrin and albumin in birds according to micro-complement fixation analysis. *Journal of Molecular Evolution*, **3**, 243–62.

Riesberg, L. H. and Soltis, D. E. (1991). Phylogenetic consequences of cytoplasmic gene flow in plants. *Evolutionary Trends in Plants*, **5**, 65–84.

Schliewen, U. K., Tautz, D., and Pääbo, S. (1994). Sympatric speciation suggested by monophyly of crater lake cichlids. *Nature*, **368**, 629–32.

12

Radiations, communities, and biogeography

Peter R. Grant

12.1 Introduction

The preceding chapters discussed causes and mechanisms of speciation. The remaining chapters extend the discussion of speciation on islands and in island-like habitats to questions about the evolution of many species from the same ancestral stock. There is a change of scale in this transition; from the evolution of two species from one to the evolution of many species from one or a few. The chapters therefore cover topics of adaptive radiations, community ecology, and biogeography. They address such questions as why are there so many species and what is the role of ecological opportunity in fostering these radiations?

The development of complex communities from simple ones depends on both ecological and evolutionary factors. Identifying the relevant factors and assigning them relative importance is a difficult task for the observational and experimental biologist (e.g. see Ricklefs and Schluter 1993; Brown 1995; Rosenzweig 1995; Cody and Smallwood 1996; Schluter 1996). The task is simplified to some extent by focusing on a species-rich assemblage of species derived from a common ancestor, which occupies the same local area where the species evolved; hence the importance of adaptive radiations. Islands and lakes are rich in such assemblages. Furthermore their endemics are more likely to have evolved where they are found now than is true of continental endemics, because the geographical distributions of species on continents often shift when the climate changes.

While focusing on the radiation of a single taxon we should remember that these clusters of species are usually embedded in yet larger and ecologically more diverse assemblages that make up the rest of local communities. Moreover the radiations are not necessarily entirely adaptive. Non-adaptive processes may have been involved. For example divergence may have occurred as a result of stochastic processes (Chapters 2, 3, and 8). Epigamic speciation driven by sexual selection (Section 6.7) may have contributed to the diversity. However, I know of no radiation suspected of being produced solely by these two processes alone or in combination that is without some role for natural selection in the radiation.

Two chapters discuss radiations: *Anolis* lizards in the Caribbean (Chapter 13) and cichlid fish in Lake Tanganyika (Chapter 14). Two chapters discuss diversification in insular habitats within extensive forested regions; habitat islands in Amazonia (Chapter 15) and *Heliconius* butterflies that may have been influenced by them (Chapter 16). This latter chapter offers the most explicit discussion of a recurring theme in this book that evolution results from an interplay of selection and drift. The chapter on plant evolution (Chapter 17) returns us to some of the major patterns displayed on islands that were discussed in Chapter 1, and considers explanations for them in the context of adaptive radiations.

12.2 Ecology of adaptive radiation

There are 139 species of lizards in the genus *Anolis* distributed throughout the Caribbean, all descended from a few mainland species at most (Williams 1983). As many as 54 species occur on a single island, and up to 12 are sympatric (on Cuba). Only *Eleuthero-dactylus* frogs among vertebrates rival this diversity in the Caribbean. It even exceeds that of *Drosophila* (Section 8.2). Other insular radiations elsewhere are more impressive in the extent of morphological diversification or in their numbers of taxa, such as the drosophilids and lobelias of Hawaii, and marsupial mammals of Australia. Nevertheless no other radiation of this magnitude is understood so well ecologically, or has been studied for so long (Williams 1969, 1983). The radiation of Darwin's finches is also well known ecologically, but is much smaller (Chapter 9; Grant 1986).

In Chapter 13 Losos asks what factors explain the variation in number of species among islands. He uses the non-evolutionary theory of island biogeography in a novel way to answer this question. He finds that ecological factors such as colonization, competition, and extinction predominate on small islands, whereas evolutionary factors are the primary determinant of species number on large and heterogeneous islands. We can think of the large islands as the main producers of species, therefore acting as evolutionary sources, and small islands acting as evolutionary sinks; lizards disperse there, or become isolated there when the islands are formed by rising sea levels, do not speciate, and in fact change little if at all. A possible exception is peripatric speciation occurring on the small offshore islands close to the large ones. Fluctuating sea levels might repeatedly separate and rejoin communities on the large islands and their fringing satellites, with a net flow of species from satellite to main island. For example this type of reconstruction has been offered, in addition to within-island allopatric speciation, to explain the surprisingly high number of endemic species (67) of flightless beetles in the genus *Miocalles* on the small and remote Pacific island of Rapa (Paulay 1985).

The evolutionary increase in species number with island area among the four large Greater Antilles islands (Puerto Rico, Jamaica, Hispaniola, Cuba) has an ecological explanation. It results from an increase in the allopatric production of new species that are ecologically similar to those which gave rise to them, and from an increase in the number of habitat niches utilized by sympatric species. The chapter is thus relevant to the question of the role of ecological opportunities in determining both within-island speciation and the number of species (formed there or elsewhere) occurring on an island. It suggests that ecological opportunities favouring diversification increase in a rather special way as island size increases. There are several possibilities for how this could happen:

(1) Diversity of structural habitats increases with island size. This is the most obvious and may be generally true (e.g. see Section 5.3 on lizards of the Canary Islands), but does not seem to be correct for variation among the largest islands of the Greater Antilles. However, food supply and predator and parasite pressure are not well enough known to be sure they are not important factors.

(2) A large pool of allopatric species enhances the likelihood that one species will diverge sufficiently to utilize a distinctive habitat niche and become sympatric; i.e. many species beget more species.

(3) A large diversity of sympatric species forces lineages (through competitive pressure) to specialize in ways not seen elsewhere.

Recalling the adaptive landscape metaphor, Losos suggests that hierarchical and sequential specialization arises either because some specialist niches are more accessible than others, or because some have higher adaptive peaks than others. I would add that occupation of one peak can lead to occupation of a more remote peak; peaks are not 'leap-frogged'. By this means a taxon may sequentially 'explore' a complex adaptive landscape, enter new regions of niche space (Sections 1.5 and 6.5) and diversify, analogous to the way in which an individual species explores a contoured fitness topography of gene combinations as envisaged by Wright (1932) in his shifting balance theory of evolution. The lizard data suggest that evolutionary divergence occurs allopatrically or parapatrically, but sympatric divergence of ecologically specialized species (ecomorphs) is presumably a viable alternative, as it is for the fish discussed in the next section. No chapter better than this exemplifies Schluter's point in Chapter 10 that a knowledge of the environment in necessary for an understanding of adaptive radiation.

12.3 Radiation in lakes

Carlquist (1974) has written that adaptive radiation is inevitable on an island or in an archipelago where a small number of immigrants meets a broad spread of ecological opportunities. This is a good starting point for considering the remarkable radiations of cichlid fishes in the Great Lakes of Africa. Three major radiations have occurred, producing more than 170 species in Lake Tanganyika, more than 300 in Lake Victoria and more than 500 in Lake Malawi. The ages of these lakes are 9–12 Ma, 0.5–0.7 Ma and 1–2 Ma respectively, but the span of time over which the radiations unfolded are currently estimated to be much shorter; 1.2, 0.2, and 0.8 Ma respectively (Meyer *et al.* 1990; Sturmbauer and Meyer 1992; Moran and Kornfield 1993). Recent evidence suggests that even one of these time spans is too long. The Lake Victoria basin may have been completely dry 12 to 15 thousand years ago (Johnson *et al.* 1996), implying that all of the more than 300 modern (monophyletic) species evolved since then at an exceptionally high rate; ignoring extinctions, if speciation rate was constant the doubling time of species number would have been less than 2000 years.

Why are there so many species in these three lakes? The answer is partly ecological. Some groups have diversified into specialized feeders on zooplankton, arthropods, molluscs, algae, and the scales and eyes of other fish (Fryer and Iles 1972). Thus the immigrants to the lakes not only encountered a broad range of ecological opportunities (cf. Carlquist 1974), their evolution helped to create new ones.

Another part of the answer lies in their appearance. Many of them display an impressive variation in colour and pattern, both within and between species. Bright colours may warn predators of unpalatability, attract mates, or both. If mate attraction is the primary function of the colours, sexual selection will have been at least as important in the diversification as natural selection, if not more so (Dominey 1984). In other words the radiation is partly adaptive (dietary diversification) and partly non-adaptive (epigamic diversification). There are many species of fish perhaps because both ecological and epigamic speciation have occurred.

Chapter 14 reviews one small part of one of the radiations using modern molecular techniques to help decipher the evolutionary history of all four nominal species in the three genera of eretmodine cichlids in Lake Tanganyika. Rüber and colleagues find that major fluctuations in the level of the lake have been important in shaping the radiation.

The geography of genetic variation in this lake reveals a high degree of within-lake endemism among three genetically well-separated lineages distributed along the inferred shore lines of three historically intermittent lake basins. In other words the lake was once split into three (Tiercelin and Mondeguer 1991), and the genetic characteristics of these fish retain the fingerprint of that separation. Remarkably, the three-clade three-basin phylogeographic pattern is repeated twice within this tribe of cichlids. Fryer and Iles (1972) had earlier hypothesized fluctuating lake levels as important in erecting and dismantling habitat barriers to gene flow, and this is now supported by, among other evidence, the presence of closely related species on opposite sides of the lake. Other barriers occur on the same shore. Most species are restricted to patchy rocky habitats along the shore, separated by sandy beaches, and a few kilometres of unfavourable coastline habitat are sufficient to effectively prevent gene flow. Ecogeographical factors are clearly important. But still, why are there so many species? Are there 170 different niches for 170 different species?

Each species has several allopatric colour morphs. Given the lengthy geographical barriers that separate them, an argument can be made for highly restricted gene flow between them. If so, each area may have been colonized by a few individuals, and divergence then occurred by a combination of drift, in the founding event and subsequently, and sexual selection. These are the conditions, coupled with the existence of a few genes with major phenotypic effects, that are believed to be conducive to founder effect speciation (Chapters 6–8). At this point it is worth recalling Barrett's caveat in Chapter 2 (p. 28): 'Some of the striking morphological differentiation that characterizes island taxa displaying "adaptive" radiation may result from stochastic processes operating during founder events and periods of small population size'.

Another interesting finding is that classifications based on mitochondrial DNA and morphology are in conflict. If the DNA classification more closely reflects history then trophic morphologies, particularly the shapes of oral teeth, have converged in some lineages and evolutionary history has repeated itself. However, the mtDNA phylogeny may not faithfully reflect history, for three reasons; random lineage sorting, retention of ancestral polymorphisms, and hybridization. The first two appear not to be a problem with this relatively old group of fish, whereas the third has not been assessed.

The problem caused by hybridization was discussed in Section 6.8, and reappears now in a new context. Although microallopatric intralacustrine speciation appears to have prevailed in Lake Tanganyika, sympatric speciation cannot be ruled out. The strongest evidence for this form of speciation in fish (see also Section 10.2) comes from another study of cichlids in two volcanic crater lakes, Barombi Mbo and Bermin, in Cameroon (Schliewen *et al.* 1994). Eleven and nine endemic species respectively coexist in these small and topographically homogeneous lakes. A mitochondrial DNA phylogeny gives evidence of monophyly within each of the lakes. Ecological (dietary) diversification has been pronounced in Barombi Mbo, and colour patterns as well as breeding times and places differ among the species to some extent, but there are no geographical barriers that might have faciltated allopatric speciation (Trewavas *et al.* 1972). These facts have given rise to the suggestion that the speciation took place sympatrically (Schliewen *et al.* 1994). The mitochondrial DNA similarity may reflect common ancestry, but alternatively it might be the product of introgressive hybridization between non-sister taxa that colonized the lake on more than one occasion. Present-day hybridization has not been discovered (Schliewen *et al.* 1994), although it may have occurred in the past; Clarke and colleagues in Section 11.5 discuss reasons why hybridization between sister taxa may have been more prevalent in the past than now. Laboratory studies have shown that cichlid species from other lakes have a strong interbreeding potential (Crapon de Caprona and Fritzsch 1991), which is to be expected for radiations as recent as these.

Regardless of the mode of origin, these African cichlids appear to have an exceptional propensity to speciate rapidly, and ecological opportunity is a part, but only a part, of the explanation. This chapter provides an example of how important it is to estimate phylogeny reliably and to reconstruct past environments correctly when attempting to understand the radiation of a taxon. Applications to other well-known lacustrine radiations immediately come to mind, such as the cyprinid fish in Lake Titicaca (Parenti 1984) and the cottids (as well as the crustacean *Gammarus*) in the ancient Siberian lake of Baikal (Brooks 1950; Kozhov 1963). Perhaps these radiations should be re-examined as possible products of sympatric speciation. It is too late to do so for the cyprinid fish in Lake Lanao on Mindanao, Phillipines (Myers 1960), as 15 of the 18 species have become extinct, apparently as a result of human activity (Kornfield and Carpenter 1984). The cichlids of Lake Victoria have suffered a similar fate (Goldschmidt 1996).

12.4 Diversification in habitat islands

The rainforest of Amazonia covers a vast area of equatorial South America. Its continuity is broken by rivers, and by four types of insular patches of non-rainforest vegetation: white sand campinas and caatingas, savannas, inselbergs (isolated granite peaks), and tepuis (sandstone table mountains). In Chapter 15 Prance compares these four types of islands as environments that foster adaptive evolution leading to speciation and radiation. The chapter is included in the group on adaptive radiation because its focus is on an unusual environment in which adaptive radiations can take place: archipelagos of island-like habitats in a 'sea' of forest. The relevance to questions about the origin of tropical biodiversity is discussed only briefly, but in more detail in Chapter 16.

The four types of Amazonian islands display different levels of endemism among plants, according to the action of at least three factors; harshness of the environment, dispersive powers of the species, and age of the habitats. Caatingas and campinas are poor in nutrients. Caatingas have sclerophyllous vegetation and a large number of endemic species. Campinas are smaller and more island-like forested or shrubby areas in rainforest. Endemism is lower than in caatingas because their areas are smaller and the species are dispersive (diaspores are mainly bird or wind dispersed). In contrast to these the llanos-type savannas are the remnants of a once widespread habitat, have only recently become restricted, do not demand special adaptations from would-be colonists, and do not have many endemics. The species that occupy them are easily dispersed. Inselbergs are crowned by low scrub forest of sclerophyllous plants, adapted to dry season aridity. Tepuis have a much higher endemism at the species level, also a moderate amount at generic level; i.e. species and genera found only on tepuis, usually on more than one. An important factor is the variety of habitats on the larger tepuis. One high tepui, the Sierra de Neblina (>3000 m), has an endemic family of insectivorous plants (Saccifoliaceae). Tepuis are therefore fairly old. In some cases speciation has occurred on different tepui islands, particularly on the older tepuis.

This is how the habitat islands are to be seen today. However, in the past expansion and contraction of the rainforest has occurred at least four times in the last 60 000 years under an oscillating climate of wet and dry regimes (Chapter 15). In past times of greater aridity and coolness associated with glacial maxima the drier vegetation has expanded in area, and it is believed the rainforests have shrunk to a point at which they themselves were the islands. The component plant species and the animals that were directly or indirectly dependent upon them were then restricted to refugia. Refugia are believed to be evolutionarily important, as inferred from the concordance of centres of endemism in different taxa including birds, insects, and plants; peaks of locally adapted species in a sea of generalists. Species would have become extinct, others would have diverged into different subspecies or species, so that each refuge would have developed a unique constellation of species. According to Prance the forest was not reduced to discrete areas, rather the areas were more or less connected by gallery rainforest along rivers, as in some modern savannas (Chapter 15). Corridors would have weakened the isolation of the 'isolates', and presumably reduced their opportunities for divergence.

The evidence for the refuge hypothesis of tropical diversity is much debated on biogeographical (e.g. Cracraft and Prum 1988) and palynological (Colinvaux *et al.* 1996) grounds; see also Section 16.5. It has also been criticized by Nelson *et al.* (1990) on the grounds that the areas of high plant endemism have also received the most collecting activity. This is not necessarily a fatal criticism, as these areas may have received the greatest attention because they were known to be botanically interesting. All these problems may be irrelevant if the span of time for understanding the evolution of tropical diversity is far greater than the few thousand years generally considered in discussions of the refuge hypothesis. Over that longer time span island-like refuges may have been of considerable importance. Certainly the global climate has oscillated repeatedly over the last million years (deMenocal 1995), producing conspicuous changes to vegetation in temperate regions and forming insular refuges at the highest latitudes (Pielou 1992; Hewitt 1996). The unanswered question for the tropics is not *has* change occurred in the forested vegetation but how much, of what type and with what ecological and evolutionary effects on the biota (e.g. see Leyden 1984; Brumfield and Capparella 1996; Colinvaux *et al.* 1996). Was the forest fragmented enough, and long enough, to promote allopatric diversification?

12.5 Radiations of mimics

Chapter16 integrates many of the features discussed in the preceding chapters: the interplay of selection and drift in microevolution; the models and metaphors of Sewall Wright; allopatric speciation; the dependence of evolving species upon other members of their communities; and the importance of correctly reconstructing phylogeny and reconstructing environments, including insular refuges, in seeking an understanding of present-day patterns of diversity.

Radiation of the neotropical butterfly genus *Heliconius* has produced 55 species. These in turn are represented by 400 geographically separated colour pattern races. Diversification has affected many morphological and life history features of these butterflies, including larval diets and the chemical defences they obtain from them, but the most conspicuous feature in which the species vary is the colour and pattern of the wings. A remarkable characteristic of the genus is the extremely close resemblance of the colour patterns displayed by sympatric species, and the extent to which the (Müllerian) mimicry is replicated in one geographical race after another. Attempts to understand the radiation in this group have thus focused largely on colour variation, especially in one mimetic pair; *Heliconius erato* and *H. melpomene*. Each has about 30 races and about 15 different colour patterns. Each race and pattern of *H. erato* except one is sympatric with a closely similar looking race of *H. melpomene*. Thus each species has radiated into different races that are convergent upon the other, with *H. erato* possibly being leader and *H. melpomene* the follower (Brower 1996). The racial divergence and convergence is paralleled by divergence and convergence between members of different heliconiine clades. The evolution of *H. erato* and *H. melpomene* diversity thus serves as a model (and not a mimic!) for the evolution of the whole genus.

All radiations pose the problem of how and why new species arise and then spread, and what limits the proliferation of species. Müllerian mimicry of *Heliconius* butterflies poses a special problem; how do novel phenotypes that arise by mutation in either member of a species pair increase in frequency when novel phenotypes should be selected against by experienced predators? Warning patterns should be subject to strong frequency-dependent and stabilizing selection. Rare, novel, patterns should be at a disadvantage, and Müllerian mimics are expected to be monotypic. In the case of *Heliconius* butterflies this is far from the truth. Why?

There are really two questions to be answered, one concerning the origin of a new race and the other concerning mimicry of that race. Since *H. erato* and *H. melpomene* only mimic each other, and do not mimic other species, the first question is how does a new race of one of the species originate? The second question is how does a population of the second species come to resemble the new race of the first one?

Wing colour patterns are governed by a small number of genes. Alternative alleles at a locus can have major effects on the patterns, and alleles at other loci modify those effects (e.g. Turner 1977; Nijhout *et al.* 1990). Thus the genetic answer to the first question is mutation at the major gene locus. The ecological circumstances under which the frequency of the new variant increases are far less clear.

The second question is best addressed in the context of mimicry rings. Up to six different assemblages (rings) of Müllerian mimics coexist. Each assemblage (ring) of Müllerian mimics can be thought of as occupying an adaptive peak determined by predators, and separated by deep valleys or troughs in the adaptive landscape from other such peaks. The valley can be crossed only when suitable major mutations arise which allow a shift between one peak and the region below another. Subsequent mutations (modifiers) selectively improve the mimicry. Ring switching occurs only when the new peak is higher than the previous one, for example as a result of better chemical defence or more abundant members. In one small area along the Rio Amazonas *H. hermathena* has apparently made the switch. A black and red pattern that closely mimics the pattern of the sympatric *H. erato* and *H. melpomene* has replaced the red band and yellow bar pattern that is typical for the species elsewhere. Mallet and Turner believe that in most cases elsewhere the valley between peaks is deeper and not so easily crossed.

The difficult question to answer is the environmental circumstances under which these changes take place. One suggested explanation invokes evolutionary change in the confines of insular refuges. The Brown–Sheppard–Turner model assumes that the conversion of one mimetic pattern to another is brought about solely by natural selection, with no contribution from genetic drift. If one of a pair of species becomes locally extinct the other may switch to another ring of co-mimicking species when the right mutation occurs. In other words there is a switch from one adaptive peak to another. This is most easily visualized occurring in a small island where by chance one of the pair of species becomes extinct, and so do some of the predators. Replicated refuge 'experiments' (in habitat patches or islands) will differ in composition of species, to a large degree stochastically, and change through time independently by a process referred to as 'biotic drift'. This is an allopatric model.

An alternative explanation, which does not depend upon insular refuges, embraces fully and explicitly Sewall Wright's shifting balance theory of evolution (Mallet 1993). A combination of random effects and natural selection causes an increase in frequency of a new variant, ultimately towards fixation, and the new variant spreads outwards from its point of origin to neighbouring populations through population pressure, transforming them through interdemic selection or by a process of moving clines, until a barrier to further spread is reached. This is a sympatric and parapatric model.

The chapter assesses the evidence for both modes of evolution, finds neither to be sufficient or entirely consistent with the facts, and develops a blend of the two. In this blend the shifting balance theory is retained and modified by combining biotic drift with genetic drift and natural selection. It can account for both the origin of a new and non-mimetic form as well as the mimicry of that form. The ecological conditions for it may not occur very often, but given enough time significant rare events may occur often enough. The time for these events to occur may be as long as 1–2 Ma (Brower 1996).

Three controversial subjects are raised but not resolved in this chapter. They concern where the evolutionary changes took place, when they occurred, and the role of genetic drift.

The first problem concerns whether or not refuges played an important, or any, role in the diversification. Ecological fluctuations could generate biotic drift and ring switching in continuous forest, but changes are much more likely and perhaps greater during periods when the biota were segregating into refuges. These refuges of rainforest islands are the same as those discussed in Chapter 15 by Prance. As pointed out earlier, the existence of refuges has been questioned. At times corresponding to glacial maxima high-altitude forest spread to low elevations, but effects of cooling on savanna vegetation are unknown (Colinvaux *et al.* 1996). Much of lowland Ecuador, that might have had refuges, was blanketed by a series of massive volcanic residues. Nonetheless, from what is known in Galápagos and Hawaii, such volcanic activity could split continuous forest into discontinuous patches around the perimeter of its zone of influence, thereby creating refuges for the somewhat sedentary butterflies. Initially their numbers may have been very low. Brower (1996) has recently found a lack of congruence between patterns of mitochondrial DNA variation and variation in the genetic factors responsible for wing patterns, and these and other facts lead him to conclude there is little evidence for or against the refuge theory.

A second and related controversial topic is when the evolutionary changes took place. Arguments over the existence and placement of refuges a little more than 10 000 years ago may be largely irrelevant to racial variation if the races of *H. erato* and *H. melpomene* began diversifying 1–2 Ma ago, continuing extensively 65 000 to 200 000 years ago (Brower 1996). A deeper knowledge of Amazonian topography through time is needed to assess the refuge theory.

A third controversial matter is whether the genetic architecture is suitable for significant changes by drift or not. A single locus displaying dominance can have major effects on wing colour patterns, modified by alleles at other loci. Following Barton (Chapter 7), whose models show that the shifting balance process works best with mutations of minor selective effect (but see Section 8.2), Mallet and Turner suggest that the genetic architecture of wing patterns in *Heliconius* supports the role of natural selection for mimicry over genetic drift as an initiator of divergence. On the other hand it is exactly this genetic architecture that Templeton (1980, 1981, 1996) has argued is favourable for change by founder effects involving drift. The theoretical debate on these issues that surfaced in 1984 (Barton and Charlesworth 1984; Carson and Templeton 1984) is still unresolved (Coyne 1992; Slatkin 1996; Templeton 1996; Chapters 6–8).

12.6 Adaptive radiations of plants

More is known about plant evolutionary diversification in the Hawaiian archipelago than anywhere else in the world, as a result of (1) extensive collecting and taxonomic description, (2) ecological fieldwork, and (3) molecular analyses that have permitted phylogenies to be reconstructed using modern (cladistic) methods (Wagner and Funk 1995). Thus students of the Hawaiian flora are in the best position to document and interpret the major and repeated features of insular evolution in plants. In Chapter 17 Givnish examines several trends manifested on islands around the world but especially clearly in the Hawaiian archipelago; trends towards arborescence, increased seed size, high incidence of endozoochory by frugivores on high islands, and the evolution of thorn-like prickles and heterophylly. He interprets these trends in the framework of adaptive radiations on islands.

Molecular work in recent years has enabled investigators to determine the origin of island forms and the direction of post-colonization evolution and speciation. For example, the 105 species of Hawaiian lobelioids are known to be monophyletic, to have dispersed and speciated predominantly in a direction from older to younger islands, and in the process to have undergone more striking morphological divergence than genetic divergence.

Putting such knowledge to work, Givnish points out that the evolution of arborescence in the Hawaiian archipelago is associated with herbaceous species that initially colonized open habitats. These, or their derivatives, subsequently moved into forested habitats. This sequence is in agreement with models which predict that optimal plant structure should increase with the density of plant coverage, and with Darwin's reasoning that competition for light was the primary cause of the evolution of arborescence. Large seed size is considered by Givnish to be an associated trait. The significance is believed to be an enhanced ability to penetrate leaf litter into mineral soil, and enhanced competitive ability in a crowded below-ground environment. The second advantage is in line with Darwin's general reasoning on the importance of competition in an environment lacking some of the constraints operating in mainland communities.

One of those missing constraints is browsing mammals. Therefore one does not expect to find defences such as prickles and thorns on the leaves of island plants, unless the species are recent colonists from a mammal-rich continental region. Nonetheless they are found on almost one-third of the 65 species in the genus *Cyanea*, the largest of six endemic lobelioid genera. In the absence of mammals, why would plants be defended up to one metre above ground in this way?

Carlquist (1962) thought the prickles were a defence against snails, but since then the sub-fossil bones of giant ducks and geese have been found, and Givnish believes a more convincing case can be made for these ground-living birds being the selective agent (see also Givnish *et al.* 1994, 1995). Through evolutionary convergence they apparently occupied the vacant browsing mammal niche. Significantly, the forest species appear to have been present, like the prickles themselves, on each of the five largest islands except Kauai.

Most of the *Cyanea* species with prickles are heterophyllous: the leaves of young plants differ from those of older plants—they are deeply divided. The herbivorous birds may have acted as a selective agent here also by favouring dissected (and protected) juvenile leaves through enhancement of the mimetic resemblance of the thorn-bearing veins of adult leaves. This idea is consistent with the fact that similar heterophylly is known on other islands where large ground-living birds occurred; on New Zealand, in the presence of moas, on Madagascar, in the presence of elephant birds, and on the

Mascarene islands, in the presence of the dodo. All these giant birds are extinct. How many other features of plants (and animals) owe their existence to selective pressures from species now extinct and perhap even unknown?

Wind pollination and associated life-history traits may be expecially common in the floras of some oceanic islands (Section 2.2), but on tall islands there is a high incidence of endozoochory, that is dispersal of fruits by birds, internally. The Hawaiian lobelioids provide a good example. The predominance of this dispersal mode is explained by both colonization and speciation. Endozoochory is effective in the long-distance dispersal that results in colonization of a remote archipelago. As mentioned above in connection with arborescence, colonization occurs largely in open habitats, reflecting more the habits of the dispersal agents (birds) than the suitability of the terrain. Several seeds may be transported at one time, which has important implications for the founding event and the likelihood of dioecious species becoming established (Section 2.3). Thus unlike arborescence, which evolves *in situ*, endozoochory is the initial (ancestral) condition.

Endozoochorous species then enter forested habitat, and multiply. Their speciation rate appears to have been substantially higher in wet forests, where wind dispersal is ineffective, than in drier habitats; higher, for example, in the genus *Cyanea* than in *Clermontia* which are found in drier habitats. Givnish attributes the higher rate to relatively low dispersal of the forest birds on which the plants depend for pollination and seed dispersal, geographical isolation of demes, and non-adaptive evolutionary change. This suggestion invites more population genetic work along the lines discussed by Barrett in Chapter 2, and more ornithological work on pollen flow and seed dispersal, although given the extinction of several of the relevant honeycreeper finches (Freed *et al.* 1987) (and lobelioids) it can have only limited success.

Givnish concludes by arguing that the cost of restricted distribution and specialized, animal-dependent, reproductive syndromes is a greater risk of extinction. Thus the work on island plants fits well with the concept of a taxon cycle developed for ants on islands (Wilson 1961) and extended to other insects (Greenslade 1969) and birds (Greenslade 1968; Ricklefs and Cox 1972; see also Section 1.5). While the argument is highly plausible, our knowledge of natural extinction to test it is unfortunately meagre. To illustrate the problem consider *Cyanea*. An exceptionally large genus, it has the greatest number of species that are threatened, endangered, or extinct in the USA. Direct and indirect human activities are responsible. Are we to assume that humans are simply one element in a natural taxon cycle that culminates in extinction, or are human effects so artificial that they distort extinction probabilities in the terminal phase of a cycle? Unless some islands are kept in a state entirely free from human influence we shall never know the answer.

References

Barton, N. H. and Charlesworth, B. (1984). Genetic revolutions, founder effects, and speciation. *Annual Review of Ecology and Systematics*, **15**, 133–64.

Brooks, J. L. (1950). Speciation in ancient lakes. *Quarterly Review of Biology*, **25**, 30–60, 131–76.

Brower, A. V. Z. (1996). Parallel race formation and the evolution of mimicry in *Heliconius* butterflies: a phylogenetic hypothesis from mitochondrial DNA sequences. *Evolution*, **50**, 195–221.

Brown, J. H. (1995). *Macroecology*. University of Chicago Press.

Brumfield, R. T. and Capparella, A. P. (1996). Historical diversification of birds in northwestern South America: a molecular perspective on the role of vicariant events. *Evolution*, **50**, 1607–24.

Carlquist, S. (1962). Ontogeny and comparative morphology of thorns of Hawaiian Lobeliaceae. *American Journal of Botany*, **49**, 413–19.

Carlquist, S. (1974). *Island biology*. Columbia University Press, New York.

Carson, H. L. and Templeton, A. R. (1984). Genetic revolutions in relation to speciation phenomena: The founding of new populations. *Annual Review of Ecology and Systematics*, **15**, 97–131.

Cody, M. L. and Smallwood, J. A. (1996). *Long-term studies of vertebrate communities*. Academic, New York.

Colinvaux, P. A., de Oliveira, P. E., Moreno, J. E., Miller, M. C., and Bush, M. B. (1996). A long pollen record from lowland Amazonia: forest and cooling in glacial times. *Science*, **274**, 85–8.

Coyne, J. A. (1992). Genetics and speciation. *Nature*, **355**, 511–15.

Cracraft, J. and Prum, R. O. (1988). Patterns and processes of diversification: speciation and historical congruence in some neotropical birds. *Evolution*, **42**, 603–20.

Crapon de Crapona, M.-D. and Fritzsch, B. (1991). African fishes. *Nature*, **350**, 467–8.

deMenocal, P. B. (1995). Plio-Pleistocene African climate. *Science*, **270**, 53–9.

Dominey, W. J. (1984). Effects of sexual selection and life history on speciation: species flocks in African cichlids and Hawaiian Drosophila. In *Evolution of fish species flocks*, (ed. A. A. Echelle and I. Kornfield), pp. 231–49. University of Maine Press, Orono.

Freed, L. A., Conant, S., and Fleischer, R. C. (1987). Evolutionary ecology and radiation of Hawaiian passerine birds. *Trends in Ecology and Evolution*, **2**, 196–9, 202–3.

Fryer, G. and Iles, T. D. (1972). *The cichlid fishes of the great lakes of Africa: their biology and evolution.* Oliver and Boyd, Edinburgh.

Givnish, T. J., Sytsma, K. J., Smith, J. F., and Hahn, W. J. (1994). Thorn-like prickles and heterophylly in *Cyanea:* adaptations to extinct avian browsers on Hawaii? *Proceedings of the National Academy of Sciences USA*, **91**, 2810–14.

Givnish, T. J., Sytsma, K. J., Hahn, W .J., and Smith, J. F. (1995). Molecular evolution, adaptive radiation, and geographic speciation in *Cyanea* (Campanulaceae, Lobelioideae). In *Hawaiian biogeography: evolution on a hot-spot archipelago*, (ed. W. L. Wagner and V. A. Funk), pp. 299–337. Smithsonian Institution Press, Washington, D. C.

Goldschmidt, T. (1996). Darwin's dreampond: drama on Lake Victoria. MIT Press, Cambridge, MA.

Grant, P. R. (1986). *Ecology and evolution of Darwin's finches.* Princeton University Press.

Greenslade, P. J. M. (1968). Island patterns in the Solomon islands bird fauna. *Evolution*, **22**, 751–61.

Greenslade, P. J. M. (1969). Insect distribution patterns in the Solomon islands. *Philosophical Transactions of the Royal Society of London* B, **255**, 271–84.

Hewitt, G. M. (1996). Some genetic consequences of ice ages, and their role in divergence and speciation. *Biological Journal of the Linnean Society*, **58**, 247–76.

Johnson, T. C., Scholz, C. A., Talbot, M. R., Kelts, K., Ricketts, R. D., Ngobi, G., Beuning, K., Ssemmanda, I., and McGill, J. W. (1996). Late Pleistocene desiccation of Lake Victoria and rapid evolution of cichlid fishes. *Science*, **273**, 1091–3.

Kornfield, I. and Carpenter, K. E. (1984). Cyprinids of Lake Lanao, Phillipines: taxonomic validity, evolutionary rates and speciation scenarios. In *Evolution of fish species flocks*, (ed. A. A. Echelle and I. Kornfield), pp. 69–84. University of Maine Press, Orono.

Kozhov, M. M. (1963). *Lake Baikal and its life*. Junk, The Hague.

Leyden, B. W. (1984). Guatemalan forest synthesis after pleistocene aridity. *Proceedings of the National Academy of Sciences USA*, **81**, 4856–9.

Mallet, J. (1993). Speciation, raciation, and color pattern evolution in *Heliconius* butterflies: evidence from hybrid zones. In *Hybrid zones and the evolutionary process*, (ed. R. G. Harrison), pp. 226–60. Oxford University Press.

Meyer, A., Kocher, T. D., Basasibwaki, P., and Wilson, A. C. (1990). Monophyletic origin of Lake Victoria cichlids suggested by mitochondrial DNA sequences. *Nature*, **347**, 550–3.

Moran, P. and Kornfield, I. (1993). Retention of ancestral polymorphism in the mbuna species flock (Teleostei: Cichlidae) of Lake Malawi. *Molecular Biology and Evolution*, **10**, 1015–29.

Myers, G. S. (1960). The endemic fish fauna of Lake Lanao, and the evolution of higher taxonomic categories. *Evolution*, **14**, 323–33.

Nelson, B. W., Ferreira, C. A. C., da Silva, M. F., and Kawasaki, M. L. (1990). Endemism centres, refugia and botanical collection density in Brazilian Amazonia. *Nature*, **345**, 714–16.

Nijhout, H. F., Wray, G. A., and Gilbert, L. E. (1990). An analysis of the phenotypic effects of certain colour pattern genes in *Heliconius* (Lepidoptera: Nymphalidae). *Biological Journal of the Linnean Society*, **40**, 357–72.

Parenti, L. R. (1984). Biogeography of the Andean killifish genus *Orestias* with comments on the species flock concept. In *Evolution of fish species flocks*, (ed. A. A. Echelle and I. Kornfield), pp. 85–92. University of Maine Press, Orono.

Paulay, G. (1985). Adaptive radiation on an isolated oceanic island: The Cryptorhynchinae (Curculionidae) of Rapa revisited. *Biological Journal of the Linnean Society*, **26**, 95–187.

Pielou, E. C. (1992). *After the ice age*. University of Chicago Press.

Ricklefs, R. E. and Cox, G. W. (1972). Taxon cycles in the West Indian avifauna. *American Naturalist*, **106**, 195–219.

Ricklefs, R. E. and Schluter, D. (1993). *Species diversity in ecological communities*. University of Chicago Press.

Rosenzweig, M. L. (1995). *Species diversity in space and time*. Cambridge University Press.

Schliewen, U. K., Tautz, D., and Pääbo, S. (1994). Sympatric speciation suggested by monophyly of crater lake cichlids. *Nature*, **368**, 629–32.

Schluter, D. (1996). Ecological causes of adaptive radiation. *American Naturalist*, **148**, Supplement, S40–S64.

Slatkin, M. (1996). In defense of founder-flush theories of speciation. *American Naturalist*, **147**, 493–505.

Sturmbauer, C. and Meyer, A. (1992). Genetic divergence, speciation and morphological stasis in a lineage of African cichlid fishes. *Nature*, **358**, 578–81.

Templeton, A. R. (1980). The theory of speciation via the founder principle. *Genetics*, **91**, 1011–38.

Templeton, A. R. (1981). Mechanisms of speciation—a population genetic approach. *Annual Review of Ecology and Systematics*, **12**, 23–48.

Templeton, A. R. (1996). Experimental evidence for the genetic-transilience model of speciation. *Evolution*, **50**, 909–15.

Tiercelin, J. J. and Mondeguer, A. (1991). The geology of the Tanganyikan Trough. In *Lake Tanganyika and its life*, (ed. G. W. Coulter), pp. 7–48. Oxford University Press.

Trewavas, E., Green, J., and Corbet, S. A. (1972). Ecological studies on crater lakes in west Cameroon fishes of Barombi Mbo. *Journal of Zoology, London*, **167**, 41–95.

Turner, J. R. G. (1977). Butterfly mimicry: the genetical evolution of an adaptation. *Evolutionary Biology*, **10**, 163–206.

Wagner, W. H. and Funk, V. A. (eds.) (1995). *Hawaiian biogeography: evolution on a hot-spot archipelago*. Smithsonian Institution Press, Washington, D. C.

Williams, E. E. (1969). The ecology of colonization as seen in the zoogeography of anoline lizards on small islands. *Quarterly Review of Biology*, **44**, 345–89.

Williams, E. E. (1983). Ecomorphs, faunas, island size, and diverse end points in island radiations of *Anolis*. In *Lizard ecology: studies of a model organism*, (ed. R. B. Huey, E. R. Pianka, and T. W. Schoener), pp. 326–70. Harvard University Press, Cambridge, MA.

Wilson, E. O. (1961). The nature of the taxon cycle in the melanesian ant fauna. *America Naturalist*, **95**, 169–93.

Wright, S. (1932). The roles of mutation, inbreeding, crossbreeding and selection in evolution. *Proceedings of the Sixth International Congress of Genetics*, **1**, 356–66.

13

Ecological and evolutionary determinants of the species–area relationship in Caribbean anoline lizards

Jonathan B. Losos

13.1 Introduction

Explaining differences in species diversity among areas has been a central goal of ecology with studies traditionally focusing either on historical or ecological factors (see recent reviews in Ricklefs and Schluter 1993; Brown 1995; Rosenzweig 1995). Studies of insular faunas and floras have played a prominent role in the development of theories of species diversity beginning with the work of the naturalists of the latter half of the nineteenth century and continuing to this day. MacArthur and Wilson's (1967) equilibrium theory of island biogeography, which explained differences in diversity in terms of the ecological processes of colonization and extinction rates, was particularly influential. Nonetheless, as MacArthur and Wilson (1967, Chapter 7) noted, ecological factors are only part of the explanation. To the extent that *in situ* evolutionary diversification contributes to the species diversity of an island, then processes operating in evolutionary, as well as ecological, time must be considered. To date, however, few studies have attempted to consider the relative roles of ecological and evolutionary factors in determining species diversity. One possible explanation for the paucity of such studies is that few groups of organisms occur in a wide enough range of situations that the role of both ecological and evolutionary factors can be examined. Caribbean *Anolis* lizards, however, are an exception to this generality.

Anolis lizards (Fig. 13.1) are arguably the dominant group of vertebrates in the Caribbean (their only rival being frogs of the genus *Eleutherodactylus*). Anoles are found on almost every island throughout the Caribbean (Williams 1969), often at extremely high densities (Schoener and Schoener 1980; Reagan 1992). There are 139 species currently recognized, with as many as 55 on a single island (Powell *et al.* 1996, plus recent species descriptions). The distribution of anole lineages among islands in the Caribbean paradoxically suggests both great dispersal ability and relatively limited dispersal success. The former is demonstrated by the presence of some Cuban species

Fig. 13.1 *Anolis vermiculatus* from Cuba (upper left), an occupant of tree trunks and rocks near streams, and three ecomorphs: *A. gundlachi* (upper right), a trunk-ground anole from Puerto Rico; *A. insolitus* (lower left), a twig anole from Hispaniola; and *A. pulchellus* (lower right), a grass anole from Puerto Rico.

of anoles on islands 250 or more miles from Cuba (e.g. Swan Island, Bay Islands; Williams 1969). These islands are relatively low-lying and must have been under water during the last glacial minimum, 120 000 years ago. Hence, colonization of these distant islands has occurred relatively recently.

Despite this great colonizing capability, however, anole faunas throughout the Caribbean are not homogeneous; rather, the vast majority (>85%) of species of *Anolis* are endemic to a single island bank (this number is only slightly inflated by the tendency to describe closely related taxa on different islands as different species). This paradox indicates that both ecological and evolutionary factors must be considered in studies of the island biogeography of Caribbean anoles. In addition, many island banks emergent 8000 years ago have been fragmented by rising sea levels (Heatwole and MacKenzie 1967; Fig. 13.2); thus, historical factors, too, must be considered to understand patterns of anole species distribution.

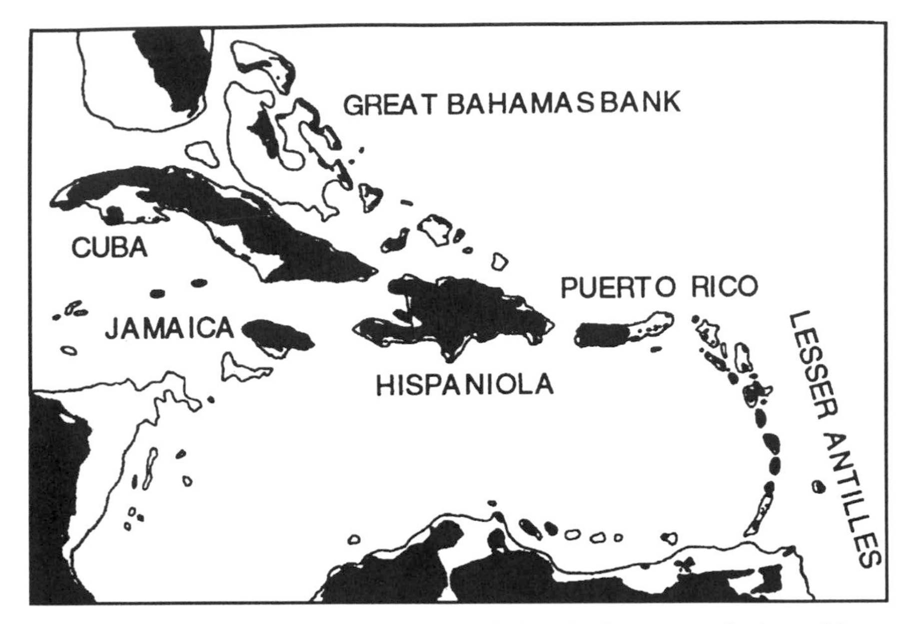

Fig. 13.2 Map of the Caribbean. Lines represent 200 m depth contours. At times of lower sea levels, these island banks were above water, thus merging currently independent islands into single large landmasses.

13.2 Data base

The primary source of data for this paper is Rand's (1969) compilation of area and species number for 145 Caribbean islands ranging in area from 0.06 to >40 000 square miles. I have updated the species counts following Powell *et al.* (1996) for the Greater Antilles and Schwartz and Henderson (1991) for islands near Hispaniola and the Isle of Pines. In addition, I have added data for two additional Hispaniolan islands, Beata and Ilé a Cabrit (Schwartz and Henderson 1991).

Rand (1969) classified the islands into four groups: Greater Antilles and fringing islands, Lesser Antilles, isolated islands (i.e. islands in the vicinity of the Greater Antilles, such as St Croix and Mona, that have never been part of a larger island), and the Bahamas. For reasons discussed below, I separate the four large Greater Antillean islands from islands on the same underwater banks; the former I refer to as the 'Greater Antilles' and the latter as 'Greater Antillean satellites.' In addition, I divide the Bahamian islands into those on the Great Bahama Bank and those on other banks; although the latter constitutes a heterogeneous group, all members of this group are similar in being isolated and occurring on relatively small island banks.

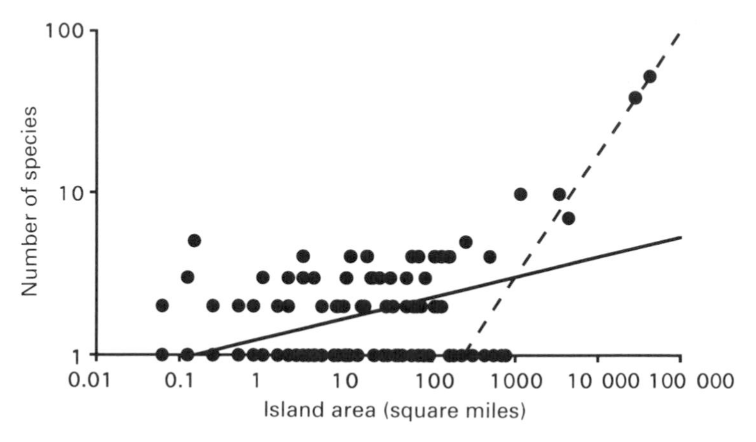

Fig. 13.3 Species–area relationship for Caribbean *Anolis*. The solid line is the best fit least squares regression line for all islands; the broken line is the line for only the major islands of the Greater Antilles (Cuba, Hispaniola, Jamaica, and Puerto Rico).

13.3 Species–area relationships in *Anolis*

A significant relationship exists between island area and species number for Caribbean *Anolis* (Fig. 13.3). MacArthur and Wilson (1967) noted that the species–area relationship could be described by the equation $s = Ca^z$, where s is the number of species, A is island area, and c and z are constants. MacArthur and Wilson further noted that the value of z normally lies in the range 0.20–0.35; subsequent work showed that many types of organisms have z-values in this range (Connor and McCoy 1979). For Caribbean *Anolis*, $z = 0.13$ (as calculated from log–log regressions), a value lower than that observed for most other taxa.

Separate analyses, however, indicate substantial heterogeneity in the species–area relationship among island classes (Fig. 13.4). No relationship exists for isolated ($r^2 = 0.00$, $F_{1,9} = 0.18$, $P > 0.65$) and 'other' (= non-Great Bahama Bank) Bahamian islands ($r^2 = 0.01$, $F_{1,36} = 1.36$, $P > 0.25$) and only a weak relationship exists for Lesser Antillean islands ($r^2 = 0.13$, $F_{1,31} = 5.89$, $P < 0.025$). By contrast, area and species number are strongly related for Greater Antillean ($r^2 = 0.92$, $F_{1,2} = 36.75$, $P < 0.03$), Greater Antillean satellite ($r^2 = 0.54$, $F_{1,44} = 53.60$, $P < 0.001$), and Great Bahamas Bank islands ($r^2 = 0.67$, $F_{1,13} = 29.30$, $P < 0.001$). Among these latter three island groups, the slope of the regression lines (z) differs between the Greater Antilles and the other two (Analysis of covariance [Ancova], Greater Antilles [$z = 0.77$] versus satellites [$z = 0.18$], $F_{1,46} = 10.75$, $P < 0.0025$; Greater Antilles versus Great Bahamas [$z = 0.19$], $F_{1,15} = 11.41$, $P < 0.005$); the regression lines for the Greater Antillean satellite and Great Bahamas Bank islands also are almost statistically different in intercept ($F_{1,58} = 3.46$, $P < 0.07$).

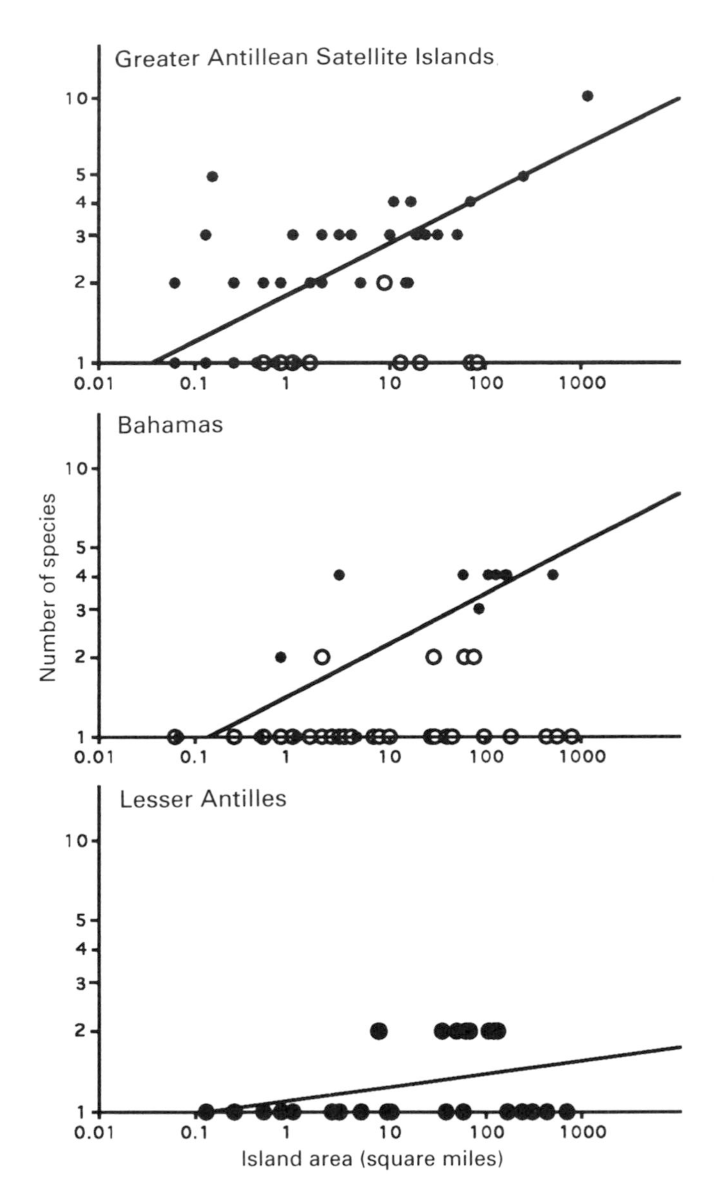

Fig. 13.4 Species–area relationship for several classes of islands. In the plots for Greater Antillean islands, closed circles are islands on the banks of the Greater Antilles, whereas open circles are isolated islands (i.e. islands that have never been connected by dry land to one of the Greater Antilles). The Greater Antilles (Jamaica, Puerto Rico, Hispaniola, and Cuba) are not included in this graph. In the plot for the Bahamas, closed circles are islands of the Great Bahama Bank and open circles are other islands. Regression lines are only provided for statistically significant relationships.

Rand (1969) previously noted this heterogeneity among island classes. He further noted that among islands that are not on the island banks of the Greater Antilles or on the Great Bahamas Bank, not only is the relationship between island area and number of species weak or non-existent, but no island has more than two species. By contrast, Greater Antillean satellites can have as many as ten species and Great Bahamian Bank islands as many as four. One possible explanation for this difference is that isolated islands are more distant from sources of colonists than are satellite islands. To test this hypothesis, I measured the distance of each island to the nearest landmass with many potential colonizing species (Cuba, Hispaniola, Jamaica, Puerto Rico, or South America). Based on these measurements, a weak negative relationship exists between species number and distance ($r^2 = 0.005$, $F_{1,91} = 6.32$, $P < 0.02$); sample sizes are smaller in this analysis than previous ones because some of the small islands reported by Rand (1969) could not be located. With the effect of island area removed by calculating residuals, relative species number is more strongly related to relative distance ($r^2 = 0.122$, $F_{1,91} = 13.73$, $P < 0.001$). Nonetheless, distance effects do not account for the difference in species number between landbridge (i.e. Greater Antillean and Great Bahama Bank) and isolated islands. With the effect of distance removed using residuals, the relationship between area and species number is stronger for landbridge islands than it is for isolated islands both in the Greater Antilles (Ancova, difference in slopes, $F_{1,33} = 4.52$, $P < 0.05$) and in the Bahamas (difference in slopes, $F_{1,27} = 11.30$, $P < 0.0025$).

The difference in number of species on isolated versus landbridge islands obviously results from how the species came to be on the islands. Islands that were connected to larger landmasses can 'inherit' great numbers of species, whereas isolated islands that must build up species diversity via over-water colonization can manage two species at best (Rand 1969). The discrepancy suggests that, although *Anolis* has great dispersal capabilities, colonists arriving on inhabited islands must have very low likelihood of becoming established and that this likelihood diminishes as a function of number of species already resident on an island (Rand 1969). Further, all two-species isolated islands contain species that differ ecologically (Schoener 1970, 1988). With one possible exception (the northern Lesser Antilles, where character displacement may have produced ecological differentiation subsequent to sympatry (Schoener 1970; Losos 1990*a*; Miles and Dunham 1996; Butler and Losos, in press; C.J. Schneider *et al.*, unpublished), the differences between species on two-species islands existed prior to sympatry; in other words, a colonizing species must be ecologically different from a resident species if it is to have any substantial hope of successful colonization.

The Great Bahamas Bank is an exception to the generalization that no more than two species can colonize an island by over-water dispersal. None of the four species found on the Great Bahamas Bank evolved there; three immigrated from Cuba and one from Hispaniola. As with other isolated islands, the species that colonized the Great Bahamas Bank are all ecologically distinctive and each had evolved specializations to use a different habitat type prior to colonization of the Great Bahamas Bank (Schoener 1968). The greater number of colonists on the Great Bahamas Bank relative to other isolated islands may be a reflection of lack of isolation; at its maximal exposure, this bank was only 10 miles from Cuba. The size of the Great Bahamas Bank may also have been important; at its maximal extent, it rivalled Cuba in total area.

13.4 The role of extinction

On the satellite islands of the Greater Antilles and within the Greater Bahamas Bank, extinction, rather than colonization, has been the primary process producing species–area patterns (Rand 1969). These islands fit the classic idea of faunal relaxation (Wilcox 1978; Richman *et al.* 1988); areas that presumably used to harbour more species experiencing a decline in species richness after a reduction in island size. Such patterns have usually been explained as differential extinction, an explanation that seems to hold for Caribbean anoles. Previous studies on Bahamian (Schoener and Schoener 1983*a*, *b*) and Puerto Rican (Roughgarden 1989; Mayer 1989) islands indicate that species extinction is not random; rather, species drop out in a pattern predictable based on island area, thus producing a nested pattern of species occurrences.

The cause of this nesting of species probably is a result of habitat specializations. Islands lose habitats in a predictable pattern as they get smaller, which would result in a deterministic pattern of species disappearance. For example, in the Bahamas, the smallest lizard-inhabited islands are scrubby, with little vegetation. Such habitat is only suitable for *A. sagrei*. Larger islands may have thicker vegetation and some trees, providing habitat for *A. carolinensis smaragdinus*. Even larger islands will have the broad-diameter trees and shade needed by *A. distichus*. Only the largest islands have the high vegetation necessary for the twig anole, *A. angusticeps* (Schoener and Schoener 1983*a*).

Colonization may play a secondary role in determining species occurrences on these islands by re-establishing or maintaining populations threatened by extinction (i.e. the 'rescue effect'; Brown and Kodric-Brown 1977). For example, in the Bahamas, distance to a large source island is negatively related to species occurrences, although the relationship is considerably weaker than that between species' occurrences and island area (Schoener and Schoener 1983*a*, *b*). This distance effect could result either because colonization occurs more frequently on less isolated islands or because more isolated islands have less suitable habitats, even when the effect of area is removed (Schoener and Schoener 1983*b*).

13.5 The role of evolutionary diversification

Thus far, discussion has focused on ecological processes—colonization, competition, extinction—and how they are affected by island area. However, this discussion does not pertain to the large islands of the Greater Antilles, for the great bulk of their species has arisen *in situ* as a result of within-island speciation (Williams 1983). Thus, the processes responsible for the species–area curve among these islands must be fundamentally different from those operating within the smaller islands of the Caribbean.

First, one must ask if there is an island size below which within-island diversification does not occur. Williams (1983; also Williamson 1981) argued that cladogenetic speciation in anoles occurs only on islands the size of Puerto Rico or larger. This hypothesis may be tested by considering the phylogenetic relationships of anoles. Within-island speciation would be suggested if sister taxa occur on the same island (or

islands); between-island speciation would be suggested if the sister taxa c
different islands (Lynch 1989; Brooks and McLennan 1991). This approac
infallible; dispersal and shifts in geographic range can obscure these patterns. Nonetheless, in this case, the method seems reasonable because dispersal and range shifts are more likely to pose problems in a continental setting than in an island situation in which dispersal appears limited. Thus, dispersal and range shifts probably do not confound interpretation of general patterns of anole distribution, although they may be important in particular situations.

Within-island speciation clearly has occurred in the Greater Antilles. Of the 18 'series' of anoles that occur in the Greater Antilles (Burnell and Hedges 1990; Hass *et al.* 1993), only four have species that occur on more than one island group. Hence, all of the diversification in the other 14 groups must have resulted from within-island speciation, rather than colonization from other islands. Even in the four species groups with species on different island groups, most of the speciation was probably within-island with one or two instances of between-island speciation. Thus, in the Greater Antilles, most speciation must have occurred by within-island diversification.

By contrast, on islands other than the Greater Antilles, only two instances exist of sister taxa occurring on the same island or group of islands, and neither provides a compelling case for within-island speciation. In the northern Lesser Antilles, the *wattsi* and *bimaculatus* groups are sister taxa. Members of the *wattsi* group only occur on six islands (and surrounding cays), all of which are occupied by members of the *bimaculatus* group (Schoener 1970; Lazell 1972). Thus, this pattern of geographic distribution is consistent with the hypothesis that the two lineages initially diverged on one of these islands; however, a more likely explanation, also consistent with the biogeographical data, is that the lineages diverged on different islands in the area and then subsequently came into sympatry. Similarly, *A. pigmaequestris* appears to be sympatric throughout its limited range in the northern Cuban fringing islands with its sister taxon, *A. equestris* (Garrido 1975). As with the case above, allopatric speciation followed by dispersal seems more likely than within-island diversification on these small islands.

Although mechanisms of speciation have been little explored in *Anolis*, the finding of a minimum island area below which within-island speciation probably does not occur is consistent with a hypothesis that geographic isolation is required for speciation. Small islands may not have the landscape necessary to produce geographic disjunction. Certainly, the landscape complexity of the Greater Antilles exceeds that of smaller Caribbean islands; in addition, the Greater Antilles have considerably more offshore islands upon which peripatric speciation might produce new species able to subsequently reinvade the larger island (Williams 1983). This explanation, however, is not entirely satisfying because some of the larger Lesser Antillean islands (e.g. Guadeloupe) are topographically diverse and have offshore islands, yet speciation has not occurred and these islands have no more than two species (note, too, that some of these islands appear as vegetationally complex as the Greater Antilles and certainly have the habitat complexity to support more than two species). More detailed studies of geographic differentiation and speciation in *Anolis* are required to more adequately address these issues.

Given that within-island speciation has been the predominant factor in the Greater Antilles, one might inquire what role colonization has played in establishing the species–area relationship within these islands. A maximal estimate for the number of possible colonization events can be derived by counting the number of monophyletic species groups on each island (from Burnell and Hedges 1990; Hass *et al.* 1993), with the assumption that each species group represents a separate colonization event. Numbers obtained in this way will surely overstate the amount of colonization because two monophyletic groups on an island may be sister taxa and thus not represent separate invasions and because the presence of some lineages may result from vicariance rather than dispersal (see discussion in Hedges *et al.*1992; Hedges 1996; Crother and Guyer 1996; Jackman *et al.*, 1997). Given these caveats, the maximal number of colonization events for each island are: Jamaica, 2 (29% of species on the island); Puerto Rico, 3 (30%); Hispaniola, 11 (28%); Cuba, 7 (13%). Thus, minimally, at least 70% of species on each of the Greater Antillean islands has arisen from *in situ* speciation.

13.6 Causes of the relationship between island area and speciation

The observed increase in species number with area in the Greater Antilles is a composite of several different factors, each of which may be affected differently by increasing island area. At the broadest level, two factors appear responsible for the increase in species number with area (Williams 1983). First, on larger islands, some species groups have produced a large number of geographically isolated species. For example, the *sagrei* series has produced 14 ecologically similar species on Cuba, only two of which are distributed island-wide and most of which have relatively small geographic ranges. A similar pattern exists among the 14 grass anoles in the *alutaceus* series in Cuba. By contrast, most species in Jamaica and Puerto Rico are geographically widespread; only one species in Jamaica and two in Puerto Rico have geographically limited ranges. Second, the maximum number of anole species that can be found sympatrically increases with island area, from six in Jamaica to 12 on Cuba (Fig. 13.5). This increase in sympatric species is accomplished by both more finely subdividing the habitat and by utilizing additional niches.

To quantify these factors, I classified Greater Antillean *Anolis* species diversity into four categories. First, I counted the number of independent habitat niches that have been occupied on each island. On Puerto Rico, for example, species adapted to five niches have evolved: trunk-ground, trunk-crown, grass, crown-giant, and twig anoles (Williams 1972). Morphological evolution was used as an index of the number of instances of evolutionary specialization to habitat niches. As a general rule, this assumption is not problematic; among all *Anolis* lizards that have been well studied, morphologically different species differ in habitat use (e.g. Moermond 1979; Estrada and Silva Rodríguez 1984; Losos 1990*b*; Irschick and Losos 1996). As an example of this procedure, the 14 grass anoles of the *alutaceus* species group on Cuba were counted as one instance of habitat specialization.

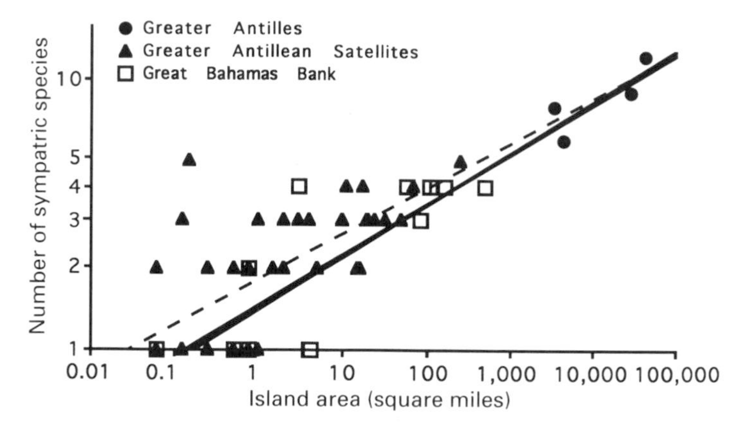

Fig. 13.5 Maximum number of sympatric species versus island area. The relationship is weak or non-existent for islands of the Lesser Antilles, isolated islands in the Greater Antilles, and non-Great Bahama Bank Bahamian islands; hence, they are not included. The broken line is the regression for the Greater Antilles.

Evolutionarily independent occupation by two lineages of the same habitat niche on the same island was counted as two events; however, this rarely occurred. The only examples are the multiple evolution of grass anoles on Cuba and Hispaniola and of twig anoles on Hispaniola. In addition, two lineages of trunk-ground anoles are present in Jamaica, but one, *A. sagrei*, is a recent colonist (Williams 1969).

Within each habitat niche, I investigated how many species have evolved the capability to be sympatric. For example, among the *alutaceus* series, as many as three species may be found sympatrically (Garrido and Hedges 1992). Hence, for this group, I scored one evolution of habitat niche specialization (grass anole) and two instances of the evolution of the ability to occur sympatrically. I broke down this latter category into two groupings, thermal differentiation and other. Coexistence of anoles using the same structural habitat by adapting to different thermal microclimates has been extensively documented in anoles as one means of permitting sympatry (reviewed in Losos 1994). Other means of attaining sympatry includes divergence in body size, which has occurred several times (e.g., *grahami* and *opalinus* on Jamaica, *evermanni* and *stratulus* on Puerto Rico; reviewed in Losos 1994). The ecology of a number of sympatric species, such as Cuban grass anoles, has not been well studied. Hence, whether these species differ ecologically cannot at this time be assessed; based on studies of other sympatric anoles, I would suspect that ecological differences will be found upon closer examination; in any case, they are placed in the 'other' category.

Finally, I categorized all remaining taxa (e.g. the remaining 11 members of the *alutaceus* series) as allopatric species, i.e. members of a species complex that are ecologically undifferentiated and allopatrically or parapatrically distributed. A potential bias exists in this procedure. First, in the absence of more detailed phylogenetic information, I assumed that the capability to achieve sympatry had evolved the minimum number of times consistent with the data. For example, many members of the *alutaceus* series occur sympatrically with one or two other members of that series, although no more than three species occur at any one site. The possibility exists that each one of these species has independently evolved the capability of coexisting with other members of its series. However, the alternative possibility is that the ability to exist sympatrically only evolved in three lineages within the *alutaceus* series, and that these lineages then produced a number of allopatrically distributed trios of species. Obviously, further phylogenetic information is needed to clarify this issue; the possibility exists that my method inflates the importance of allopatric diversification and diminishes the importance of ecological differentiation permitting coexistence among species adapted to the same habitat niche.

Given this categorization, Fig. 13.6 yields several insights. First, almost all species in Jamaica and Puerto Rico differ in habitat use, but allopatric species are the largest component of the Cuban fauna and a major component in Hispaniola. Second, the increase in species number with area primarily results from increases in number of habitat niches utilized and in the number of ecologically similar but allopatrically distributed species; the slope of increase for number of allopatric species is substantially greater than for habitat types (Ancova, $F_{1,4} = 13.59$, $P < 0.025$). Third, the number of species that coexist by using different thermal microclimates is approximately equal on three of the islands, but anomalously low for Jamaica.

Another point, not evident from the figure, is that only one of the habitat types found on Jamaica or Puerto Rico (the montane species *A. reconditus* from Jamaica) is not also

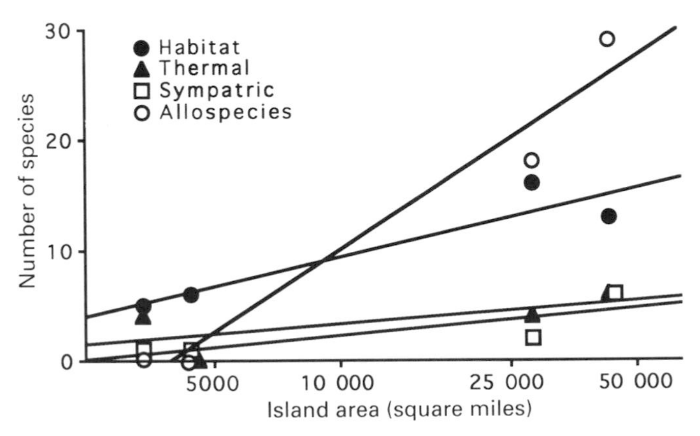

Fig. 13.6 Species–area relationship for several aspects of evolutionary diversification in the Greater Antilles. The species on each island are divided into four categories. Habitat refers to the number of different habitats occupied on an island. In the few instances in which two species (or clades) have adapted to use the same habitat on an island, both species are counted in the habitat category. When members of a monophyletic group use the same habitat, then the species are divided into three categories: allospecies refers to allopatrically distributed species, thermal refers to sympatric species that use differ thermal microhabitats, and sympatric includes all other sympatric species. See text for more details.

found on both Hispaniola and Cuba. By contrast, many of the habitat specialists on Hispaniola (e.g. *Chamaelinorops barbouri, A. etheridgei*) and Cuba (e.g. *Chamaeleolis* species, *A. vermiculatus*) have no ecological counterpart on any of the other islands (note that *Chamaeleolis* and *Chamaelinorops* are phylogenetically part of the *Anolis* radiation (Hass *et al.* 1993, Jackman *et al.*, in press)).

These conclusions lead me to present the following hypothesis for the species–area relationship within the Greater Antilles. First, a core set of six types of habitat specialists, termed ecomorphs (Williams 1972), is present on most or all of the Greater Antilles (the exceptions being the lack of grass anoles on Jamaica and trunk anoles on Jamaica and Puerto Rico). Within each of these islands, most or all of the ecomorph types are present at most localities and generally are more abundant than other types of habitat specialists (although twig anoles and crown-giants are often rarer than the other ecomorphs). Further, as already noted, some of the ecomorph types can be represented by many species on a single island. Second, other than ecomorph species, few other types of habitat specialist are represented by more than one species on an island. The only exceptions are three rock-dwelling members of the *monticola* series on Hispaniola (two sympatric), three bush-inhabiting species of the *argillaceus* series on Cuba (two probably sympatric), two species of the *lucius* group on Cuba (allopatric), and four species of *Chamaeleolis* on Cuba (only two of which can be found at any given locality).

Consequently, the increase in species number on large Greater Antillean islands reduces to two factors: first, lineages of the six common ecomorphs diversify by producing many allopatrically distributed species and a number of species capable of coexisting in sympatry. Second, many unique types of habitat specialists have evolved on the larger islands. The first factor may simply be a consequence of larger islands affording greater opportunities for allopatric speciation, as discussed above. Certainly, the existence of these allopatric species indicates that allopatric speciation occurs on these islands. In turn, this larger pool of reproductively isolated species enhances the possibility that some of these species will evolve ecological differences sufficient to permit sympatry.

The increased number of habitat types on the larger islands is less readily explained. Certainly, habitat availability would not seem to differ among the islands, all of which have wet and dry forests and high mountain ranges. One possible explanation is that the greater pool of species on larger islands enhances the possibility that one species will diverge sufficiently to utilize a distinctive habitat niche. This hypothesis makes the prediction that species utilizing unique habitats should have arisen within a lineage of a particular ecomorph type and should render such lineages paraphyletic. Although the phylogenetic affinities of many of these species are not clear-cut, the only evidence to date that supports this hypothesis is from the Jamaican *A. reconditus*, which may have evolved from within the Jamaican trunk-ground lineage (D. Irschick, personal communication). In contrast, most of the data seem to suggest that once a lineage has specialized to a particular habitat niche, it rarely evolves away from this specialization.

A third possibility is that the greater diversity of sympatric assemblages on larger islands forces lineages to specialize in ways not seen on small islands. Perhaps a hierarchy of habitat niches exists such that species evolve to occupy certain niches only after others have been occupied. This hierarchy, or deterministic pathway, might result ecause specialization for some niches is evolutionarily more accessible than zation for other niches, or because the adaptive peaks (*sensu* Simpson 1953) of so iches are higher than others. This hypothesis will prove difficult to test and does not accord with the observation that some unique habitat species occur in assemblages containing few other anole species (Williams 1983). Better phylogenetic information will be necessary to provide insight into how, and possibly why, these unique habitat species evolved.

13.7 Conclusions

The species–area relationship among anoline lizards is more complicated than it appears on first inspection. Only by detailed examination of differences in island types and relationships among taxa was it possible to discover the heterogeneity in the importance of different factors in determining species number on different islands. Broad-scale studies involving many taxa are important for documenting general patterns and suggesting general hypotheses, but such studies must be complemented by more detailed studies of particular groups to fully understand patterns of species diversity (Williams 1969, 1983).

13.8 Summary

Species–area relationships were studied for anoline lizards on 147 islands in the Caribbean. The relative importance of ecological and evolutionary factors in determining species number varied with island size. On small islands, only ecological factors effect the species–area relationship. Further, the importance of different ecological factors such as colonization, competition, and extinction, varied among different types of islands. On landbridge islands, differential extinction as a function of island area appears to play a key role in producing a species–area relationship. By contrast, limited colonization success generally prevents oceanic islands from accumulating more than two species. Among the larger islands, evolutionary factors are the primary determinant of species number. Detailed examination of several components of evolutionary diversification indicates that the species–area relationship among the Greater Antilles anoles primarily results because larger islands have increased number of habitat niches occupied and a greater number of closely related species that are ecologically similar and allopatrically distributed; increased subdivision of certain habitats plays a lesser role.

Acknowledgements

This research was supported by the National Science Foundation (DEB 9318642 and 9407202), the David and Lucile Packard Foundation, and the National Geographic Society. I thank B. Hedges, M. Leal, D. Pepin, and T. Schoener for helpful comments and assistance.

References

Brooks, D. R. and McLennan, D. A. (1991). *Phylogeny, ecology, and behavior: A research program in comparative biology*. University of Chicago Press.

Brown, J. H. (1995). *Macroecology*. University of Chicago Press.

Brown J. H. and Kodric-Brown, A. (1977). Turnover rates in insular biogeography: effect of immigration on extinction. *Ecology*, **58**, 445–9.

Burnell, K. L. and Hedges, S. B. (1990). Relationships of West Indian *Anolis* (Sauria: Iguanidae): an approach using slow-evolving protein loci. *Caribbean Journal of Science*, **26**, 7–30.

Butler, M. A. and Losos, J. B. Testing evolutionary hypotheses that require reconstruction of the past: A simulation study comparing squared-change and linear parsimony methods for a continuous character. *Evolution*, in press.

Connor, E. F. and McCoy, E. D. (1979). The statistics and biology of the species–area relationship. *American Naturalist*, **113**, 791–833.

Crother, B. I. and Guyer, C. (1996). Caribbean historical biogeography: Was the dispersal-vicariance debate eliminated by an extraterrestrial bolide? *Herpetologica*, **52**, 440–65.

Estrada, A. R. and Silva Rodríguez, A. (1984). Análisis de la ecomorfología de 23 especies de lagartos cubanos del género *Anolis*. *Ciencias Biológicas*, **12**, 91–104.

Garrido, O. H. (1975). Nuevos reptiles del archipiélago Cubano. *Poeyana*, **141**, 1–58.

Garrido, O. H. and Hedges, S. B. (1992). Three new grass anoles from Cuba (Squamata: Iguanidae). *Caribbean Journal of Science*, **28**, 21–9.

Hass, C. A., Hedges, S. B., and Maxson, L. R. (1993). Molecular insights into the relationships and biogeography of West Indian anoline lizards. *Biochemical Systematics and Ecology*, **21**, 97–114.

Heatwole, H. and MacKenzie, F. (1967). Herpetogeography of Puerto Rico. IV. Paleogeography, faunal similarity and endemism. *Evolution*, **21**, 429–38.

Hedges, S. B. (1996). Vicariance and dispersal in Caribbean biogeography. *Herpetologica*, **52**, 466–73.

Hedges, S. B., Hass, C. A., and Maxson, L. R. (1992). Caribbean biogeography: molecular evidence for dispersal in West Indian terrestrial vertebrates. *Proceedings of the National Academy of Sciences USA*, **89**, 1909–13.

Irschick, D. J. and Losos, J. B. (1996). Morphology, ecology, and behavior of the twig anole, *Anolis angusticeps*. In *Contributions to West Indian herpetology: A tribute to Albert Schwartz*, (ed. R. Powell and R. Henderson), pp. 291–301. Society for the Study of Amphibians and Reptiles, Ithaca, NY.

Jackman, T., Losos, J. B., Larson, A., and de Queiroz, K. (1997). Phylogenetic studies of convergent adaptive radiations in Caribbean *Anolis* lizards. In *Molecular evolution and adaptive radiation*, (ed. T. Givnish and K. Sytsma), pp. 535–57. Cambridge University Press.

Lazell, J. D., Jr. (1972). The anoles (Sauria, Iguanidae) of the Lesser Antilles. *Bulletin of the Museum of Comparative Zoology, Harvard University*, **143**, 1–115.

Losos, J. B. (1990*a*). A phylogenetic analysis of character displacement in Caribbean *Anolis* lizards. *Evolution*, **44**, 558–69.

Losos, J. B. (1990*b*). Ecomorphology, performance capability, and scaling of West Indian *Anolis* lizards: an evolutionary analysis. *Ecological Monographs*, **60**, 369–88.

Losos, J. B. (1994). Integrative approaches to evolutionary ecology: *Anolis* lizards as model systems. *Annual Review of Ecology and Systematics*, **25**, 467–93.

Lynch, J. D. (1989). The gauge of speciation: on the frequencies of modes of speciation. In *Speciation and its consequences*, (ed. D. Otte and J. A. Endler), pp. 527–53. Sinauer, Sunderland, MA.

MacArthur, R. H. and Wilson, E. O. (1967). *The theory of island biogeography*. Princeton University Press.

Mayer, G. C. (1989). Deterministic patterns of community structure in West Indian reptiles and amphibians. Unpubl. Ph.D. thesis. Harvard University, Cambridge, MA.

Miles, D. B. and Dunham, A. E. (1996). The paradox of the phylogeny: Character displacement of analyses of body size in island *Anolis*. *Evolution*, **50**, 594–603.

Moermond, T. C. (1979). The influence of habitat structure on *Anolis* foraging behavior. *Behaviour*, **70**, 147–67.

Powell, R., Henderson, R. W., Adler, K., and Dundee, H. A. (1996). An annotated checklist of West Indian amphibians and reptiles. In *Contributions to West Indian herpetology: A tribute to Albert Schwartz*, (ed. R. Powell and R. Henderson), pp. 51–93. Society for the Study of Amphibians and Reptiles, Ithaca, NY.

Rand, A. S. (1969). Competitive exclusion among anoles (Sauria: Iguanidae) on small islands in the West Indies. *Breviora*, **319**, 1–16.

Reagan, D. P. (1992). Congeneric species distribution and abundance in a three-dimensional habitat: the rain forest anoles of Puerto Rico. *Copeia*, **1992**, 392–403.

Richman, A. D., Case, T. J., and Schwaner, T. D. (1988). Natural and unnatural extinction rates of reptiles on islands. *American Naturalist*, **131**, 611–30.

Ricklefs, R. E. and Schluter, D. (1993). *Species diversity in ecological communities: Historical and geographical perspectives*. University of Chicago Press.

Rosenzweig, M. L. (1995). *Species diversity in space and time*. Cambridge University Press.

Roughgarden, J. (1989). The structure and assembly of communities. In *Perspectives in ecological theory*, (ed. J. Roughgarden, R. M. May, and S. A. Levin), pp. 203–26. Princeton University Press.

Schoener, T. W. (1968). The *Anolis* lizards of Bimini: Resource partitioning in a complex fauna. *Ecology*, **49**, 704–26.

Schoener, T. W. (1970). Size patterns in West Indian *Anolis* lizards. II. Correlations with the size of particular sympatric species—displacement and convergence. *American Naturalist*, **104**, 155–74.

Schoener, T. W. (1988). Testing for non-randomness in sizes and habitats of West Indian lizards: Choice of species pool affects conclusions from null models. *Evolutionary Ecology*, **2**, 1–26.

Schoener, T. W. and Schoener, A. (1980). Densities, sex ratios and population structure in four species of Bahamian *Anolis* lizards. *Journal of Animal Ecology*, **49**, 19–53.

Schoener, T. W. and Schoener, A. (1983*a*). Distribution of vertebrates on some very small islands. I. Occurrence sequences of individual species. *Journal of Animal Ecology*, **52**, 209–35.

Schoener, T. W. and Schoener, A. (1983*b*). Distribution of vertebrates on some very small islands. II. Patterns in species number. *Journal of Animal Ecology*, **52**, 237–62.

Schwartz, A. and Henderson, R. W. (1991). *Amphibians and reptiles of the West Indies: Descriptions, distributions, and natural history*. University of Florida Press, Gainesville, FL.

Simpson, G. G. (1953). *The major features of evolution*. Columbia University Press, New York.

Wilcox, B. A. (1978). Supersaturated island faunas: a species-age relationship for lizards on post-Pleistocene land-bridge islands. *Science*, **199**, 996–8.

Williams, E. E. (1969). The ecology of colonization as seen in the zoogeography of anoline lizards on small islands. *Quarterly Review of Biology*, **44**, 345–89.

Williams, E. E. (1972). The origin of faunas. Evolution of lizard congeners in a complex island fauna: a trial analysis. *Evolutionary Biology*, **6**, 47–89.

Williams, E. E. (1983). Ecomorphs, faunas, island size, and diverse end points in island radiations of *Anolis*. In *Lizard ecology: studies of a model organism*, (ed. R. B. Huey, E. R. Pianka, and T. W. Schoener), pp. 326–70. Harvard University Press, Cambridge, MA.

Williamson, M. (1981). *Island populations*. Oxford University Press.

14

Lake level fluctuations and speciation in rock-dwelling cichlid fish in Lake Tanganyika, East Africa

Lukas Rüber, Erik Verheyen, Christian Sturmbauer, and Axel Meyer

14.1 Introduction

The East African Lakes Tanganyika, Malawi, and Victoria each harbour hundreds of endemic invertebrate and vertebrate species. The endemic cichlid fish faunas of the East African Lakes are biologically astonishingly diverse and each of the lakes contains its distinct species flock of cichlid fishes. These species flocks are viewed as the most spectacular example among living vertebrates for evolutionary phenomena termed adaptive radiation, and explosive speciation (Fryer and Iles 1972; Futuyma 1986; Coulter 1991; Martens *et al.* 1994) and provide ample opportunity of the study of the evolutionary mechanisms that might be responsible for the formation of these species flocks (Fryer and Iles 1972). Inferences about the ecological and evolutionary processes responsible for the origin of these species flocks will only be possible when they are based upon explicit phylogenetic hypotheses of the studied species linked to pathways of morphological and ecological diversification (Avise 1994). To determine the relative importance of intrinsic characteristics versus extrinsic factors for the intralacustrine evolution of these faunas may offer information about the processes that resulted in the diversification and speciation in these species flocks.

14.2 The Tanganyikan cichlid species flock

Probably due to its greater age and its probable polyphyletic origin, the Tanganyikan cichlid species flock is morphologically and behaviourally more diverse than the flocks of Lakes Malawi and Victoria (Fryer and Iles 1972) even if—in terms of numbers—it harbours the lowest number of endemic cichlid species (more than 170) (but see Snoeks *et al.* 1994). With its estimated age of 9 to 12 million years (Ma), Lake Tanganyika is considerably older than Lakes Malawi and Victoria (Cohen *et al.* 1993). Most species are therefore probably on average older and hence more genetically distinct. This makes the utilization of DNA sequences for molecular phylogenetic work more feasible than in the younger species flocks of Lakes Malawi and Victoria (Meyer *et al.* 1990; Meyer 1993*a*). Questions about the evolution of cichlid fishes can be addressed with molecular phylogenetic techniques which avoid the potential pitfalls caused by parallel evolution of similar morphologies, in the reconstruction of the evolutionary relationships among these fishes (e.g. Meyer *et al.* 1990; Sturmbauer and Meyer 1992, 1993; Klein *et al.* 1993; Kocher *et al.* 1993; Moran and Kornfield 1993; Sturmbauer *et al.* 1994; Sültmann *et al.* 1995).

The great majority of the species of Tanganyika cichlids are confined to the patchy rocky habitats which are discontinuously distributed (Brichard 1989). Most cichlids have restricted geographic distributions within their respective lakes and only a very small number of species are found lake-wide (Fryer and Iles 1972; Brichard 1989; Snoeks *et al.* 1994; Kohda *et al.* 1996). Molecular studies indicate that rock-dwelling cichlid species in Lakes Malawi and Tanganyika are usually strongly subdivided in genetically distinguishable populations (e.g. Sturmbauer and Meyer 1992; Bowers *et al.* 1994; Moran and Kornfield 1995). Their typically high habitat specificity, site fidelity, and low capacity for dispersal are all expected to reduce gene flow between populations and be at least partially responsible for the extensive intralacustrine allopatric speciation in cichlid fishes. The high speciation rates in rock-dwelling cichlids are believed to be the result of intralacustrine speciation caused by both intrinsic (e.g. stenotopy, sexual selection) and extrinsic factors such as vicariant biogeographical processes that restrict gene flow between (micro)allopatric populations (Sturmbauer and Meyer 1992; Meyer 1993*a*; Ribbink 1994; Sturmbauer *et al.* 1997; see Chapter 7).

14.3 Lake basin subdivision and allopatric speciation

Intralacustrine allopatric speciation involving spatial isolation, either by basin subdivision or intralacustrine microallopatric segregation, has been invoked to be the most important mode of speciation in cichlid species flocks (Brooks 1950; Poll 1951; Ribbink 1986; Coulter 1991; Meyer 1993*b*). Yet the occurrence of some locally restricted sister taxa could also indicate sympatric speciation, as suggested for two small cichlid species flocks endemic to crater lakes in Cameroon (Schliewen *et al.* 1994).

Geological evidence indicates that periods of aridity that persisted for several thousands of years have caused dramatic drops in water level—of up to 600 m—that split Lake Tanganyika into three separate lakes approximately 200 000 years ago (Tiercelin and Mondeguer 1991). So far, only a few studies support the hypothesis that lake-wide phylogeographic patterns and possibly the process of speciation in lacustrine cichlids are often associated with such abiotic, historical events as lake level fluctuations (Greenwood 1964; Owen *et al.* 1990; Sturmbauer and Meyer 1992; Rossiter 1995; Sturmbauer *et al.* 1997); for geological data for the timing of speciation see Chapter 8. A recent study of the Tanganyikan rock-dwelling cichlid genus *Tropheus* showed that the amount of genetic differentiation among neighbouring *Tropheus* populations can be either considerable or quite small, depending on the sampling localities in the lake (Sturmbauer and Meyer 1992). The intralacustrine mitochondrial DNA (mtDNA) distribution of haplotypes of *Tropheus*, provides some evidence that major lake level fluctuations may have played a dominant role in determining population genetic structure and possibly speciation in rock-dwelling cichlids of Lake Tanganyika (Sturmbauer and Meyer 1992). However, biological characteristics of species, such as their capacity to disperse, to defend breeding and feeding territories, the size of their broods, and other life-history characteristics may also influence the genetic population structure, and thus be of importance for determining modes of speciation (Fryer and Iles 1972; Meyer *et al.* 1996).

14.4 Testing evolutionary hypotheses on eretmodine cichlids

In order to further test the relative importance of biotic and abiotic factors more comparative phylogeographic data are required. Here, we investigate the variation in the mtDNA control region of cichlids of the tribe Eretmodini (Poll 1986) (Fig. 14.1). This tribe comprises four species, assigned to three genera: *Eretmodus cyanostictus* Boulenger 1898, *Spathodus erythrodon* Boulenger 1900, *Spathodus marlieri* Poll 1950 and *Tanganicodus irsacae* Poll 1950. As a unique feature among lake cichlids, the eretmodines have a reduced swimbladder that allows them to live in the uppermost littoral zone in gravel and rocky shores in the surge zone of Lake Tanganyika. The limited dispersal ability of these cichlids probably caused to the formation of several allopatric colour morphs of these morphologically very similar taxa (Konings 1988; Brichard 1989). The shape of the mandibular teeth is the most important taxonomic character for the Eretmodini and reflects ecology and feeding behaviour of each species. The teeth of *Eretmodus* are spatula shaped with a slender neck region, those of *Spathodus* are cylindrically shaped, and those of *Tanganicodus* are slender and pointed. These differences in dental morphology (e.g. also the position of the mouth and the morphology of the dental arcade) are causally linked to trophic differences, for example *Tanganyicodus* is an invertebrate 'picker' whereas *Eretmodus* and *Spathodus* mainly scrape algae off rocks (Yamaoka *et al.* 1986; Yamaoka 1987).

Populations studied and gene sequenced

A total of 43 specimens from 32 localities were studied (for the importance of such fine-grained sampling see Chapter 5). They were collected during two expeditions in 1991 and 1992 along the Burundian and Tanzanian coastline of Lake Tanganyika (Fig. 14.2). All voucher specimens have been deposited in the Africa Museum at Tervuren (Belgium). The specimens were identified based on Poll (1986). However, it is clear from our study that the taxonomy might need to be modified (Rüber *et al.* in preparation). Altogether 336 base pairs (bp) of the mt control region were determined (see Verheyen *et al.* 1996) (EMBL accession numbers are X90593–X90635). Based on a phylogenetic analysis of these data we examined the evolutionary history of these fishes and attempted to evaluate the importance of major lake level fluctuations on the intralacustrine speciation patterns. Only one 2 bp insertion/deletion event was found. A total of 81 positions (24%) contain variation. Even among the most diverged lineages

Fig. 14.1 Two cichlid fish from Lake Tanganyika: *Eretmodus cyanostictus* from Kapampa, above, and *Spathodus erythrodon* from Masanza, below. (Photos H. H. Büscher.)

within the Eretmodini transitions outnumber transversions, indicating that transitions still contain phylogenetic information (DeSalle *et al.* 1987).

Age estimates for the Eretmodini

The maximum corrected (Kimura 1980) sequence divergences within the Eretmodini were compared with those that had been found within other Lake Tanganyika cichlids (recalculated for the published sequences): the Ectodini, the Lamprologini, and the genus *Tropheus* (Sturmbauer and Meyer 1992, 1993; Sturmbauer *et al.* 1994). The calculated divergences are based on all substitutions and also on transversions only (given in parentheses): Lamprologini, 22.7% (11.0%); Ectodini, 15.7% (7.4%);

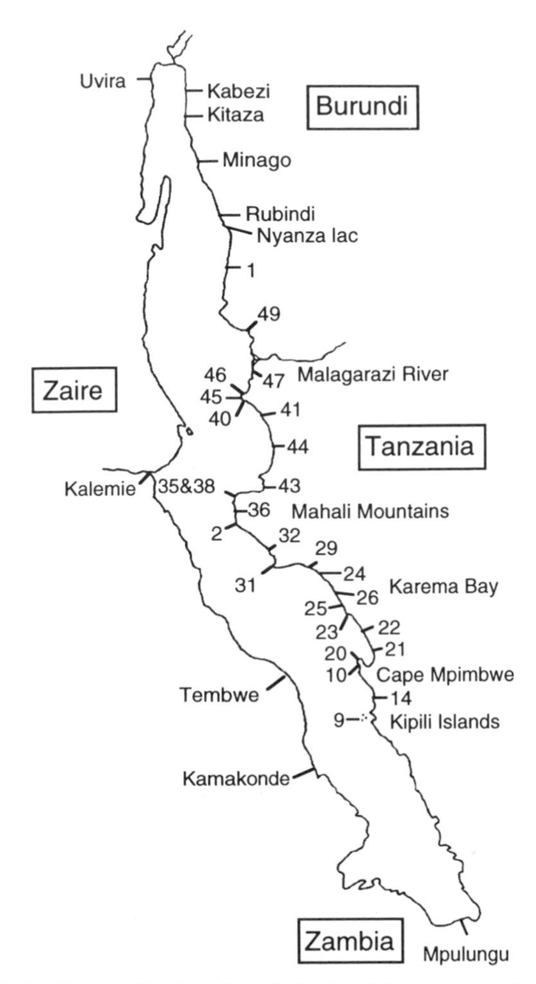

Fig. 14.2 Map of Lake Tanganyika showing all the localities and sample sites mentioned in the text. Lake Tanganyika is about 650 km in length and maximally 80 km wide.

Eretmodini, 12.5% (5.2%); *Tropheus*, 16.1% (4.7%). Assuming a comparable rate of molecular divergence among these lineages, the maximum observed corrected sequence divergence within three other Tanganyikan cichlid lineages indicates that the Eretmodini are approximately 0.5 to 0.8 times as old as the Lamprologini and the Ectodini, and approximately the same age as the genus *Tropheus* (Sturmbauer and Meyer 1992; Sturmbauer *et al.* 1997).

Phylogeny of the Eretmodini

Two major mt lineages within this tribe (A and B in Figs. 14.3 and 14.4) were identified. Lineage A contains three clades A1–A3 whereas lineage B contains four clades B1–B4. The evolutionary relationships within the Eretmodini are corroborated by both parsimony (Swofford 1993) and neighbour-joining (Saitou and Nei 1987) phylogenetic methods, and most branches defining the major clades are supported with high bootstrap confidence (Felsenstein 1985) (Figs. 14.3 and 14.4). Differences between the parsimony and the neighbour-joining method were only observed within clade B (Figs. 14.3 and 14.4). In the parsimony trees subclade B1 is sistergroup to B2+B3+B4, whereas in the neighbour-joining tree B4 is sistergroup to B1+B2+B3. However, the relative positions of the four clades within the B-lineage are not supported by high bootstrap values (Fig. 14.3).

The current generic classification within the Eretmodini is in partial conflict with the mtDNA phylogeny obtained and may be in need of revision (Figs. 14.3 and 14.4). Clades A1 and B4 both contain mtDNA haplotypes from specimens of all three genera. The taxonomic and phylogenetic implications and the evolution of dentitional differences within the Eretmodini will be discussed elsewhere (Rüber *et al.* in preparation).

Sequence divergence and relative age estimates for the eretmodine radiations

Seven genetically distinct clades within two major eretmodine lineages were identified (Figs. 14.3 and 14.4). The estimated genetic divergences within the three clades from lineage A are small and similar (averages about 1.3%), suggesting that each of these radiations originated relatively recently and simultaneously in a 'secondary radiation' (Fig. 14.3). The average corrected sequence divergences between the clades within lineage A are also rather similar (about 3.1% to 3.4%) and also indicate that the A-clades are about of the same age and might have arisen in a 'primary radiation' of the A-lineage. The estimated genetic divergences within clades from lineage B are, however, higher and more variable (means range from 1.3% to 3.2%) (Fig. 14.3) hinting that those clades are of different evolutionary ages and might not have radiated within as short a time span as the clade within the A-lineage. The considerably higher average corrected sequence divergence (4.5% to 7.7%) observed between the B-clades indicate that the B-lineage is considerably older than the A-lineage.

Intralacustrine distribution of mtDNA clades

The three clades (A1–A3) show a non-overlapping phylogeographic pattern along the eastern shore of Lake Tanganyika (Fig. 14.5a; for another example of extensive within-island geographic variation see Chapter 5). Clade A1 is widely distributed and ranges from Burundi to the northern edge of the Mahali mountain area. The distribution of clade A2 is restricted to the Mahali mountain area in the central part of the lake. Clade A3 ranges from the southern edge of the Mahali mountain area to Cape Mpimbwe.

The four clades within the B-lineage also show restricted and non-overlapping geographic distributions (Fig. 14.5b). Clade B1 only contains specimens of *S. erythrodon* and ranges from Burundi to north of the Malagarazi river delta. Clade B2, which consists exclusively of *T. irsacae*, is only found south of the Malagarazi river delta. The southernmost range of the distribution of this clade is the southern edge of the Mahali

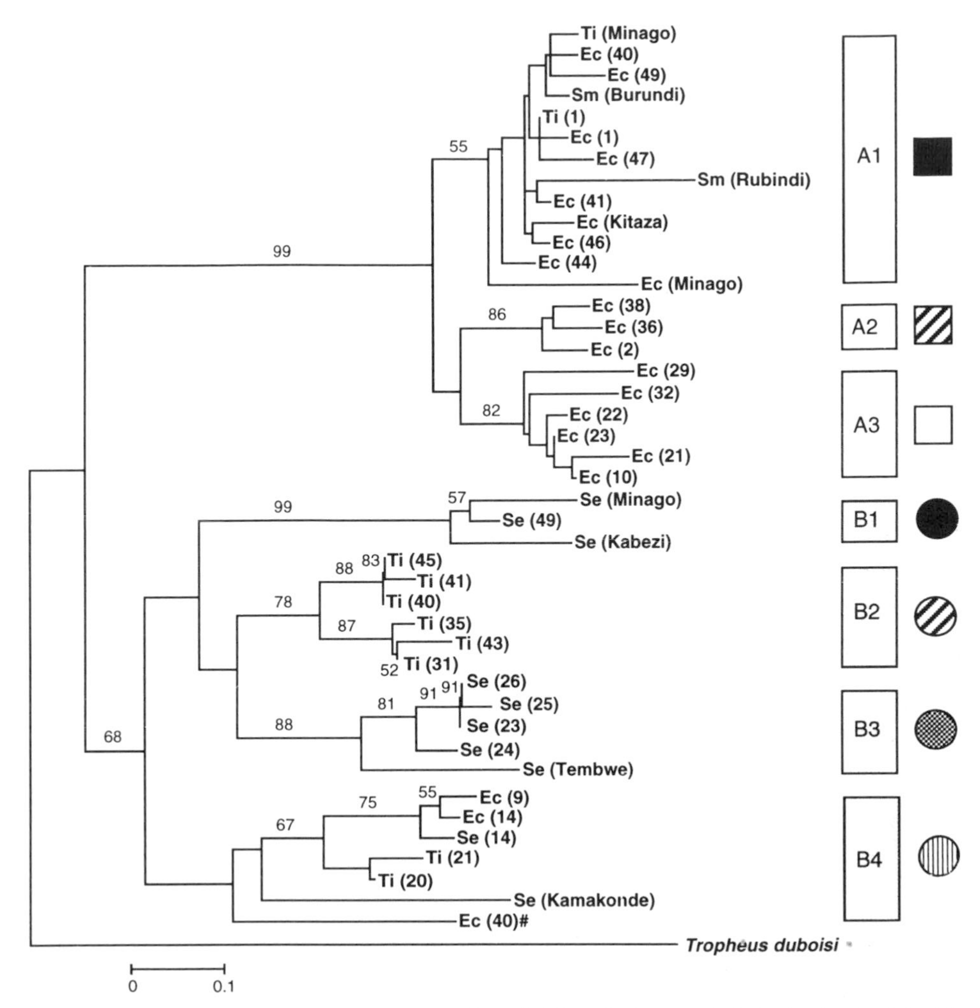

Fig. 14.3 Neighbour-joining tree (Saitou and Nei 1987) of the Eretmodini obtained using MEGA (version 1.01, Kumar *et al.* 1993). *Tropheus duboisi* was declared outgroup based upon a phylogenetic analysis of the major mouthbrooding lineages of Lake Tanganyika (Sturmbauer and Meyer 1993). Genetic distances were corrected for multiple substitutions (Kimura 1980). Gap sites and missing information (insertions and deletions or indels) are ignored in distance estimation; the option 'pairwise-deletion' was used to analyse sequences that contain such sites. Bootstrap values are given on those branches that were obtained in > 50% of the 1000 replications. Branches are drawn to scale, with the bar representing per cent divergence. The species names are given according to the current taxonomic assignments: Ec = *Eretmodus cyanostictus*, Ti = *Tanganicodus irsacae*, Se = *Spathodus erythrodon*, Sm = *Spathodus marlieri*. Locality names and numbers are given in parentheses. Ec (40)# indicates a morphologicaly distinct *Eretmodus* from Ec (40). Clade designations (see text) are based on neighbour-joining and parsimony analyses.

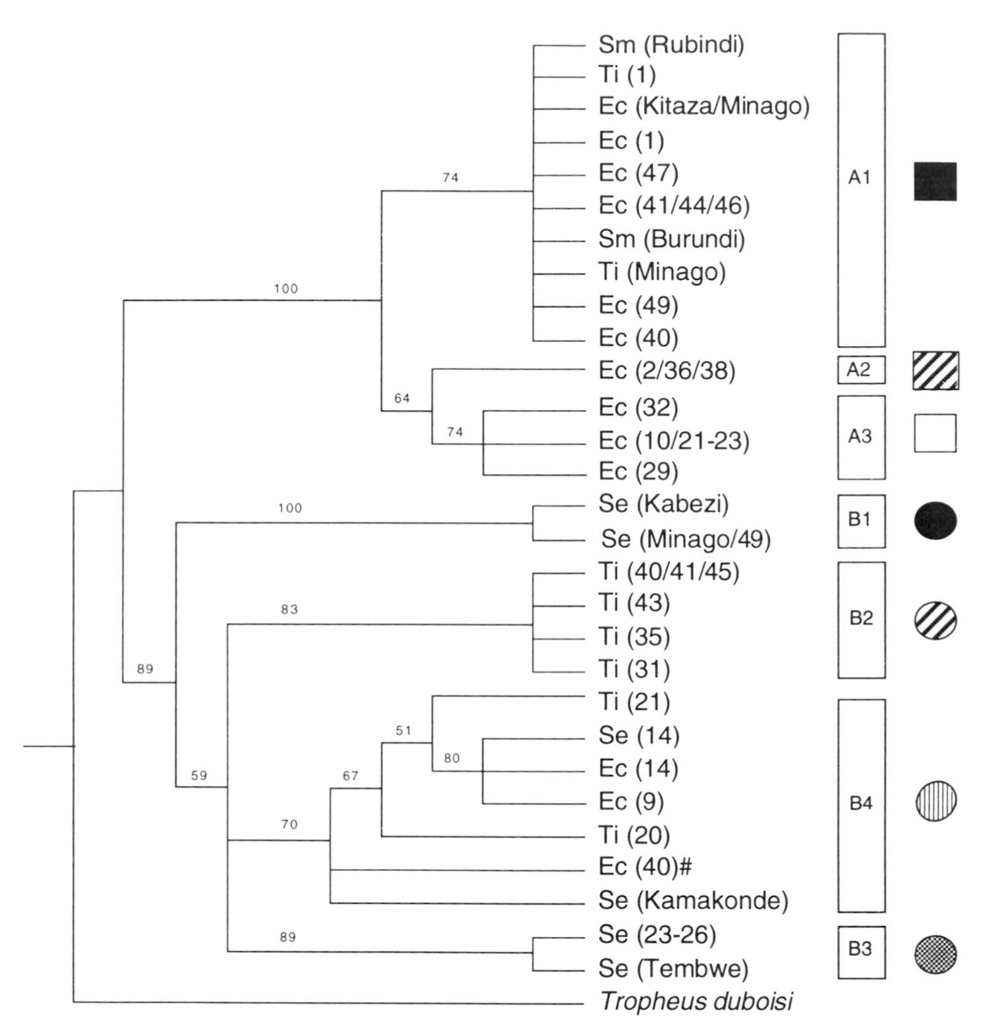

Fig. 14.4 The sequences were analysed by means of the parsimony method using PAUP (version 3.1.1; Swofford 1993). Confidence estimates were obtained using the bootstrap method (Felsenstein 1985). Strict consensus tree constructed from 275 equally parsimonious trees (tree length of 208 steps, consistency index (CI) 0.62, rescaled consistency index (RC) 0.53. Heuristic search with random addition of taxa (10 replications). Bootstrap (Felsenstein 1985) values are given on those branches that were obtained in >50% of the replications (heuristic search, simple addition of taxa, 100 bootstrap replications). The shown bootstrapped parsimony tree was not constructed using all specimens in our data set. Several of the used OTUs are represented by a consensus control region sequences of conspecific eretmodines that show identical cytochrome b sequences (authors' unpublished data). The species names are given according to the current taxonomic assignments: Ec = *Eretmodus cyanostictus*, Ti = *Tanganicodus irsacae*, Se = *Spathodus erythrodon*, Sm = *Spathodus marlier.* Locality names and numbers are given in brackets. Ec (40)# indicates a morphologicaly distinct *Eretmodus* than Ec (40). Clade designations (see text) are based on neighbour-joining and parsimony analyses.

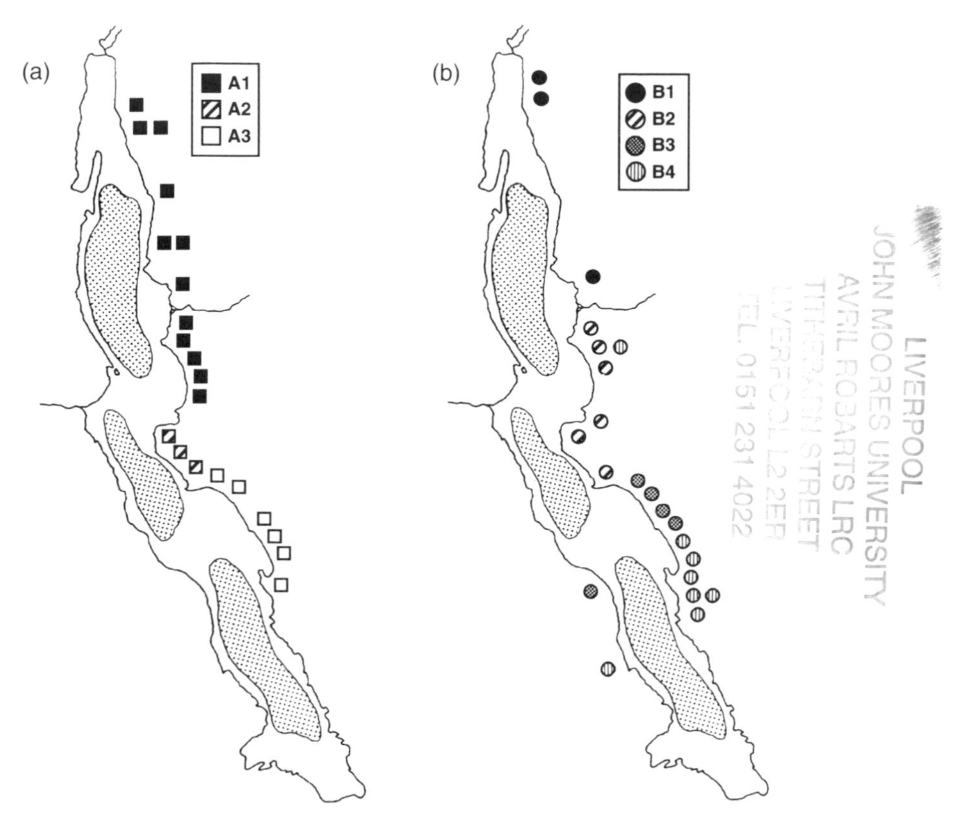

Fig. 14.5 Maps of Lake Tanganyika showing the mtDNA distribution of the studied Eretmodini. The symbols indicate genetically distinct lineages based upon the phylogenetic analyses (see Figs. 14.2 and 14.3). (a) Intralacustrine distribution of mtDNA haplotypes belonging to lineages A1, A2, and A3. (b) Intralacustrine distribution of mtDNA haplotypes that belong to lineages B1, B2, B3, and B4. Each map shows the three separate paleolakes that follow the present 600 m depth contour. The lake level dropped by almost 600 m from its current level (Tercelin and Monteguer 1991). Localities where two identical symbols (see Figs. 14.2 and 14.3) appear (e.g. locality 1) (Figs. 14.1, 14.5a) are meant to indicate that individuals with different tooth morphology and hence different generic assignment have been collected and sequenced.

mountain area. Another clade (B3) containing *S. erythrodon* is found around the Karema Bay and in Tembwe on the opposite western shore line. Clade B4 is morphologically heterogeneous and contains individuals classified as *Spathodus*, *Eretmodus*, and *Tanganicodus*. Its distribution ranges from south of Cape Mpimbwe to the Kipili Islands, the southern end of the sampling area and Kamakonde on the western shore of the lake (Fig. 14.5b).

Specimens that belong to the two most inclusive phylogenetic lineages, lineages A and B, live sympatrically along nearly the whole length of the regions collected. It seems worth pointing out that where the members of the two lineages co-occur they generally have different tooth shapes and hence different trophic ecologies and are also identified as belonging to different genera. The geographic distributions of each clade (e.g. A1 and B1, etc.) between the two lineages are similar (Figs. 14.5a and 14.5b).

14.5 The influence of Lake Tanganyika's history on adaptive radiation

Allopatric speciation, either through geographic isolation or appropriate habitat type within the same water body, seems to be the most important mode of intralacustrine speciation for lacustrine cichlids (reviewed by Meyer 1993*b*; but see Schliewen *et al.* 1994). The age estimates for various endemic Tanganyikan lineages including the Eretmodini suggest that during the geological history of the lake, when the actual rift formation occurred, cichlids rapidly filled the available niches over the entire lake (e.g. Liem and Osse 1975; reviewed in Coulter 1991; Cohen *et al.* 1993; Meyer 1993*b*; Snoeks *et al.* 1994). When much more recent (in the Pleistocene) climatic changes resulted in lowered water levels, the single Lake Tanganyika basin became subdivided into three paleolakes for probably many thousands of years (Scholz and Rosendahl 1988; Tiercelin and Mondeguer 1991). These lake level fluctuations effectively isolated populations of cichlids, including eretmodines, into northern, central, and southern basin populations (Figs. 14.2 and 14.5). It has been suggested that not only basin subdivision but also minor fluctuations in lake level probably influenced the evolution of the littoral cichlid fauna (Fryer and Iles 1972; Coulter 1991). Sandy beaches or estuaries that separate rocky shores are supposed to act as effective barriers to gene flow, thereby influencing the distribution of genetic variation and probably speciation in these fishes (Coulter 1991). In particular, stenotopic and philopatric species like the eretmodine cichlids that seem to be adapted to living in shallow gravel and rocky shores could have been isolated by small continuous changes in lake levels. This results in physical changes of distinct habitats, such as rocky shores, that are patchily distributed along the coastline. This life-history characteristic ties these fishes strongly to their habitat and would seem to make these cichlids particularly sensitive to lake level fluctuations which will affect the availability of habitat and thereby facilitate or inhibit gene flow by creating or destroying rocky habitat. If these periods of isolation existed long enough, genetic differences between populations are likely to arise by drift and thus may or may not result in the formation of new species, depending on whether the degree of differentiation is enough to arise as a reproductive barrier (see Chapter 7). However, as shown earlier, on the basis of intralacustrine species distributions of a number of Lamprologini (Snoeks *et al.* 1994; Kohda *et al.* 1996) even the huge Malagerazi delta (Fig. 14.2) did not seem to be a strict barrier for some taxa (Figs. 14.5a, b); however, these phenotypic data for lamprologine cichlids need to be matched with future genetic data (Verheyen *et al.* in preparation).

14.6 Phylogeographic patterns in Tanganyikan rock-dwelling cichlids

The evolutionary history of the Eretmodini is strongly connected to the geological history of Lake Tanganyika. The data presented here clearly highlight the effects of lake level fluctuations on the present distribution of genetic variation in these cichlid fishes. Indeed, the distribution of recent Eretmodini mtDNA clades matches quite closely the now inundated shorelines of the three intermittant Lake Tanganyika paleolakes (Figs. 14.5a, b). The within-lake distribution of all clades, but in particular those of clades A1, A2, and A3, is restricted to the northern, central, or southern intermittent lake basins which existed about 200 000 to 75 000 years ago (Scholz and Rosendahl 1988). Also the occurrence of genetically distinct clades that show restricted distributional patterns and, furthermore, the presence of closely related populations on both sites of the lake (e.g. *Spathodus* from localities [23–26] and Tembwe, and lineage B from around the Kipili Islands and Kamakonde) suggests that major water level fluctuations in Lake Tanganyika had pronounced effects on the speciation and the distribution of the cichlid fauna from rocky littoral habitats.

Similar to what was observed for the genus *Tropheus* (Sturmbauer and Meyer 1992), but in contrast with findings in *Simochromis* (Meyer *et al.* 1996), our sequence divergence data indicate that at least two consecutive periods of rapid diversification occurred in the evolutionary history of Eretmodini clades (Fig. 14.3). This seems in particular to be the case for the A-lineage (Fig. 14. 3), whereas the ages of the B-clades seem to be varied and are in general older. Hence B-lineage eretmodine cichlids may have originated somewhat earlier and due to different causes than the members of the A mtDNA eretmodine lineage. Just as in *Tropheus* populations (Sturmbauer and Meyer 1992), the Eretmodini populations occurring over some stretches of the Tanganyika coastline appear to be effectively isolated from each other, even if they are separated by only a few kilometres (Figs. 14.3–14.5). However, one mtDNA lineage of Tropheus was found lake-wide (Sturmbauer and Meyer 1992). In contrast to this one lineage of *Tropheus*, all the genetically distinct Eretmodini lineages seem to have a restricted distribution along the rocky littoral coastline of Lake Tanganyika (Fig. 14.5). Our phylogeographic analysis reveals only two cases in which two allopatric populations share an identical haplotype (*Tanganicodus* from localities 40 and 45; *Spathodus* from localities 23 and 26). Since these localities are only about 20 km apart, the amount of gene flow of mtDNA haplotypes might be low even among geographically near populations. Since only a few individuals per locality have been analysed so far, we do not know if these two populations are fixed for these mtDNA haplotyes.

14.7 Morphology based taxonomy versus mtDNA phylogeny of the eretmodines

Morphology based taxonomy places the Eretmodini in their respective genera mainly on the basis of their dental features (Poll 1986). The major mtDNA clades contain mostly morphologically homogeneous groups. Our mtDNA phylogeny is in partial conflict with the current generic classification of the Eretmodini. For example lineage A is mainly constituted of specimens which morphologically correspond to *E. cyanostictus* as defined by Poll (1986). However, clade A1 also contains *T. irsacae* (from localities 1 and Minago) and *S. marlieri* (Rubindi and another locality in Burundi). In addition, unpublished *S. marlieri* sequences (Sturmbauer unpublished results) were also placed within the A1 clade. Clade B4 contains the three genera, as they are currently defined. *Tanganicodus irsacae* are found in several of the mtDNA clades. These *Tanganicodus* can be differentiated on the basis of morphological features. *Tanganicodus* populations found in Minago and locality 1 resemble the fishes from the type locality (Uvira, north Zaire) and can be distinguished from the *Tanganicodus* populations found south of the Malagarazi delta by the presence of a dark spot in the soft-rayed part of the dorsal fin and by their colour pattern. Also the occurrence of two distinct *Eretmodus cyanostictus* mtDNA haplotypes that belong to clades A1–A3 and B4 at locality 40 is supported by morphological characters (Rüber *et al.* in preparation). Since the type locality of *E. cyanostictus* is Kinyamkolo (=Mpulungu) at the southern edge of the lake in Zambia, the *Eretmodus* specimens from localities 9, 14, and 40 may represent the genuine *Eretmodus cyanostictus*. The isolated occurrence of southern genuine *E. cyanostictus* at locality 40 which is separated by more than 200 km from the other members of clade B4 may represent a remnant population of a previously more widespread clade, although a translocation by the aquarium trade cannot be ruled out.

Our data suggest that the species originally assigned to three different genera represent several more genetically and morphologically distinct lineages. Since several genetically distinct lineages are found within each of the studied genera, the conclusion that *Spathodus* and *Tanganicodus* are derived monophyletic lineages with *Eretmodus* as their ancestral sister lineage (Liem 1979), needs to be re-examined. Therefore it is our intention to study other molecular markers as well as morphological characters to establish a phylogenetically based generic classification of the Eretmodini. Our results also suggest that the shape of the oral jaw teeth, which is the main morphological feature used for the present classification of the Eretmodini, may be highly variable, homoplasious, and thus not a reliable character for taxonomic purposes. These differences in the dental morphology (*e.g.* position of the mouth and the morphology of the dental arcade) of these fishes and their relative gut length are related to differences in feeding behaviour (Yamaoka 1985, 1987; Yamaoka *et al.* 1986).

In most localities where two species of the Eretmodini occurred in sympatry these were assigned to one of the two most basal branches within the Eretmodini. Sympatric taxa also seemed to differ morphologically and hence ecologically, pointing to the possibility that ecological diversification may be important for the coexistence of two eretmodine lineages. In the majority of the cases that we examined, these species pairs consisted of one species with a typical *Eretmodus*-like dentition whereas the second taxon is usually characterized by a *Tanganicodus*- or *Spathodus*-like dentition. The phylogenetic and geographic distribution of these dental characteristics might indicate that competition (niche partitioning through competition avoidance during periods of low lake stands) for food between members of the Eretmodini may have been a cause for the multiple occurrence of similar trophic specializations (Rüber *et al.* in preparation; for other examples of parallel evolution see Chapter 10).

MtDNA phylogenies do not necessarily reflect the true species phylogeny; for example lineage sorting and the retention of ancestral polymorphisms can result in the occurrence of mitochondrial poly- and paraphyly between biological species (Moran and Kornfield 1993, 1995). Although this reasoning may be valid for the extremely young mbuna cichlids from Lake Malawi, it is not a likely explanation for the Eretmodini since they are considerably older than the Lake Malawi and Lake Victoria cichlids (Meyer *et al.* 1990; Meyer 1993*b*). Another explanation for the occurrence of mtDNA polymorphisms across species boundaries is introgressive hybridization after secondary contact (e.g. Dowling and DeMarais 1993). Parental-care patterns and mating systems differ between *Spathodus marlieri* and the representatives of the two other genera. However, no relevant information is available on the breeding biology of *Spathodus erythrodon* (Kuwamura *et al.* 1989) and there are no studies that allow us to refute the possibility of introgressive hybridization. To test this hypothesis it will be necessary to investigate nuclear markers and conduct breeding experiments with *Eretmodus*, *Spathodus*, and *Tanganicodus* from the different clades characterized by distinct mtDNA sequences (for examples of breeding experiments see Chapter 10).

14.8 Summary

Geographic patterns of genetic variation reveal a high degree of within-lake endemism among genetically well-separated lineages which are distributed along inferred shore lines of three historically intermittent lake basins. These facts have important implications for taxonomists, and future taxonomic work should account for them. The phylogeographic pattern of eretmodine cichlids suggests that eretmodine cichlids are poor dispersers, and that major and minor lake level fluctuations have been important in shaping the adaptive radiation and speciation in these fishes. The mitochondrially defined clades are in conflict with the current taxonomy of this group of species. Taxonomy needs to be revised in the light of apparently extensive convergent evolution in trophic morphology.

Acknowledgements

We are grateful to the Governments of the Republic of Burundi (Professor Dr Ntakimazi, University of Burundi) and Tanzania (Professor Dr Bwathondi, Tanzanian Fisheries Research Institute, Dar es Salaam) for granting us research permits to collect fishes on Lake Tanganyika. We also thank Heinz H. Büscher for providing us with specimens from Zaire (Tembwe and Kamakonde) and Luc 'Tuur' De Vos for the samples from Rubindi. Jos Snoeks, Thierry Backeljau, Ole Seehausen, and Koen Martens provided helpful comments. During this study, L.R. was supported by an ERASMUS grant, A.M. was supported by National Science Foundation grants (BSR-9119867 and

BSR-9107838) and the Max-Planck-Society, Germany; A.M. and E.V. by a NATO collaboration grant (CRG-910911) and E.V. was funded by grants of the Belgian Fund for Joint Basic Research (F.K.F.O. program 2.0004.90 and F.K.F.O.M.I.30–35).

References

Avise, J. C. (1994). *Molecular markers. Natural history and evolution*. Chapman and Hall, London.

Bowers, N., Stauffer, J. R., and Kocher, T. D. (1994). Intra- and interspecific mitochondrial DNA sequence variation within two species of rock-dwelling cichlids (Teleostei: Cichlidae) from Lake Malawi, Africa. *Molecular Phylogenetics and Evolution*, **3**, 75–82.

Brichard, P. (1989). *Cichlids of Lake Tanganyika*. TFH Publications, Neptune City.

Brooks, J. L. (1950). Speciation in ancient lakes. *Quarterly Review of Biology*, **25**, 30–60, 131–76.

Cohen, A. S., Soreghan, M. J., and Scholz, C. A. (1993). Estimating the age of formation of lakes: an example from Lake Tanganyika, East African Rift system. *Geology*, **21**, 511–14.

Coulter, G. W. (1991). Zoogeography, affinities and evolution with special regard to the fishes. In *Lake Tanganyika and its life*, (ed. G. W. Coulter), pp. 275–305. Oxford University Press.

DeSalle, R., Freedman, T., Prager, E. M., and Wilson, A. C. (1987). Tempo and mode of sequence evolution in mitochondrial mtDNA of Hawaiian *Drosophila*. *Journal of Molecular Evolution*, **26**, 157–64.

Dowling, T. E. and DeMarais, B. D. (1993). Evolutionary significance of introgressive hybridization in cyprinid fishes. *Nature*, **362**, 444–6.

Felsenstein, J. (1985). Confidence limits on phylogenies: an approach using the bootstrap. *Evolution*, **39**, 783–91.

Fryer, G. and Iles, T. D. (1972). *The cichlid fishes of the Great Lakes of Africa: their biology and evolution*. Oliver and Boyd, Edinburgh.

Futuyma, D. J. (1986). *Evolutionary biology*. Sinauer, Sunderland, MA.

Greenwood, P. H. (1964). Explosive evolution in African lakes. *Proceedings of the Royal Institution of Great Britain*, **40**, 256–69.

Kimura, M. (1980). A simple method for estimating the evolutionary rate of base substitutions through comparative studies of nucleotide sequences. *Journal of Molecular Evolution*, **16**, 111–20

Klein, D., Ono, H., O'hUigin, C., Vincek, V., Goldschmidt, T., and Klein, J. (1993). Extensive MHC variability in cichlid fishes of Lake Malawi. *Nature*, **364**, 330–4.

Kocher, T. D., Conroy, J. A., McKaye, K. R., and Stauffer, J. R. (1993). Similar morphologies of cichlid fishes in Lake Tanganyika and Malawi are due to convergence. *Molecular Phylogenetics and Evolution*, **2**, 158–65.

Kohda M., Yanagisawa Y., Sato, T., Nakaya, K., Niimura Y., Matsumoto, K., and Ochi, H. (1996). Geographic colour variation in cichlid fishes at the southern end of Lake Tangayika. *Environmental Biology of Fishes*, **45**, 237–48.

Konings, A. (1988). *Tanganyika cichlids*. Verduijn Cichlids and Lake Fish Movies, Zevenhuizen, Holland.

Kumar, S., Tamura, K., and Nei, M. (1993). *Molecular evolutionary genetics analysis*, (MEGA version 1.01). Pennsylvania State University, University Park, PA.

Kuwamura, T., Nagoshi, M., and Sato, T. (1989). Female-to-male shift of mouthbrooding in a cichlid fish, *Tanganicodus irsacae*, with notes on breeding habits of two related species in Lake Tanganyika. *Environmental Biology of Fishes*, **24**, 187–98.

Liem, K. F. (1979). Modulatory multiplicity in the feeding mechanism in cichlid fishes, as exemplified by the invertebrate pickers of Lake Tanganyika. *Journal of Zoology (London)*, **189**, 93–125.

Liem, K. F. and Osse, J. W. M. (1975). Biological versatility, evolution, and food resource exploitation in African cichlid fishes. *American Zoologist*, **15**, 427–54

Martens, K., Coulter, G., and Goddeeris, B. (1994). Speciation in ancient lakes—40 years after Brooks. *Advances in Limnology*, **44**, 75–96.

Meyer, A. (1993*a*). Evolution of mitochondrial DNA in fishes. In *Biochemistry and molecular biology of fishes*, Vol. 2, (ed. P. W. Hochachka and T. P. Mommsen), pp. 1–38. Elsevier, London.

Meyer, A. (1993*b*). Phylogenetic relationships and evolutionary processes in East African cichlid fishes. *Trends in Ecology and Evolution*, **8**, 279–84.

Meyer, A., Kocher, T. D., Basasibwaki, P., and Wilson, A. C. (1990). Monophyletic origin of Lake Victoria cichlids suggested by mitochondrial DNA sequences. *Nature*, **347**, 550–3.

Meyer, A., Morrissey, J. M., and Schartl, M. (1994). Recent origin of sexually selected trait in *Xiphophorus* fishes inferred from a molecular phylogeny. *Nature*, **368**, 539–42.

Meyer, A., Knowles, L., and Verheyen, E. (1996). Widespread distribution of mitochondrial haplotypes in rock-dwelling cichlid fishes from Lake Tanganyika. *Molecular Ecology*, **5**, 341–50.

Moore, J. E. S. (1903). *The Tanganyika problem*. Hurst and Blackett, London.

Moran, P. and Kornfield, I. (1993). Retention of ancestral polymorphism in the mbuna species flock (Teleostei: Cichlidae) of Lake Malawi. *Molecular Biology and Evolution*, **10**, 1015–29.

Moran, P. and Kornfield, I. (1995). Were population bottlenecks associated with the radiation of the mbuna species flock (Teleostei: Cichlidae) of Lake Malawi? *Molecular Biology and Evolution*, **12**, 1085–93

Owen, R. B., Crossley, R., Johnson, T. C., Tweddle, D., Kornfield, I., Davison, S., Eccles, D. H., and Engstrom, D. E. (1990). Major low levels of Lake Malawi and their implications for speciation rates in cichlid fishes. *Proceedings of the Royal Society of London* B, **240**, 519–53.

Poll, M. (1951). Histoire du peuplement et origine des espèces de la faune ichthyologique du Lac Tanganika. *Annales de la Societe Royal Zoologique de Belgique*, **81**, 111–40

Poll, M. (1986). Classification des Cichlidae du lac Tanganika: Tribus, Genres et Espèces. *Acad. Royale de Belgique-Mém. Sci.* Coll. 8°-2°série, **XLV**, fascicule 2, 1–163.

Ribbink A. J. (1986). Species concept, sibling species and speciation. *Annales du Musée Royal de l'Afrique Centrale (Tervuren), Sciences Zoologiques*, **251**, 109–16.

Ribbink, A. J. (1994). Alternative perspectives on some controversial aspects of cichlid fish speciation. *Advances in Limnology*, **44**, 101–25.

Rossiter, A. (1995). The cichlid fish assemblages of Lake Tanganyika: Ecology, behaviour and evolution of its species flock. *Advances in Ecological Research*, **26**, 187–252.

Saitou, N. and Nei, M. (1987). The neighbor-joining method: a new method for constructing trees. *Molecular Biology and Evolution*, **4**, 406–25.

Schliewen, U., K. Tautz, D., and Pääbo, S. (1994). Sympatric speciation suggested by monophyly of crater lake cichlids. *Nature*, **368**, 629–32.

Scholz, C. A. and Rosendahl, B. R. (1988). Low lake stands in Lake Malawi and Tanganyika, East Africa, delineated with multifold seismic data. *Science*, **240**, 1645–48.

Snoeks, J., Rüber, L., and Verheyen, E. (1994). The Tanganyika problem: taxonomy and distribution patterns of its ichthyofauna. *Advances in Limnology*, **44**, 357–74.

Sturmbauer, C. and Meyer, A. (1992). Genetic divergence, speciation and morphological stasis in a lineage of African cichlid fishes. *Nature*, **358**, 578–81.

Sturmbauer, C. and Meyer, A. (1993). Mitochondrial phylogeny of the endemic mouthbrooding lineages of cichlid fishes from Lake Tanganyika in Eastern Africa. *Molecular Biology and Evolution*, **10**, 751–68.

Sturmbauer, C., Verheyen, E., and Meyer, A. (1994). Mitochondrial phylogeny of the Lamprologini, the major substrate spawning lineage of cichlid fishes from Lake Tanganyika in Eastern Africa. *Molecular Biology and Evolution*, **11**, 691–703.

Sturmbauer, C., Verheyen, E., Rüber, L., and Meyer, A. (1997). Phylogeographic patterns in populations of cichlid fishes from rocky habitats in Lake Tanganyika. In *Molecular systematics of fishes*, (ed. T. Kocher and C. Stepien), pp. 93–107. Academic Press, San Diego, California.

Sültmann, H., Meyer, W. E., Figuero, F., Tichy, H., and Klein, J. (1995). Phylogenetic analysis of cichlid fishes using nuclear DNA markers. *Molecular Biology and Evolution*, **12**, 1033–47.

Swofford, D. L. (1993). *Phylogenetic analysis using parsimony* (PAUP version 3.1.1). Illinois Natural History Survey, Champaign, IL.

Tiercelin, J. J. and Mondeguer, A. (1991). The geology of the Tanganyika Trough. In *Lake Tanganyika and its life*, (ed. G. W. Coulter), pp. 7–48. Oxford University Press.

Verheyen, E., Rüber, L., Snoeks, J., and Meyer, A. (1996). Mitochondrial phylogeography of rock-dwelling cichlid fishes reveals evolutionary influence of historical lake level fluctuations of Lake Tanganyika, Africa. *Philosophical Transactions of the Royal Society London* B, **351**, 797–805

Yamaoka, K. (1985). Intestinal coiling pattern in the epilithic algal-feeding cichlids (Pisces, Teleostei) of Lake Tanganyika, and its phylogenetic significance. *Zoological Journal of the Linnean Society*, **84**, 235–61.

Yamaoka, K. (1987). Comparative osteology of the jaw of algal-feeding cichlids (Pisces, Teleostei) from Lake Tanganyika. *Reports of the Usa Marine Biological Institute, Kochi University*, **9**, 87–137.

Yamaoka, K., Hori, M., and Kuratani, S. (1986). Ecomorphology of feeding in goby-like cichlid fishes in Lake Tanganyika. *Physiology and Ecology Japan*, **23**, 17–29.

15

Islands in Amazonia

Ghillean T. Prance

15.1 Introduction

It may seem surprising to have a chapter about the Amazon basin, a continental area *par excellence*, in a book about evolution on islands. In fact there are hundreds of islands within Amazonia varying from Marajó Island in the Amazon delta, the size of Switzerland, to tiny islands that come and go all along the principal rivers. These islands are not of particular interest in the evolution of species, although recent work has shown their importance in creating a succession of many different types of flood plain vegetation types (Tuomisto *et al.* 1995), and thereby increasing the vegetational diversity of the region. I am not, however, going to treat the actual river islands, but rather insular patches of vegetation that are in situations which are comparable to islands within the forested regions of Amazonia. In the lowlands there are many areas of savanna and of white sand caatinga or campina that are in effect small isolated island habitats of a different type of vegetation surrounded by rainforest. Where higher ground exists in Amazonia, principally on the Guiana shield to the north and the Brazilian shield to the south, there are isolated granite peaks (inselbergs) and a large number of sandstone table mountains (tepuis), the tops of which are also isolated islands of vegetation. It is these four areas each with their own vegetation type that are discussed here.

The vegetational cover of Amazonia has not been stable through the geological time scale and variations in climate have resulted in variations in the balance between the drier types of vegetation such as savanna, deciduous forest, and caatinga and the wetter tropical moist forest. At times the current situation has been reversed and patches of rainforest have been islands within savanna or semi-deciduous forest (Haffer 1969; Prance 1973, 1983; Van der Hammen 1974; Thomson *et al.* 1995). These changes have all had a profound effect on the evolution and distribution of plant species. This too has created islands of rainforest surrounded by savanna, and other types of vegetation adapted to a drier climate acted as refugia for the species of the more humid rainforest vegetation.

It is, therefore, worthwhile considering the vegetation of Amazonia in terms of islands as well as in terms of being the largest continental mass of rainforest. These continental islands have certainly increased the total diversity of the region because they offer different habitats into which adaptation has caused the evolution of new species.

15.2 Amazonian savannas

The Amazon basin is bordered by the two major savanna regions of South America the Llanos of Colombia and Venezuela to the north and the cerrado of the Planalto of central Brazil to the south. However, within the rainforest region there are a considerable number of smaller isolated savannas and some quite large ones such as the Roraima–Rupununi savanna of Roraima Territory, Brazil and the Rupununi district of Guyana, and the Llanos de Mojos in Bolivia. (Figs. 15.1 and 15.2).

The savanna islands within Amazonia have been known and discussed for many years (for example Huber 1900, 1902; Ducke 1907, 1913) but only more recently have sufficient phytosociological studies been made to compare the distribution of their component species (for example Takeuchi 1960; Medina 1969; Eiten 1972; Eden 1974; Huber 1982; Gottsberger and Morawetz 1986). Extensive work has been carried out on the cerrados of central Brazil (for example Eiten 1972, 1982; Prance and Schaller 1982; Ratter and Dargie 1992, Guarim Neto *et al.* 1994; Ratter *et al.* 1996). Although there are some species of plants common to almost all the South American savannas from Mexico to Paraguay, such as *Curatella americana* L. and *Roupala montana* Aubl., most workers agree that the Amazonian savanna islands are considerably different from those of the cerrado and the llanos (Eiten 1978; Kubitzki 1979, 1983). Huber (1982) divided the savannas of Amazonas Territory in Venezuela into three distinct types:

1. Savanna in seasonally flooded areas where the absence of woody vegetation is clearly due to edaphic factors. The large flooded savannas of Amapá and Marajó Island in Brazil also fall into this category although they are quite different in their species composition.

2. Savannas on white sand soil, with extremely poor nutrient availability and rapid drainage. In these savannas there is a considerable amount of endemism and Huber (1982) suggested that these are 'highly selected and specialised plant communities surviving best on these sites where they evolved'. These are different from the campinas and Amazonian caatingas discussed below.

3. The llanos-type savanna which has a more or less continuous grass layer and a sparse covering of trees and shrubs. Unlike the white-sand savannas endemism is rare and fire is a common feature. Many of the smaller savanna patches within Amazonia fall into this category.

The existence of the first two categories of savanna is easy to explain, the environmental factors of flooding and of white sand control their existence. They are likely to remain savanna even during periods of the most favourable climate for rainforests. In the case of the white-sand savannas it is not surprising that there is a high degree of endemism because many white-sand specialist species appear to remain on white sand even when there is more savanna on clay soil surrounding them. Once adapted to white sand and its specific low nutrient availability these species do not readily migrate to the richer clay soil habitats.

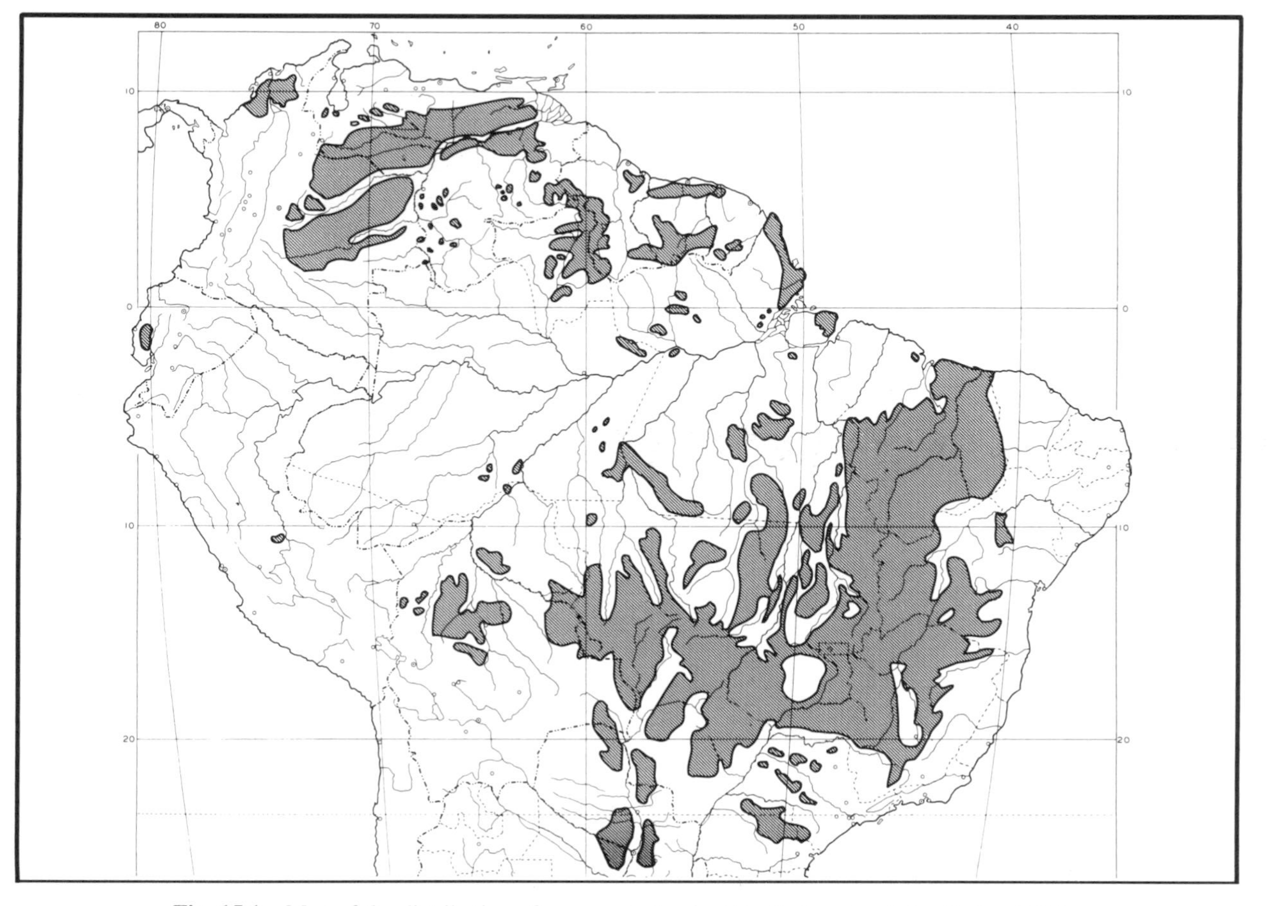

Fig. 15.1 Map of the distribution of savanna in tropical South America (after Huber 1974).

Fig. 15.2 Top, an Amazonian savanna island at Humaitá, Amazonas one of the larger areas of savanna; bottom, general aspect of a white-sand campina near to Manaus, Amazonas.

Table 15.1 Some wide-ranging neotropical savanna species

Species	Family	Distribution
Arrabidaea corallina (Jacq.) Sandw.	Bignoniaceae	Mexico to Argentina
Bowdichia virgilioides H.B.K.	Leguminosae	Venezuela to S. Brazil
Byrsonima crassifolia (L.) H.B.K.	Malpighiaceae	Cuba to S. Brazil
Byrsonima verbascifolia Juss.	Malpighiaceae	Cuba to S. Brazil
Curatella americana L.	Dilleniaceae	Mexico to Paraguay
Hirtella glandulosa Spreng.	Chrysobalanaceae	Venezuela to São Paulo
Palicourea rigida H.B.K.	Rubiaceae	Colombia to São Paulo
Polystachya estrellensis Rehb.f.	Orchidaceae	Trinidad to Paraguay
Roupala montana Aubl.	Proteaceae	Mexico to Brazil
Xylopia aromatica (Lam.) Mart.	Annonaceae	Panama, Cuba to Paraguay

The non-flooded grassland savannas which occur scattered throughout Amazonia (llanos-type of Huber (1982)) occur on both sandy and lateritic soils and it is notable that there is no difference between the soil under savanna vegetation and the adjacent forest areas (Eden 1974; Blancaneaux *et al.* 1977; Prance unpublished studies). These savannas are probably relicts of formerly widespread vegetation that developed during the drier climatic conditions of the late Pleistocene (Ducke and Black 1953; Prance 1973; Eden 1974; Huber 1982). That so many of these patches of savanna remain and did not return to a rainforest cover during the Holocene is probably partially due to human intervention. Early hunter-gatherer peoples used repeated burning of natural islands of savanna to flush game and to kill small animals, thereby maintaining or even expanding the savannas (Aubréville 1961; Hills 1969; Rivière 1972; Smith 1995). The occurrence of a large number of widespread species that are in most Amazonian savannas and beyond supports the theory that these savanna patches are relicts of a more extensive savanna (Table 15.1).

As would be expected the northern Amazonian savannas show a closer link to the llanos and the southern ones a closer link to the cerrados. For example, the Humaitá–Puciari savannas of Amazonia, Brazil contain such cerrado species as *Orthopappus angustifolius* (Sw.) Gleason, *Cyperus diffusus* Vahl, *Axonopus aureus* Beauv., *Vochysia haenkeana* Mart., *Didymopanax distractiflorum* Harms (Gottsberger and Morawetz 1986), *Keilmeyera coriacea* Mart. (Kubitzki 1979; Janssen 1986), and *Hirtella ciliata* Mart. (Prance 1972; Fig. 15.3). Various northern species extend southwards through the savannas. For example, the Savanna de Amelia on the south bank of the Rio Negro near Manas contains such species as *Ouratea spruceana* Engl. (Ochnaceae); *Ruizterrania retusa* (Spruce ex Warm.) Marcano Berti (Vochysiaceae). It is also interesting that this most isolated savanna is dominated by the ubiquitous savanna species *Antonia ovata* Pohl (Asteraceae).

There is also little morphological differentiation between populations of the common savanna species in different savanna islands which is an indication of more recent origin. In the widespread savanna species which I have studied such as *Hirtella ciliata*

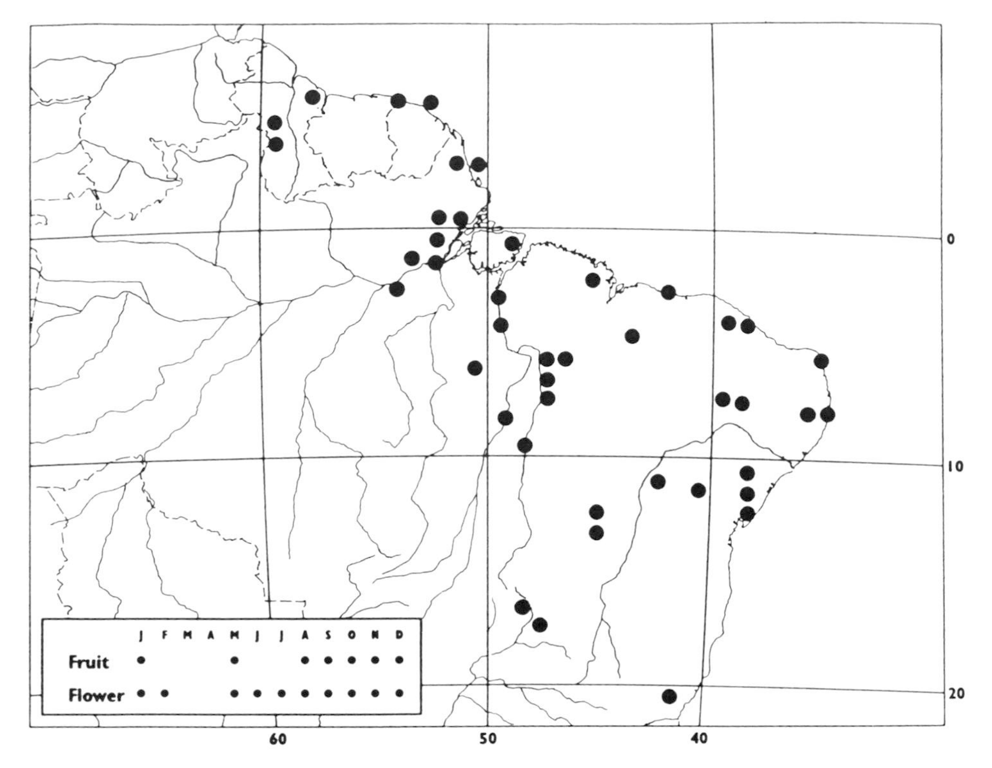

Fig. 15.3 Distribution of *Hirtella ciliata* Mart., a widespread savanna species.

Mart. and Zucc. (Fig. 15.3) (Chrysobalanaceae) no morphological differences can be found in individuals in such isolated places as the Humaitá Savanna or the Savanna de Amelia near to Manaus. The only example of taxonomic differentation between patches of savanna that has been published is that of Gentry in the Bignoniaceae occurring in the savannas of Peru (Gentry 1979, 1982). Gentry studied the species of Bignoniaceae that occur in isolated patches of dry forest along the Andes. Some species such as *Tabebuia impetiginosa* (Mart. ex DC.) Standl. showed no differentation between isolated populations; others, for example *Tabebuia ochracea* (Cham.) Standl., were divided into taxonomic subspecies because of the variation in different patches of dry forest. Two species of *Tecoma* seem to be in an intermediate situation where the populations in adjacent Andean valleys are perceptably different, but 'their taxonomic circumscriptions and states remain murky' (Gentry 1979). This is exactly what one might expect in more recently separated populations where evolution towards taxonomic isolation is actively progressing. This topic of contemporary refugia would make a very interesting molecular study where differences between populations might be revealed.

Many savanna species are easily dispersed (Kubitzki 1979, 1983) and this explains their ability to migrate rapidly during times of drier climate. Kubitzki showed that several cerrado species with large diaspores such as *Caryocar brasiliense* Camb. do not occur in most savannas.The only isolated savanna in which I have found that species is at Serra do Cachimbo in Pará, which is only separated from the cerrado by about 200 km of forest and was evidently connected to the cerrado during periods of drier climate.

15.3 Campinas and Amazonian caatingas

These two Brazilian terms are applied to areas of vegetation over white sand which differ considerably in their physiognomy from the surrounding rainforest on oxisols. The Amazonian caatinga is a large area of white sand which lies between the Rio Branco and the Rio Negro in north-western Amazonian Brazil and which extends into Venezuela. The campinas are similar 'island' patches of white-sand vegetation throughout Amazonia and are equivalent to the heath forests of tropical Asia. The large area of caatinga of north-western Amazonia is formed on the sand leached from the sandstone highlands of Venezuela and most of the campinas are formed from uplifted former river beaches. The effect of both is the same, a nutrient-poor sandy soil (spodsol). Caatingas have been described by Spruce (1908), Ducke and Black (1953), Rodrigues (1961), Takeuchi (1962), Klinge *et al.* (1977), and Klinge and Medina (1979) and Campinas by Takeuchi (1961), Anderson *et al.* (1975), and Anderson (1981). The caatinga and the campinas have some species in common (e.g. *Glycoxylon inophyllum* (Miq.) Ducke, *Humiria balsamifera* St. Hil., and *Lissocarpa benthami* Guerke), but are noteably different in species composition. Some of the characteristic species of the caatinga are given in Table 15.2.

The Amazonian white-sand caatingas should not be confused with the caatingas of north-eastern Brazil where the same term is used for semi-arid scrubland. The Amazonian caatinga is dominated by species of Leguminosae, especially the genus *Eperua*. A similar white-sand vegetation called Wallaba forest occurs in the Guianas (Richards 1952, 1996), and is dominated by *Eperua falcata* Aubl. (Wallaba). In spite of dominance by two or three species, both the caatinga and the Wallaba forest is rich in tolul species and in endemic species (Ducke and Black 1953) most of which are from the common genera of the Amazonian rainforest (for example, *Caryocar gracile* Wittm. and *Couepia racemosa* Benth.). The poor soil and the low sclerophyllous vegetation of the caatinga is a separate formation, and species have adapted to it and separated through niche specialization. The caatinga formation can be defined both by its characteristic sclerophyllous vegetation with tortuous branched trees and by the presence of an extremely large number of endemic species.

More island-like are the smaller patches of campina within the rainforest. These small areas have a physiognomy similar to that of the caatinga and vary from a continuous forest cover to open areas with scattered shrubs. As would be expected from this small area, these areas have less endemism than the large caatinga area, but many of the species which occur are those which are confined to the white-sand areas. An analysis of a campina near to Manaus (Macedo and Prance 1978) showed that the majority of species (75.5%) have the capacity for long-distance dispersal (59.5% birds, 13.5% wind, 2.5% bats) which accounts for the related low endemism in campinas. The species are mainly either those occurring in the caatingas, the sandy black-water river beaches, or the white-sand forests of the Guianas. Their vagility has tended to reduce

Table 15.2 Some characteristic woody species of Amazonian caatinga, Amazonian Campina,and Wallaba forest

Caatinga	Campina	Wallaba forest
Aldina discolor Spruce ex Benth. (Leguminosae)	*Aldina heterophylla* Spruce ex Benth.(Leguminosae)	*Aspidosperma excelsum* Benth. (Apocynaceae)
Aspidosperma album (Vahl) Benoist (Apocynaceae)		*Catostemma fragrans* Benth. (Bombacaceae)
Bactris cuspidata Mart.(Arecaceae)	*Clusia insignis* Mart. (Clusiaceae)	*Dicymbe corymbosa* Benth. (Leguminosae)
Barcela odora Mart. (Arecaceae)	*C. columnaris* Engl.	*D. altsoni* Sandw.
Carocar gracile Wittm. (Caryocaraceae)	*Eperua purpurea* Benth. (Leguminosae)	*Dimorphandra conjugata* (Splitg.) Sandw. (Leguminosae)
Clusia insignis Mart. (Clusiaceae)	*Humiria balsamifera* St. Hil. (Humiriaceae)	*D. davisii* Sprague and Sandw.
Compsoneura debilis (A.DC.) Warb. (Myristicaceae)	*Macrolobium arenarium* Ducke (Leguminosae)	*Eperua falcata* Aubl. (Leguminosae)
Eperua leucantha Benth. (Leguminosae)	*Pagamea duckei* Standl. (Rubiaceae)	*E. grandiflora* (Aubl.) Benth.
E. purpurea Benth.	*Protium heptaphyllum* (Aubl.) March (Burseraceae)	*Licania buxifolia* Sandw. (Chrysobalanaceae)
E. rubiginosa Miq.	*Sandemania hoehnei* (Cogn.) Wurdack (Melastomataceae)	
Hevea rigidifolia (Benth.) M.Arg. Euphorbiaceae)	*Swartzia dolichopoda* Cowan (Leguminosae)	
Lissocarpa benthamii Guerke (Ebenaceae)		
Lucuma sp. (Sapotaceae)		
Micrandra crassipes (M.Arg.) R.E. Schult. (Euphorbiaceae)		
Micrandra crassipes (M.Arg.) R.E. Schult. (Euphorbiaceae)		
M. spruceana (M.Arg.) R.E. Schult.		
Peltogyne catingae Ducke (Leguminosae)		
Retiniphyllum chloranthum Ducke (Rubiaceae)		

both endemism and the evolution of new species in each of these isolated islands of campina.

There are a large number of white-sand savannas and savanna forests in the Guianas. Although termed savannas there, they are much more comparable to the white-sand campinas and caatingas than to the grass-dominated savannas. A good description of an area of white-sand vegetation in Surinam was given by Heyligers (1963).

That the different isolated areas of white-sand vegetation have given the opportunity for radiation of certain genera is seen from the data in Table 15.2 which lists some of the dominant tree species of caatinga, campina, and wallaba forest. Such woody genera as *Aspidosperma* and *Eperua* have speciated into each of the major white-sand areas.

15.4 Inselbergs

Scattered throughout the Guianas and northern Amazonian Brazil there are a series of isolated granitic outcrops which rise above the rainforest to a height of 300–800 m. In some cases these hills are covered by dense forest (e.g. Palunlouiméempeu and Mitraka), but most of them are characteristic inselbergs (Grabert 1976) with low scrub forest and open areas of exposed rocks (Granville and Sastre 1974; Hurault 1974; Sastre 1977). These are obviously islands of a special type of vegetation surrounded by rainforest. The top of the inselbergs are well drained and become very dry in the dry season so that the vegetation is often dominated by sclerophyllous plants or any with other adaptations to drought such as the orchid *Cyrtopodium andersonii* R. Brown with large pseudobulbs that store water or various cacti (*Epiphyllum* and *Melocactus*). The dominant shrub is usually a species of *Clusia* and other terrestrial orchids include *Epidendrum nocturnum* Jaq. and *Encyclia ionosma* (Lindl.) Schlecht.; the Bromeliad *Pitcairnia geyskesii* L. B. Smith is also common. Sastre (1977) found that about 55% of the plant species on the inselbergs of the Guianas are confined to that habitat. Unlike the white-sand areas it is not generally the genera of lowland rainforest that have adapted to this habitat, although a few epiphytic genera such as *Pitcairnia* are found there. Many of the genera of the inselbergs have their origin in the more arid regions of the neotropics. There has been considerable speciation in adaptation to the summit of inselbergs with their arid dry season conditions and very humid rainy season conditions. However, in contrast to tepuis, there has been much less speciation between the different inselbergs.

15.5 Tepuis

Tepui is the Venezuelan term for the sandstone mountains of the Guayana Highland (Fig. 15.4). They extend from the easternmost tepui Tafelberg in Surinam through Guyana and Venezuela into Amazonian Colombia and south just into Brazil. Mount Roraima is on the corner of Venezuela, Guyana, and Brazil and two tepuis occur within the Brazilian Amazon, Tepequém and Araca. The highest one is Mount Neblina (3045 m) on the Brazil–Venezuela frontier. The lowland area between the tepuis is either rainforest

Fig. 15.4 Top, view of a typical tepui, Kirun-tepui near to Canaima, Venezuela; bottom, the plateau vegetation of Serra Araca a Brazilian tepui with a granitic dome in the background.

or savanna, depending on the locality. These spectacular mountains which tower above the surrounding forest are from 800–3000 m and are islands of montane vegetation surrounded by lowland forest.

The Guyana Highlands have now been well studied, largely through the work of Bassett Maguire of the New York Botanical Garden (Maguire 1953–84,1956, 1970), Julian Steyermark (1967, 1974, 1979, 1984, 1987), and Otto Huber (1987, 1990), and the *Flora* of which the first two volumes have recently been published (Steyermark *et al.* 1995). This work has demonstrated that there is a great deal of endemism on the tepuis and that many less vagile genera have speciated extensively through the isolation of one tepui from another. Maguire (1970) almost certainly overestimated the level of endemicity on the tepuis when he estimated it as between 90 and 95% of the species, and this was challenged by Steyermark (1979) who showed that 39 out of the 459 genera known to occur on the summits of tepuis (8.5%) were endemic to the summits. However, as an additional 40 genera are found mainly on the summits but also occur on the lower talus slopes of the surrounding lower altitudes of contiguous areas, making a total of 79 or 17.2% of the genera largely confined to tepuis. That differentiation has taken place to such an extent at the generic level as well as the specific level would argue for the tepuis being ancient formations rather than recent upheavals of the Neogene or Pleistocene as argued by Kubitzki (1989). Both Steyermark (1979) and Maguire (1956) argue from an analysis of summit species of the tepuis that dispersal was centripetal rather than centrifugal, i.e. migration has taken place from the lowlands to the highlands rather than the reverse.

Table 15.3 confirms that there is considerable endemism on tepuis; the table lists the 16 new species that were described as a result of the author's two expeditions to Serra Araca, a rather low (1000–1400 m) outlier of Guayana Highland well into Brazil. This

Table 15.3 New species and subspecies from Serra Araca, a sandstone tepui in Brazil, from two expeditions there by the author (see Prance and Johnson 1992 for further details)

Caraipa aracaensis Kubitzki (Clusiaceae)
Diacidia aracaensis Anderson (Malpighiaceae)
Gleasonia prancei Boom (Rubiaceae)
Gongylolepis oblanceolata Pruski (Asteraceae)
Licania aracaensis Prance (Chrysobalanaceae)
Meriana aracaensis Wurdack (Melastomataceae)
Pitcairnia pranceana L.B. Smith (Bromeliaceae)
Podocarpus aracaensis De Laubenfels & Silba (Podocarpaceae)
Raveniopsis aracaensis Kallunki & Steyerm. (Rutaceae)
Stenopadus aracaensis Pruski (Asteraceae)
Styrax tepuiensis Steyerm. & Maguire subsp. nov. (Styracaceae)
Tepuianthus aracaensis Steyerm. (Tepuianthaceae)
Ternstroemia aracae Boom (Theaceae)
Ternstroemia prancei Boom (Theaceae)
Tetrapterys cordifolia Anderson (Malpighiaceae)
Vaccinium pipolyi Luteyn (Ericaceae)
Xyris brachyfolia Kral & Wanderley (Xyridaceae)

Table 15.4 The Phytogeographic affinities of species of Serra Araca plateau flora which are not endemic with a few examples of each category (from Prance and Johnson 1992, where further details are given)

Phytogeographic affinty	Example
Widespread	*Cyrilla racemiflora* L. (Cyrillaceae)
	Viburnum tinoides L. f. (Rosaceae)
	Burmannia bicolor Mart. (Burmanniaceae)
Lowland Amazonia	*Pochota amazonica* (Robyns) (Bombacaceae)
	Acmanthera parviflora W. Anderson (Malpighiaceae)
Confined to Guyana Highland but widespread	*Cyathea demissa* (C. Morton) A.R. Smith ex Lel. (Cyatheaceae)
	Ouratea roraimae Engl. (Ochnaceae)
	Schefflera duidae Steyerm. (Araliaceae)
Restricted Guyana Highland	
1. Neblina Massif	*Gongylolepis oblanceolata* Pruski (Asteraceae)
2. Duida Complex	*Phyllanthus neblinae* Jabl. (Euphorbiaceae)
	Remijia maguirei Steyerm. (Rubiaceae)
3. Roraima Complex	*Pagamea montana* Gleason & Standl. (Rubiaceae)
	Sipaneopsis maguirei Steyerm. (Rubiaceae)
	Stegolepis membranacea Maguire (Rapateaceae)
	Symbolanthus elisabethae (Schomb.) Gilg (Gentianaceae)
	Bactris simplicifrons Mart. (Arecaceae)

tepui was little explored until the expeditions of 1985 and 1986 and so the endemics were only recently described. Like other tepuis, Araca also contains a large number of more widespread species characteristic of the Guayana Highland such as *Oedematopus duidae* Gleason, *Saxofridericia spongiosa* Maguire, *Retiniphyllum scabrum* Benth., *Perissocarpa steyermarkii* (Maguire) Steyerm. & Maguire, and *Macairea duidae* Gleason. The phytogeographic affinities of the species that are not endemic to Araca are given in Table 15.4 with a few selected examples. It can be seen that the affinity is strongly within the Guayana Highland with a few more widespread and lowland Amazonian species also occurring there.

There are many genera that have differentiated at the species or subspecies level from one tepui to another. Details of two such genera which also occur in the lowland savannas, *Tepuianthus* (Tepuianthaceae a Guayana endemic family) and *Heliamphora* (Sarraceniaceae) are given in Table 15.5. It can be seen that in both genera there are several species confined to one tepui. There is sufficient isolation for speciation to have taken place. The higher tepuis such as Sierra de la Neblina (over 3000 m) have a greater number of endemic species. Neblina even has an endemic family, the insectivorous

Table 15.5 The species of *Tepuianthus* (Tepuianthaceae) and *Heliamplora* (Sarraceniaceae) showing the distribution which is largely on summits of tepuis, in Venezuela unless otherwise indicated

Species	Locality
Tepuianthus auyantepuiensis Mag. & Steyerm.	Auyán-tepui
T. yapacanensis Mag. & Steyerm.	Cerro Yapacana
T. savannensis Mag. & Steyerm.	White sand savannas of SW Amazonas, Venezuela
T. colombianus Mag. & Steyerm.	Cerro Isibukuri (Colombia)
T. sarisariñamensis Mag. & Steyerm	
subsp. *sarisariñamensis*	Sarisariñama tepui
subsp. *duidensis*	Cerro Duida
T. aracaensis Steyerm.	Serra Araca (Brazil)
Heliamphora nutans Benth.	Mount Roraima to Ilu-tepui
H. minor Gleason	Auyán-tepui and Chimantá-tepui
H. heterodoxa Steyerm.	Eastern Gran Sabana from Ptari-tepui northward
H. ionasi Maguire	Ilu-tepui
H. tatei Gleason	Cerro Duida
H. neblinae Maguire	Cerro Neblina

Saccifoliaceae. The tepuis can certainly be treated as islands in terms of phytogeographic analysis.

Another reason for the species diversity of the tepuis is the large range of habitat which they offer. Steyermark (1979) recognized five major habitats on the summits of tepuis: forest associations on soils; forest associations of epiphytes; shaded bluffs, ledges and crevices; savannas; exposed rocks, sands, and rocky open areas. There is also a lot of scrubland on top of many tepuis and the habitats can obviously be much more finely divided. This accounts for some genera which have more than one species on a single tepui. The lower mountains have a considerable amount of forest on their summits. For example, Steyermark and Dunsterville (1980) showed that Cerro Guaiquinima, one of the largest tepuis, has 40% of the summit covered by forest. The species in this forest are generally closely related to those of the surrounding lowlands.

15.6 Radiation of forest groups into vegetational islands

Many of the species which occupy the various types of vegetational islands described above, particularly the lowland ones, belong to predominantly forest families and genera of plants and these have adapted and radiated into savanna, white-sand, and to a lesser extent to montane habitats. For example, the genus *Carocar* (Caryocaraceae) is predominantly a forest group with rainforest species. However, *C. brasiliense* Camb. is a common species of the Brazilian cerrados, *C. cuneatum* Wittm. is a species of the arid

Table 15.6 Some species of the predominantly lowland rainforest genus *Licania* (Chrysobalanaceae) which have radiated into different vegetation types

Habitat	Species
White-sand caatinga and campina	*L. boyanii* Tutin
	L. buxifolia Sandw. (Wallaba forest)
	L. crassivenia Spruce ex Hook.f.
	L. joseramosii Prance
	L. stewardii Prance
Amazonian lowland Savanna isands	*L. cordata* Prance
	L. foldatsii Prance
	L. lanceolata Prance
	L. hebantha Mart. ex Hook.f.
Tepuis	*L. aracaensis* Prance
	L. pakaraimensis Prance
	L. tepuiensis Prance
Cerrado of Central Brazil	*L. dealbata* Hook.f. (geoxylic suffrutex)
	L. humilis Cham. & Schlecht. (thick corky bark)
	L. nitida Hook.f.
Widespread throughout northern savannas	*L. incana* Aubl
Caatinga of north-east Brazil	*L. rigida* Benth.

caatinga of north-eastern Brazil, *C. gracile* Wittm. is a species of the Amazonian white-sand caatingas of the upper Rio Negro region, and *C. montanum* Prance and M. F. Silva is a montane species found on the slopes of some Venezuelan tepuis. Table 15.6 shows species of the large predominantly rainforest genus (205 species) *Licania* (Chrysobalanaceae) which has radiated into some of the island habitats discussed here.

15.7 Rainforest islands in the history of Amazonia

Today's Amazonian landscape is a forest with small islands of the vegetation types already discussed. There are now extensive data from pollen, charcoal, and phytolyths to show that the present forest cover of the region has by no means been constant (for example Van der Hammen 1974; Absy and Van der Hammen 1976; Absy 1979; Brown and Ab'Saber 1979; Absy *et al.*1993; Colinvaux 1993; Thompson *et al.* 1995). Corresponding to the lower temperatures and drier climates of the periods of glaciation there was a much more widespread distribution of savanna and caatinga within the areas covered with rainforest (Prance 1973, 1982, 1987). These changes in vegetational cover over history have resulted in islands of rainforest which have been termed refugia. In many cases the refugia correspond with centres of endemism (Fig. 15.5) of forest organisms, for example birds (Haffer 1969); insects (Brown 1976, 1987), and plants (Prance 1973, 1983, 1987). The reduction of the forest in this way has lead to both extinctions and speciation through isolation in refugia. This expansion and contraction of forest has occurred four times over the last 60 000 years (Absy *et al.* 1991). Both the processes of isolation and of the dynamics of disturbance caused by changing vegetation have contributed to the biological diversity of lowland Amazonia. The forest, however, was certainly not reduced into neat discrete areas as depicted in many papers on refugia. It would have been much more like some of the present-day savanna landscapes where there are also many small islands of forest and large areas of galley forest along the waterways. In most work on refugia the importance of these corridors of galley forest has often been ignored. The data presented here on the differences in species composition of dry forest types show that it would not take the presence of grassland savanna to isolate patches of rainforest. It is far more likely that much of the intervening spaces between well preserved areas of rainforest were occupied by semi-deciduous forest such as is found today in the woody cerrado (cerradão) and the savanna forests of the Guianas and Venezuela. Colinvaux *et al.* (1996) criticizes refugia due to the absence of grass pollen in his pollen profiles. However, it would not necessarily take the presence of grass to isolated true humid evergreen rainforest into discrete patches. The study of speciation and of the distribution of species (Fig. 15.6) in Amazonia must take into account the fact that the vegetation cover has oscillated several times between savanna and rainforest and that the rainforest has been in an island situation prior to its current wide distribution.

15.8 Conclusions

The various islands of different types of vegetation described above offer a diversity of habitats which greatly increase the total plant species diversity of Amazonia. These islands offer new niches for adaptation and many genera of plants have diversified by extending into each of the habitats. For example, in the genus *Caryocar* (Caryocaraceae) which is predominantly distributed in lowland Amazonia rainforest *C. brasiliense* Camb. is a savanna species, *C. gracile* Wittm. is a species of Amazonian caatinga, and *C. montanum* Prance occurs on the slopes of tepuis in the Guayana Highland.

Conservation planning for the region needs to focus not only on the rainforests but also on the small island habitats in order to preserve the maximum number of species. These areas are of particular interest because of the way in which new species have evolved in adaptation to these continental island habitats.

15.9 Summary

There are a number of habitats within the lowland Amazonian rainforest ecosystem that are functional islands. Those that are discussed are savannas, white-sand campinas and caatingas, inselbergs, and sandstone table mountains or tepuis. Each of these habitats is a series of isolated islands of vegetation, and adaptive evolution has lead to speciation of forest groups into these islands. In some cases speciation has taken place between

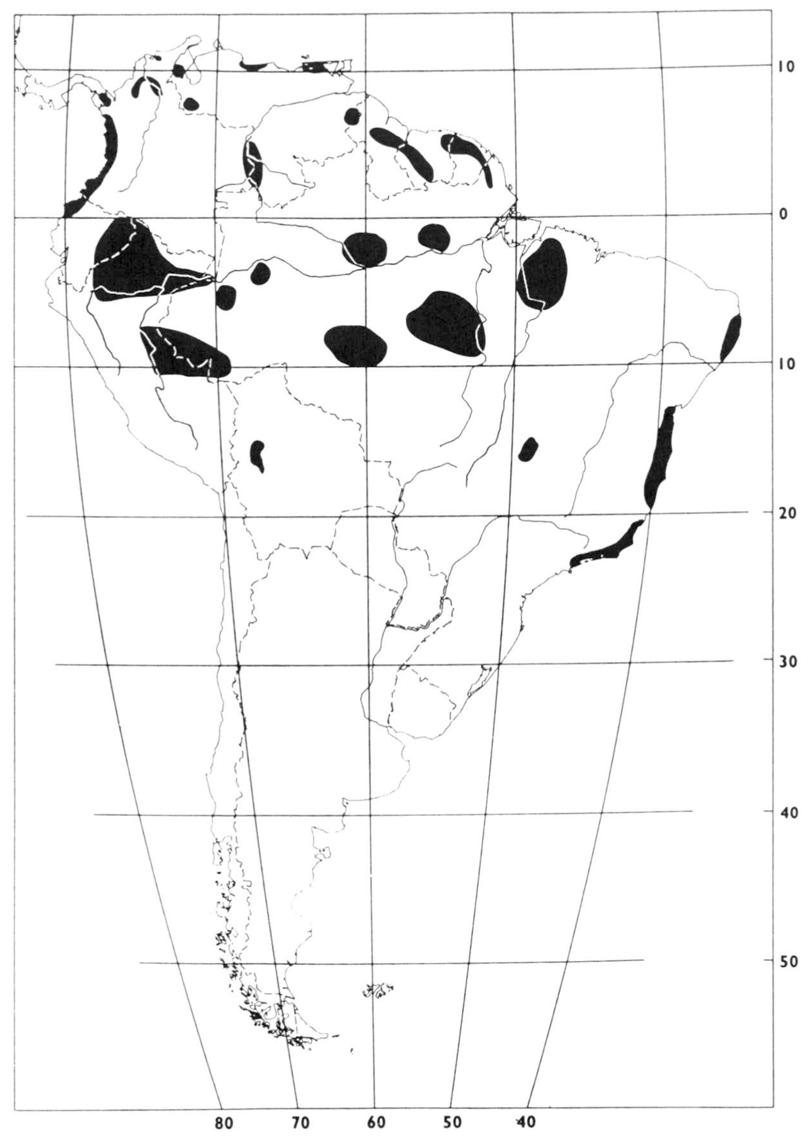

Fig. 15.5 Lowland centres of plant endemism in tropical South America (adapted from Prance 1982).

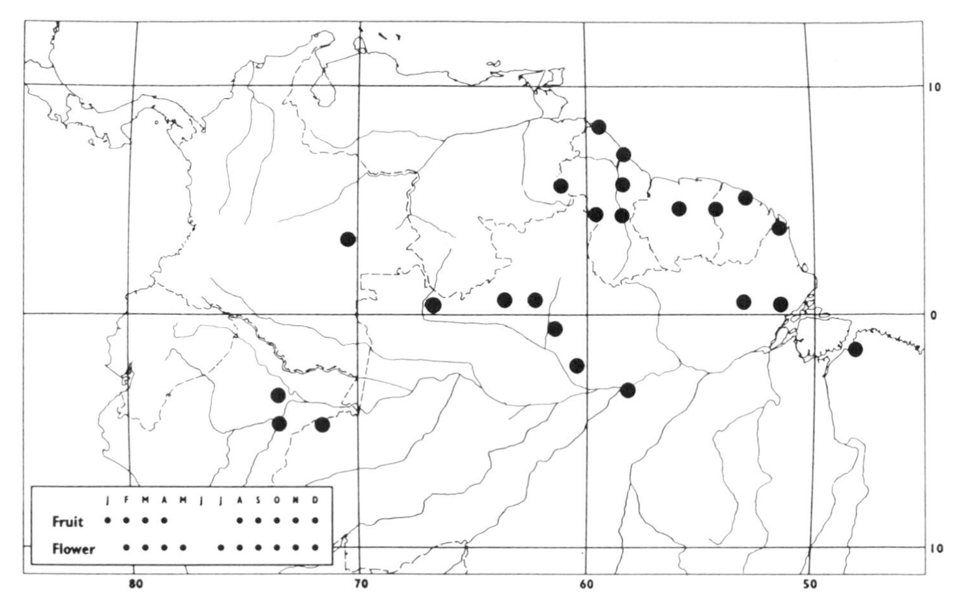

Fig. 15.6 Distribution of *Couepia parillo* DC. with an eastern and western population probably separated by Pleistocene climate changes.

islands, especially the older tepuis, and in other cases there is still little differentiation due to more vagile diaspores and recent climatic changes. The effects of Pleistocene climatic changes that altered the balance between the distribution of savanna and rainforest is discussed.

References

Absy, M. L. (1979). A palynological study of holocene sediments in the Amazon Basin. Unpubl. Ph.D thesis, University of Amsterdam.

Absy, M. L. and Van der Hammen, T. (1976). Some paleoecological data from Rondônia, southern part of Amazon basin. *Acta Amazonica*, **6**, 293–9.

Absy, M. L., Cleef, A., Fournier, M., Martin, L., Servant, M., Sifeddine, A., Ferreira da Silva, M., Soubies, F., Suguio, K., Turcq, B., and Van der Hammen, T. (1991). Mise en évidence de quatre phases d' ouventur de la forêt dense dans le sud-est de l'Amazonie an cours des 60,000 dervières années. Première comparaison avec d'autres régions tropicales. *Compte Rendus Hebdominaires des Séances de l'Académie des Sciences, Paris*, **312**, 673–8.

Absy, M. Lucia, Servant, M., and Absy, M. Laila. (1993). A historia do clima e da vegetação pelo estudo do pólen. *Ciência Hoje*, **16**, 26–30.

Anderson, A. B. (1981). White-sand vegetation of Brazilian Amazonia. *Biotropica*, **13**, 199–210.

Anderson, A. B., Prance, G. T., and de Albuquerque, B. W. P. (1975). Estudos sobre a vegetação das campinas amazonicas III. *Acta Amazonica*, **5**, 225–46.

Aubréville, A. (1961). *Etude écologique des principales formations végétales du Brésil et contribution à la connaissance des forêts de l'Amazonie Brésilienne.* Centre Technique Forestier, Nogent-sur-Marne.

Blancaneux, P., Hernandez, S., and Araujo, J. (1977). *Estudio edafológico preliminar del sector Puerto Ayacucho, Territorio Federal Amazonas, Venezuela.* MARNR, Serie Informes Científicas, Caracas.

Brown, K. S. Jr. (1976). Geographical patterns of evolution in neotropical lepidoptera: systematics and derivation of known and new Heliconiini (Nymphalidae: Nymphalinae). *Journal of Entomology* B, **44**, 201–42.

Brown, K. S. Jr. (1987). Biogeography and evolution of neotropical butterflies. In *Biogeography and quaternary history in tropical America*, Oxford Monographs on Biogeography 3, (ed. T. C. Whitmore and G. T. Prance), pp. 64–104. Clarendon Press, Oxford.

Brown, K. S. and Ab'Saber, A. N. (1979). Ice-age forest refuges and evolution in the Neotropics: correlation of paleoclimatological geomorphological and pediological data with modern biological endemism. *Paleoclimas*, **5**, 1–30.

Colinvaux, P. (1993). Pleistocene biogeography and diversity in tropical forests of South America. In *Biological relationships between Africa and South America*, (ed. P. Goldblatt), pp. 473–99. Yale University Press, New Haven, CT.

Colinvaux, P. E., de Oliveira, P. E., Moreno, J. E., Miller, M. C., and Bush, M. B. (1996). A long pollen record from Lowland Amazonia: Forest and cooling in glacial times. *Science*, **274**, 85–8.

Ducke, A. (1907). Voyage aux campos de l'Ariramba. *Bulletin de la Société de Géographie, Paris*, **16**, 19–26.

Ducke, A. (1913). Explorações científicas no Estado do Pará. *Boletim do Museo Goeldi de Historia Natural e Etnographia*, **7**, 100–97.

Ducke, A. and Black, G. A. (1953). Phytogeographical notes on the Brazilian Amazon. *Anais Academia Brasileira de Ciências*, **25**, 1–46.

Eden, M. J. (1974). Palaeoclimatic influences and the development of savanna in southern Venezuela. *Journal Biogeography*, **1**, 95–109.

Eiten, G. (1972). The cerrado vegetation of Brazil. *Botanical Review*, **38**, 201–341.

Eiten, G. (1978). Delimitation of the cerrado concept. *Vegetatio*, **36**, 169–78.

Eiten, G. (1982). Brazilian 'savannas'. In *Ecological Studies Vol. 42: Ecology of tropical savannas*, (ed. B. J. Huntley and B. H. Walker), pp. 25–47. Springer, Berlin.

Gentry, A. H. (1979). Distribution patterns of neotropical Bignoniaceae. Some phytogeographical implications. In *Tropical botany*, (ed. K. Larsen and L. B. Holm-Nielsen), pp. 339–54. Academic Press, London.

Gentry, A. H. (1982). Phytogeographic patterns as evidence for a Chocó refuge. In *Biological diversification in the tropics*, (ed. G. T. Prance), pp. 112–36. Columbia University Press, New York.

Gottsberger, G. and Morawetz, W. (1986). Floristic, structural and phytogeographical analysis of the savannas of Humaitá (Amazonas). *Flora*, **178**, 41–71.

Grabert, H. (1976). Die inselberglandschaft des Roraima in Venezolanisch-Guayana. *Die Erde*, **107**, 57–69.

Granville, J. J. and Sastre, C. (1974). Aperçu sur la végétation des inselbergs du sud-ouest de la Guyane Française. *Compte Rendu Société de Biogéographie*, **439**, 54–8.

Guarim Neto, G., Guarim, V. L. M. S., and Prance, G. T. (1994). Structure and floristic composition of the trees of an area of cerrado near Cuiabá, Mato Grosso, Brazil. *Kew Bulletin*, **49**, 499–509.

Haffer, J. (1969). Speciation in Amazonian forest birds. *Science,* **165**, 131–7.

Heyligers, P. C. (1963). *Vegetation and soil of a white-sand savanna in Suriname.* Amsterdam, Elsevier.

Hills, T. L. (1969). *The savanna landscapes of the Amazon Basin*, Savanna Research Series 14. McGill University, Montreal.

Huber, J. (1900). *Arboretum amazonicum*. Iconographia dos mais importantes vegetães espontaneos e cultivados da região amazônica. Belém, Instituto Poliographico. Decade 1:9.

Huber, J. (1902). Vur Entstehungesgeschichte der brasilischen Campos. *Petermanns Geographische Mitteilungen*, **48**, 92–5.

Huber, O. (1974). *Le savane neotopicali.* Istituto dei Italo-Latino Americano, Rome.

Huber, O. (1982). Significance of savanna vegetation in the Amazon Territory of Venezuela. In *Biological diversification in the tropics*, (ed. G. T. Prance), pp. 221–44. Columbia University Press, New York.

Huber, O. (1987). Algunas consideraciones sobre el concepto de Pantepui. *Pantepui*, **1**, 2–10.

Huber, O. (1990). Estado actual de los conocimientos sobre la flora y vegetación de la region Guayana, Venezuela. In *El Río Orinoco como ecosistema*, (ed. F. H. Weibzahn, H. Alvarez, and W. M. Lewis Jr.), pp. 337–86. Caracas.

Hurault, J. (1974). Les inselbergs rocheux des régions tropicales humides, témoins de paléoclimats. *Compte Rendu sommaire des Séances de la Société de Biogeographie*, **439**, 49–54.

Janssen, A. (1986). Flora and vegetation der Savannen von Humaitá und ihre Standortbedingungen. *Dissertationes Botanicae* 13. Cramer, Berlin.

Klinge, H. and Medina, E. (1979). Rio Negro caatingas and campinas, Amazonas states of Venezuela and Brazil. In *Heathlands and related shrublands of the world. A. descriptive study*, (ed. R. L. Specht), pp. 483–8. Elsevier, Amsterdam.

Klinge, H., Medina, E., and Herrera, R. (1977). Studies on the ecology of Amazonian Caatinga forest in southern Venezuela 1. *Acta Científica Venezolana*, **28**, 270–6.

Kubitzki, K. (1979). Ocorrência de *Kielmeyera* nos 'Campos de Humaitá' e a natureza dos 'campos'—Flora da Amazônia. *Acta Amazonica*, **9**, 401–4.

Kubitzki, K. (1983). Dissemination biology in the savanna vegetation of Amazonia. In *Dispersal and distribution*, (ed. K. Kubitzki). *Sonderbände des Naturwissenshaftlichen Vereins in Hamburg*, **7**, 353–7.

Kubitzki. (1989). Amazon lowland and Guayana Highland—historical and ecological aspects of the development of their floras. *Amazoniana*, **11**, 1–12.

Macedo, M. and Prance, G. T. (1978). Notes on the vegetation of Amazonia II. The dispersal of plants in Amazonia white sand campinas. The campinas as functional islands. *Brittonia*, **30**, 203–15.

Maguire, B. (1956). Distribution, endemicity, and evolution patterns among Compositae of the Guayana Highland of Venezuela. *Proceedings of the American Philosophical Society*, **100**, 467–75.

Maguire, B. (1970). On the flora of the Guayana Highland. *Biotropica*, **2**, 85–100.

Maguire, B. (1978). Sarraceniaceae. In *The botany of the Guayana Highland. Part X*, (ed. B. Maguire and Collaborators). *Memoirs of The New York Botanical Garden*, **29**, 36–61.

Maguire, B. and collaborators (eds.) (1953–84). *The botany of the Guayana Highland. Part I. Memoirs of The New York Botanical Garden*, **8** (2): 87–160. 1953. Part II (ibid.) **9** (3): 235–392. 1957. Part III. (ibid.) **10** (1): 1–156. 1958. Part IV-1. (ibid.) **10** (2): 1–37. 1960. Part IV-2 (ibid.) **10** (4): 1–87. 1961. Part V (ibid.) **10** (5): 1–278. 1964. Part VI. (ibid.) **12** (3): 1–185. 1965. Part VII. (ibid). **17** (1): 1–439. 1967. Part VIII (ibid.) **18** (2): 1–290. 1969. Part IX. (ibid.) **23**: 1–832. 1972. Part X. (ibid.) **29**: 1–288. 1978. Part XI. (ibid.) **32**: 1–391. 1981. Part XII. (ibid.) **38**: 1–84. 1984. Part XIII. (ibid.) **51**: 1–127. 1989.

Maguire, B. and Steyermark, J. A. (1981). Tepuianthaceae. In *The botany of the Guayana Highland. Part XI*, (ed. B. Maguire and collaborators). *Memoirs of The New York Botanical Garden*, **32**, 4–21.

Medina, E. (1969). Expedición Aso VAC al Alto Orinoco. *Acta Científica Venezolana*, **20**, 9–13.

Prance, G. T. (1972). Monograph of Chrysobalanaceae. *Flora Neotropica*, **9**, 1–406. Hafner, New York.
Prance, G. T. (1973). Phytogeographic support for the theory of pleistocene forest refuges in the Amazon Basin, based on evidence from distribution patterns in Caryocaraceae, Chrysobalanaceae, Dichapetalaceae and Lecythidaceae. *Acta Amazonica*, **3**, 5–28.
Prance, G. T. (ed.) (1982). *Biological diversification in the tropics*. Columbia University Press, New York.
Prance, G. T. (1983). A review of the phytogeographic evidences for pleistocene climate changes in the Neotropics. *Annals of the Missouri Botanical Garden*, **69**, 594–624.
Prance, G. T. (1987). Biogeography of neotropical plants. In *Biogeography and quaternary history in tropical America*, Oxford Monographs on Biogeography 3, (ed. T. C. Whitmore and G. T. Prance), pp. 46–65. Clarendon Press, Oxford.
Prance, G. T. and Schaller, G. B. (1982). Preliminary study of some vegetation types of the Pantanal, Mato Grosso, Brazil. *Brittonia*, **34**, 228–51.
Ratter, J. A. and Dargie, T. C. D. (1992). An analysis of the floristic composition of 26 cerrado areas in Brazil. *Edinburgh Journal of Botany*, **49**, 235–50.
Ratter, J. A., Bridgewater, S., Atkinson, R., and Ribeiro, J. F. (1996). Analysis of the floristic composition of the Brazilian cerrado vegetation II: Comparison of the woody vegetation of 98 areas. *Edinburgh Journal of Botany*, **53**, 153–80.
Richards, P. W. (1952). *The Tropical rainforest: an ecological study*. Cambridge University Press. (2nd edn 1996.)
Rivière, P. (1972). *The forgotten frontier: ranchers of North Brazil*. Holt Rinehart and Winston, New York.
Rodrigues, A. (1961). Aspectos fitssociológicos das catingas do Rio Negro. *Boletim do Museo Paraense Emílio Goeldi, Botânica*, **15**, 1–41.
Sastre, C. (1977). Quelque, aspects de la phytogéographie des milieux ouverts guyanais. *Publication des Laboratoires d'École Normal Superieure, Paris*, **9**, 67–74.
Smith, N. J. H. (1995). Human-induced landscape changes in Amazonia and implications for development. In *Global land use change: A perspective from the Columbian encounter*, (ed. B. L. Turner II), pp. 221–51. Consejo Superior de Investigaciones Científicas, Madrid.
Spruce, R. (1908). *Notes of a botanist on the Amazon and Andes*, 2 Vols. Macmillan, London.
Steyermark, J. A. (1967). Flora del Auyan-tepui. *Acta Botanica Venezolana*, **2**, 5–370.
Steyermark, J. A. (1974). The summit vegetation of cerro Autana. *Biotropica*, **6**, 7–13.
Steyermark, J. A. (1979). Flora of the Guayana Highland: endemicity of the generic flora of the summits of the Venezuelan tepuis. *Taxon*, **28**, 45–54.
Steyermark, J. A. (1984). Flora of the Venezuelan Guayana 1. *Annals of the Missouri Botanical Garden*, **71**, 297–340.
Steyermark, J. A. (1987). Flora of the Venezuelan Guayana II. *Annals of the Missouri Botanical Garden*, **74**, 85–116.
Steyermark, J. A., Berry, P. E., and Holst, B. K. (1995). *Flora of the Venezuelan Guayana. Vol. 1: Introduction; Vol. 2: Pteridophytes, Spermatophytes, Acanthaceae-Araceae*. Timber Press, Portland, OR.
Steyermark, J. A. and Dunsterville, G. C. K. (1980). The lowland floral element on the summit of Cerro Guaiquinima and other cerros of the Guyana Highland of Venezuela. *Journal of Biogeography*, **7**, 285–303.
Takeuchi, M. (1960). A estrutura da vegetação na Amazonia III. A mata de campina na região do Rio Negro. *Boletim do Museo Paraense Emílio Goeldi, Botânica*, **8**, 1–13.
Takeuchi, M. (1961). The structure of the Amazon vegetation II. Tropical rain forest. *Journal of the Faculty of Science, University of Tokyo, Section III, Botany*, **8**, 1–26.

Takeuchi, M. (1962). The structure of the Amazonian vegetation IV. High campina forest in the Upper Rio Negro. *Journal of the Faculty of Science, University of Tokyo, Section III, Botany*, **8**, 279–88.

Thompson, L. G., Mosley-Thompson, E., Davis, M. E., Lin, P.-N., Henderson, K. A., Cole-Dai, J., Bulzan, J. F., and Liu, K.-G. (1995). Late glacial stage and holocene tropical ice core records from Huascarán, Peru. *Science*, **269**, 46–50.

Tuomisto, H., Ruokolainen, K., Kalliola, R., Linna, A., Danjoy, W., and Rodriguez, Z. (1995). Dissecting Amazonian biodiversity. *Science*, **269**, 63–6.

Van der Hammen, T. (1974). The Pleistocene changes of vegetation and climate in tropical South America. *Journal of Biogeography*, **1**, 3–26.

16

Biotic drift or the shifting balance—did forest islands drive the diversity of warningly coloured butterflies?

James L. B. Mallet and John R. G. Turner

16.1 The problem

According to Stephen Hawking 'God does throw dice'. Our question will be 'When are the dice thrown?' It has been customary to think of Wright's shifting balance as a stochastic theory of evolution and Fisherian natural selection as deterministic; contrariwise both R. A. Fisher and Sewall Wright believed in fundamental stochasticity (Turner 1987, 1992). We propose a different dichotomy: that divergence can take place as an adaptive response either to stochastic changes in the genetic structure of the population, or to stochastic changes in the biotic environment, or of course, both. We will explore this through the spectacular evolutionary divergence of the colour patterns of South American butterflies, particularly *Heliconius*.

16.2 The facts

Geographical variation in South American long-winged butterflies (*Heliconius*, ithomiines, and others) is among the most striking examples of evolutionary radiation. 'In tropical South America, a numerous series of gaily-coloured butterflies and moths are found all to change their hues and markings together, as if by the touch of an enchanter's wand, at every few hundred miles' (Bates 1879). The geographical diversity of colour patterns, coupled with the close similarity of colour patterns between species within any one area led to Bates' (1862) theory of mimicry. Turner (1968, 1976*a*) showed that there has been a pattern of convergent evolution within the Heliconiini, such that individual clades have radiated into many of the mimetic patterns, and each of these patterns in turn is shared by members of separate clades. This pattern of divergence and convergence within the genus is repeated as a pattern of divergence and convergence between races within species: it is a fair conjecture that this racial divergence/convergence is what underlies the pattern at the higher taxonomic level. The most spectacular example is the now well-known parallelism of the mutual Müllerian mimics *Heliconius melpomene* and *H. erato*, which each have around 30 identifiable races and around 10 distinct colour patterns. Each race and pattern of *H. melpomene* is sympatric with a closely similar looking and presumably mimetic race of *H. erato* (there is one exception) (Fig. 16.1).

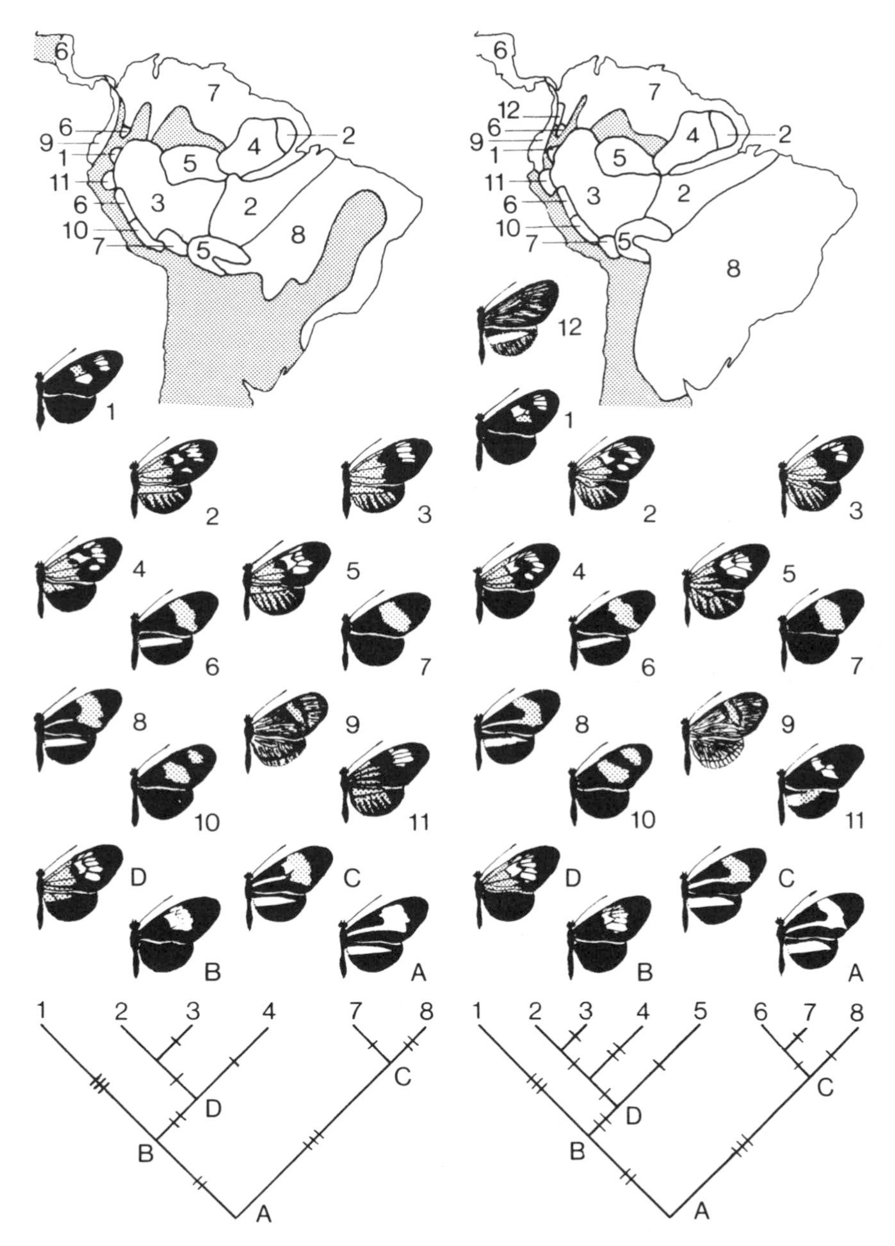

Fig. 16.1 Parallel mimicry and geographical variation between (a) *Heliconius melpomene* and (b) *Heliconius erato*. From Sheppard *et al.* (1985).

The rampant geographical diversification within these species is paradoxical. Predators' memories ensure that warning patterns are subject to strong frequency-dependent and stabilizing selection: rare patterns are at a disadvantage and a new pattern will be favoured only once it has evolved to a high frequency, selection usually favouring the commoner and best recognized form (e.g. Mallet and Singer 1987; Speed and Turner submitted). The same problem arises with the origin of warning colour itself: a rare pattern may gain little advantage of its own, and may sacrifice fitness by being conspicuous compared with the original cryptic pattern. Hence there are difficulties in explaining how warning colours increase from a low frequency within a population (Fisher 1930; Turner 1971*a*; Harvey and Greenwood 1978) without resorting to a group selectionist argument that aposematism is 'an adaptation involving only a small sacrifice of life' (Poulton 1890). Thus there are three related evolutionary puzzles: the origin of warning colour itself, the origin of new warning colours from old, and the divergence of existing warning patterns, as seen in *Heliconius*.

16.3 The biotic drift model

The Brown–Sheppard–Turner model (Brown *et al.* 1974; Sheppard *et al.* 1985; Turner 1983, 1984*a*) assumes that the conversion of these butterflies from one mimetic pattern to another is brought about by natural selection but driven by stochastic changes in the ecological community; there is no contribution from random genetic drift.

Because species are usually distributed patchily, as now adumbrated in the 'metapopulation model' (Gilpin and Hanski 1991), any small (or indeed large) isolated part of a 'continent' of species contains at any one time only a stochastic subsample of the species which might be present: thus the islets of the Gulf of Finland have sharply differentiated floras as a result of more or less random accidents in their colonization history (Halkka *et al.* 1975). The species composition of the flora and fauna undergoes biotic drift ('faunal' and 'floral drift' are alternatives). The classic case of mammals on the mountain tops that contain the fragments of a previously continuous moist forest now split by the arid conditions of the Great Basin (Brown 1971) shows what happens during biotic drift (Fig. 16.2): there is a significant stochastic element. The pattern is, however, one of 'disorderly extinction' (Turner 1977*a*, 1984*b*) rather than 'random extinction': some islands have a greater loss of species than others, and some species have a higher probability of extinction than others; the variance in the number of mountains occupied by a species is 4.7 times greater than expected if each species had an equal probability of extinction. Such changes must surely have considerable effects on evolution: the form which natural selection takes, for all the constituent species, will differ from island to island or patch to patch. This introduces a strong stochastic element into the process of adaptation itself.

It is suggested that the patterns of warningly coloured butterflies evolve primarily by 'ring switching' (Turner 1976*b*, 1977*b*, 1984*a*, *c*, *passim*; Sheppard *et al.* 1985). Natural selection favours the convergence of the patterns of warningly coloured butterflies onto a common pattern, so that they all become Müllerian mimics: such an

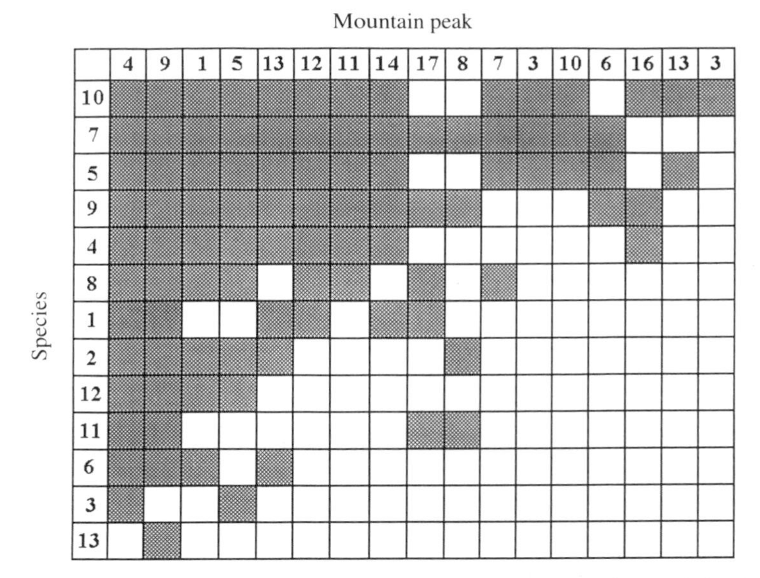

Fig. 16.2 Biotic drift or disorderly extinction of the boreal mammal fauna of the Great Basin (western USA) following its fragmentation by desert conditions: each mountain peak contains a subset of the total fauna. Shading indicates presence; numbers are mountain peaks or mammal species. After Brown (1971), redrawn by Turner (1977*a*, 1984*b*). See Brown (1971) for the names of the peaks and the mammals.

assemblage is called a mimicry ring. In the tropics it is normal for around six such rings to coexist within one life-form (bees, long-winged butterflies), indicating that something prevents total convergence on a single ring. This barrier is held to be the considerable differences between the patterns of the rings, such that the cognitive generalization of predators never confuses one ring with another; probably this is enhanced by microhabitat differences between both mimicry rings (Mallet and Gilbert 1995; Beccaloni 1995) and the predatory birds. There is thus an 'adaptive valley' between the rings. Individual species can cross the valley, and thus 'switch rings' or undergo 'mimetic capture' if they generate a mutation that gives a passable resemblance to the pattern of another ring. The resemblance does not have to be perfect, only sufficiently good to confuse enough of the predators enough of the time; higher levels of mimetic resemblance are achieved later through gradual convergent evolution.

To effect the capture of a species, the capturing ring has to be better protected than the ring from which the capture occurs: better protection arises either from more effective chemical defence, or from being more numerous. Thus capture is expected to be ongoing, with species exchanged between different rings at different times or even whole rings totally captured, as rings change in their overall abundance or in their composition: a ring which contains a particularly distasteful species or which has become more numerous (as a result of containing a few common species or a very large number of species) is the ring that will effect the captures from other rings. As rings change their abundance over time, species will be captured and maybe recaptured, and if rings differ in their capturing power in different geographical areas, the recaptured species will undergo racial divergence. The theory is thus capable of explaining racial differentiation of the kind seen so spectacularly in *H. melpomene* and *H. erato*.

The Brown–Sheppard–Turner model therefore requires nothing more than changes in the abundance of species in Müllerian mimicry rings that last for a sufficient time to enhance mimetic captures, the time involved being the delay before a suitable mutation occurs, and the time taken to substitute one allele for another. But Turner (1982) argued that although the biotic drift model could indeed generate racial divergence in mimetic *Heliconius* purely by ecological fluctuations in continuous forest, the changes were much more likely and possibly larger during periods when the fauna and flora were fragmented into refuges undergoing extensive and long-term biotic drift. Local extinction of species is universal, but in a continuous habitat is likely to be relatively short-term, as the vacated patches of habitat are recolonized from neighbouring populations. In a habitat fragmented into islands, such an extinction is permanent except through infrequent long-distance recolonization, or until the islands coalesce again. In such island refuges, there is stochastic change in the composition of the flora and fauna which is profound and long lasting. In general, the extinction of species affects the ecology and hence evolution of the remaining species, by removing host plants, predators, parasites, or competitors, and leaving open ecological niches that were previously occupied. Specifically the mimicry rings will change their species composition by the loss of individual species, but also their abundance because of the loss of parasites, competitors, host plants, etc. Such long-term changes would have great power to generate ring capture and geographical divergence in mimetic butterflies.

The biotic drift model has therefore been allied with the 'forest refuge model' for evolution in Amazonia (Fox 1949; Turner 1965; 1976*b*; Haffer 1969; Prance 1973, and Chapter 15; Brown *et al.* 1974; Brown 1979, 1987*a*, *b*), which proposes that the patterns of race formation seen in the biota of much of tropical America are to be explained by a model of allopatric race formation on 'island' refuges; the islands being formed by the shrinkage of the now continuous humid forests during the cool dry periods that accompanied the glacial maxima of the Pleistocene.

16.4 The shifting balance model

The refuge model of divergence is hard to test, in part because of a lack of alternative models. Mallet has therefore suggested random drift as a counter-hypothesis to biotic drift: that the evolution and diversification of warning colours in *Heliconius* might constitute a good example of Wright's (1932, 1977) shifting balance theory occurring in parapatry (Mallet 1986*a*, *b*, 1993; Mallet and Singer 1987). There are three phases:

Phase I: random factors such as mutation, drift, kin-founding, or fluctuations in selection might occasionally cause a novel colour pattern rise to a high frequency in a local population; if by chance this frequency is high enough to exceed the point of unstable equilibrium imposed by the positive frequency-dependent selection on warningly coloured forms (say around 50% in a species which does not belong to a mimicry ring), this will trigger

Phase II: this positive frequency-dependent selection (advantageous when common) raises the frequency of the new colour pattern, until it becomes locally predominant; once it approaches fixation in a local area this leads to

Phase III: the new pattern spreads to other populations at the expense of the older patterns. This might occur by group or interdemic selection via population pressure (see Wright 1977) or by cline movement (Barton 1979; Mallet and Barton 1989*a*; Mallet 1993), or both.

Once a new form has evolved in a local area—whether by drift or by selection—contact with the original form will not lead to swamping: frequency-dependent selection will ensure that the two warning colour patterns remain apart with a sharp cline between them (Mallet and Barton 1989*a*). The cline that forms in this way will be similar to clines formed through selection against heterozygotes, as in chromosomal evolution (Bazykin 1969; Barton 1979); because the spatial location of the cline depends only on predators' labile memories, and is not rooted to any environmental gradient, these clines can move. If there are any asymmetries in selection or migration between the two patterns—for instance, if one form is better at warning away predators—or even if there is genetic dominance for one of the patterns, the clines will move at constant velocity in the direction of the fitter or of the dominant form. Briefly this is because at the centre of the cline the allele frequency will be 0.5 but the frequency of the dominant phenotype at 0.75 will be above the unstable equilibrium point (see Fig. 16.3; also Mallet 1986*a*; Mallet and Barton 1989*a*). In phase III therefore the new form spreads by cline movement in favour of better-adapted or genetically dominant colour patterns. Phase III is perhaps the most important feature of the shifting balance, since it preserves and amplifies the rare events of phases I and II; this is analogous to the natural selection of randomly produced mutations in simple selective processes, except that it can spread novel adaptations even when they are disfavoured at low frequency.

There are many limits to spread. Moving clines will become trapped at a variety of local barriers. Novel, spreading colour patterns may encounter abiotic conditions under which they are no longer advantageous, whereupon their advance will be halted (Haldane 1948; Slatkin 1973; May *et al.* 1975; Endler 1977). Alternatively, the fronts between invading colour patterns may become trapped in regions of low density (Bazykin 1969; Barton 1979; Hewitt 1988), because more individuals flow into the centre of a density trough than flow out from the low population in the centre. Finally, any actual breaks in the distribution (which are equivalent to extreme density troughs) will also stop clines moving. Density trough trapping could explain why so many hybrid zones between colour pattern races are located at the major Amazonian rivers, in low mountain ranges, and at the isthmus of Panama. The biogeographical result of multiple shifting balances is therefore expected to be a dynamic equilibrium consisting of a patchwork of different colour pattern races exactly like that seen in *Heliconius* today.

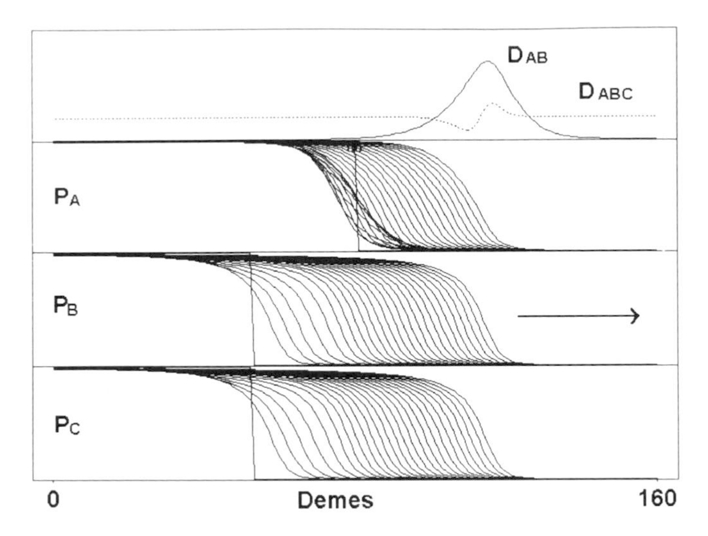

Fig. 16.3 Cline movement as phase III of Wright's shifting balance. The three lower panels show 250 generations of the movement of a three-locus (A, B, C) diploid warning colour cline system (a model of the *Heliconius* three-locus hybrid zone in Tarapoto—see Mallet and Barton (1989*b*) and Mallet *et al.* (1990)) in a linear environment of 160 demes. Each curve represents the gene frequency at ten-generation intervals. These clines are moving to the right under the influence of dominance alone, although adaptive pressures will produce very similar types of cline movement. The simulation starts with two secondary contacts between three patches, a patch on the left fixed for alleles A, B, and C; a central patch fixed for A, b, and c; and a patch on the right fixed for a, b, and c. Initially the identically selected clines at dominant loci B, C move at a constant rate of 0.37 demes per generation towards the symmetrical stationary cline at the codominant locus A. As the dominant clines approach, the A cline is briefly attracted back towards these moving systems. After about 30 generations of cline coalescence, all three clines move off together in the same direction at a reduced velocity of about 0.20 demes per generation. Clines will 'stick together' because of gametic disequilibrium only under conditions of high selection, as here, or if there is epistasis. Selection against foreign phenotypes at each locus is set at 50%, equivalent to heterozygous disadvantage of $s = 0.25$, and is similar to selection observed in *H. erato* of $s \approx 0.23$ per locus (Mallet *et al.* 1990). Gaussian migration is set to $\sigma^2 = 15$ demes^2 per generation. Two-locus gametic disequilibrium under these conditions (top panel) reaches about $D_{AB} \approx +0.09$ (D_{AC} and D_{BC} have similar values), giving a gametic correlation coefficient of $R_{AB} \approx 0.36$ at the centre, and the three-locus disequilibrium varies sigmoidally in the hybrid zone, $D_{ABC} \leq |0.016|$. For further details of the model see Mallet and Barton (1989*a*).

16.5 The evidence

Genetic architecture

Because mimicry produces a highly rugged adaptive surface, ring switching requires mutations of relatively large effect to cross the adaptive trough between two mimicry rings (Turner 1977*b*; Sheppard *et al.* 1985). Phase I of the shifting balance works best with mutations of minor selective effect (Barton and Rouhani 1987, 1991, 1992; Whitlock 1995; see also Chapter 7). Colour patterns in *Heliconius* are inherited at loci of major effect, with measured selection coefficients reaching 0.2 per locus (Sheppard *et al.* 1985; Mallet 1989; Mallet *et al.* 1990). Apparently then the genetic architecture of warning colours in *Heliconius* supports natural selection for mimicry over genetic drift as an initiator of divergence, unless there are initial periods of less intense selection.

Mimetic capture

Brown, Benson, and Sheppard found direct evidence for the capture of a species by a mimetic ring, caught in the act as it were (Brown and Benson 1977; Sheppard *et al.* 1985). *H. hermathena* is confined to isolated colonies in dry, open fields within the Amazonian rainforest. Through almost the whole of its range it has its own distinctive pattern of a red band with yellow bars, resembling none of the mimicry rings which fly with it. But in one small area along the Rio Amazonas it flies with the black and postman race of *H. melpomene* and *H. erato*, which it encounters along the edge where field and forest meet. This postman pattern differs only in lacking the yellow bars and spots of *hermathena*. In one of the half dozen locations where it flies with the postman pattern *H. hermathena* quite accurately mimics the other two species. From Hardy–Weinberg frequencies it is likely that the bars were removed by a single mutation.

This case is considered to provide rather strong evidence in support of the theory of ring capture: *H. hermathena* has crossed the gap between two mimicry rings (or been captured by a mimicry ring) only where this gap is narrow enough to be bridged by a single mutation, in this case one that suppresses the extensive yellow bars on the wings. In the remainder of its range, the ancestral pattern of *H. hermathena* is very different from the black, yellow, and red patterns of *Heliconius* species that fly with it (including further races of *H. melpomene* and *H. erato*), and it can reasonably be supposed that more than one large mutation would be required to convert one pattern to the other.

Biotic drift and mimicry in present-day refuges

The evidence for disorderly extinction in heliconiines is still equivocal. As with the mammals (Fig. 16.2), local extinction is often non-random (Mallet 1993, p. 252). In two currently isolated forest refuges, the Serra Negra in north-east Brazil (Sheppard *et al.* 1985, p. 588) and the Serranía de la Macuira, in the Guajira peninsula, Colombia, only *H. erato* remains. *Heliconius erato* is also commonly found, together with *H. melpomene* and one or two other common *Heliconius* species such as *H. sara*, in dry forest and the gallery forest of savannah regions (e.g. in parts of Darién, the Rupununi and coastal savannahs in Guyana, and parts of the Urubamba and Huallaga valleys in Peru), from which the rarer species are absent. In none of these areas are the colour patterns different from those in the adjacent wet forest. In general, when extinctions happen, it is the rarer species that disappear first, not abundant ‘weed’ species like *H. erato* and *H. melpomene*. It then becomes extremely difficult to explain the geographical divergence of *H. erato* and *H. melpomene*, whose most noticeable feature is that they mimic each other, by means of mimicry of an entirely different set of species in each refuge. Within the heliconiines, there are not enough different species to explain the observed diversification within each species.

Phases I and II: random drift and local change

The shifting balance model would be in severe doubt if the conditions for random drift did not occur in *Heliconius* populations: such conditions can be investigated via studies of population structure, which show that $F_{st} \approx 0.012–0.025$ at allozyme loci in local populations (Silva and Araujo 1994; Jiggins *et al.* 1996) and $F_{st} \approx 0.06$ at colour pattern loci in a hybrid zone (Mallet 1993): a substantial fraction of individuals in a population can be caught, so these values of F_{st} are close to their actual values, rather than depending as is usual on unreliable estimates of gene frequency from much larger populations. Assuming equilibrium between migration and drift, these values imply local effective population sizes of $Nm \approx 4–21$. Shifting balances in underdominant chromosomal rearrangements, which have near-identical dynamics to warning colours (Mallet and Barton 1989a), are likely to occur when $Nm \approx 1$ (Barton and Rouhani 1987, 1992), suggesting that *Heliconius* populations are generally too large; however, most studies of population structure are done where samples are easy to find, and will therefore usually overestimate the true average population size.

It can be argued against this model that because of mimicry the ecological conditions for the operation of phases I and II will only very infrequently be met: the population has to be small for the initial drift of the allele to occur, but if the species is rare, then establishment of a new and by definition non-mimetic pattern (phase II) will be resisted by natural selection imposed by the other species in the mimicry ring, which will strongly favour the old pattern. Thus either (1) the evolving species has to pass through a very brief bottleneck, with the population expanding to the size where it can then overcome selection imposed by the other species (and afterwards, perhaps, capture them) or (2) the species has to be in possession of a unique pattern already, that does not belong to a mimicry ring. This could readily happen if the other species in the ring had become extinct, perhaps as a result of the same forces that reduced the population size of the evolving species. This suggests that both biotic and genetic drift will be required, and perhaps even produced by the same ecological crisis, in order to cause a shifting balance.

Phase III: cline movement

Movement of *Heliconius* hybrid zones would provide good evidence for phase III. Barton and Hewitt (1989) have argued that cline movement will mostly be prevented by trivial population structural fluctuations, and that adaptations will spread much faster by simple natural selection or population spread from refuges than by cline movement; this would constitute a major limitation on the effectiveness of the shifting balance. Indeed chromosomal and warning colour clines are often found at partial barriers to gene flow like Amazonian rivers and the tops of mountains (Nichols and Hewitt 1986; Hewitt 1988; Mallet 1993). But clines will only become associated with such major barriers to gene flow if they can overcome slight perturbations in population density elsewhere in their range. In theory clines can move relatively quickly (Bazykin 1969; Barton 1979; Mallet and Barton 1989*a*; Johnson *et al.* 1990): in *Heliconius*, potential cline movement speeds can be predicted from selection pressures and migration rates using cline theory, assuming that dominance is the only force causing movement. These predicted speeds range from 50–200 km per century (Mallet 1986*a*; Mallet *et al.* 1990).

Many hybrid zones are known to be mobile: the hooded crow/carrion crow zone has moved tens of kilometres in this century (Perrins 1987). Some peculiar disjunct patterns of *Heliconius* and of ithomiine distribution strongly suggest hybrid zone movement in the past (Brown 1979; Lamas 1976; Mallet 1993). Many other hybrid zones in *Heliconius* are apparently stationary, or occur at geographical or environmental barriers from which they are unlikely to move (Turner 1971*b*; Benson 1982; Mallet 1993). But the data on cline stability are very weak; the Guiana hybrid zone studied by Turner (1971*b*, 1976*b*) could have moved 50 km in several hundred years without detection because early data are inaccurate. The only good historical data we know of show major (100 km) changes in the distributions of *H. hecale* and *H. ismenius* races in Panama where accurate locality data were collected in 1916 during the construction of the Panama Canal (G. Small, personal communication; Mallet 1993). Another butterfly hybrid zone, between *Anartia fatima* and *A. amathea*, has moved about 20 km to the east in Darién in the last two decades (Davies *et al.* 1997).

Centres of endemism

A tight correlation of patterns of endemism between different taxa has always been considered strong evidence for differentiation in refuges. However, the data for most groups have either not been analysed, or are equivocal. Nelson *et al.* (1990) showed that centres of plant endemism discovered by Prance (1973; see also Chapter 15), and used by him as evidence that Pleistocene refuges caused speciation among Amazonian plants, were also areas from which most collections had been taken. It seems that plant centres of endemism may be merely artefacts of poorly documented distributions.

The distribution of some butterflies is much better known. Brown (1979, 1987*a*) collected an enormous data set for Heliconiini and the better-known Ithomiinae in order to find areas with high probabilities for Pleistocene refuges. However, Brown's analysis has a number of problems. Using the data, Brown postulated preliminary areas of endemism. Locally endemic taxa were then assigned to one of these areas. Maps of the taxa assigned to a particular area of endemism were overlaid to produce a contour map of the endemic region. After correction for hybridization and soil types, the centres of such contour maps were assumed to be areas of high probability for the existence of Pleistocene refuges.

There are two inherent circularities in this system. First, some of Brown's earlier taxonomic papers used refuge theory as a justification for naming weakly differentiated taxa as geographical races; these new taxa were then themselves used to identify refuges. Second, the whole analysis depends on the preliminary decisions as to which endemic regions exist where, and on the assignment of taxa to endemic regions. If the races and species that form Brown's data set were actually distributed randomly, it is easy to see that he could have picked areas with weak, randomly produced peaks of endemism and defined them as endemic regions. Any endemic taxa whose distributions centred on this endemic region would be assigned to that region; other taxa with edges running through that endemic region would be assigned to adjacent endemic regions.

The assigned taxa, when overlaid on a map would then produce peaks of endemism near the centre of the endemic region first thought of. Brown's knowledge of the geography and taxa is second to none, and we believe that many of the patterns he has identified may be real; however, his analysis should not be interpreted as an independent statistical test for the existence of non-random peaks in endemism or refuges.

Further, Brown's analysis assumes the distributions of co-mimics to be independent. The geographical races of *H. erato* and *H. melpomene* may have similar patterns of distribution only because the two species are highly mimetic: where one changes its pattern, the other follows. One could even argue that species in different mimicry rings might affect each other in the competition for signalling space, so explaining some correlations between taxa in different rings. We must conclude that there is no clear evidence for refuges from the distribution of centres of endemism, at least without a proper statistical analysis of biogeographical patterns.

Range edges

Beven *et al.* (1984) tested the bird data of Haffer (1981) using simulations and cut-out models of bird distribution data randomly thrown onto maps of the Amazon basin to find how concordance between the edges of randomly distributed taxa compared with actual concordance of range edges: the random distributions were no more discordant than the natural bird distributions except that the natural distributions tended to be broken significantly more often at major rivers. Many of the butterflies studied by Brown (1979) also show distribution edges on major rivers or low mountain ranges. Such correlations between range edges unfortunately do not discriminate clearly between the refuge and parapatric models. Whereas the original zone of contact after expansion from a refuge is not expected to be at a barrier, under both theories hybrid contact zones or tension zones would move to such density troughs. No similar studies have been performed on *Heliconius* but we believe the patterns to be similar.

Fossil evidence of refuges

Much recent work on the Quaternary of South America could have provided support for the refuge theory. There were indeed major climatic disturbances, especially drying, associated with glaciation of temperate regions (Brown 1987*b*; Haffer 1987; Iriondo and Latrubesse 1994). What is unclear is whether these turbulent conditions caused the formation of forest refuges. Pollen studies of cores in the lowlands of Ecuador and Panama suggest that cooling of the order of 5–7 °C occurred in the late Pleistocene. This cooling may have precluded the use of mountainous regions as refuges by lowland forest taxa, and at the same time may have allowed major *friagems* accompanied by killing frosts to sweep throughout much of the lowland Amazon basin where a number of proposed refuges are sited (Bush *et al.* 1990, 1992; Latrubesse and Ramonell 1994). The existence of the Pantepui refuge in the endemic centre of the Guiana Shield is similarly in doubt; recent geological studies indicate that peat deposition under humid conditions started only in the Holocene, and there is evidence for arid conditions of the Pantepui region during the last glacial maximum (Schubert *et al.* 1994). Eastern

Ecuador is now home to a variety of endemic taxa, particularly *Heliconius* subspecies, thought to have originated in Pleistocene refuges, yet between 10 000 and 8000 years ago a series of massive volcanic eruptions coated most of the area with ash whose residue is even now 1.5–4 m thick; this would have had a devastating effect on the biota (Iriondo 1994). These studies show that, while there is plenty of evidence for catastrophic environmental changes, these may have been very destructive to all lowland tropical organisms.

It is not inconceivable that many of the endemic races we now see scattered around Amazonia have evolved in the last few tens of thousands of years: the island of Marajó is known to have been completely submerged in the Amazon embayment 5000 years ago, yet there are three endemic subspecies of heliconiine and ithomiine butterflies (Brown 1979). In contrast, there are still no examples of forest refuges known to have provided lowland wet forest conditions throughout the late Pleistocene.

Molecular data

Molecular phylogenies have the potential to reveal the historical branching processes of populations, races, and species (Avise 1994). However, there are no fixed allozyme differences between *Heliconius* races (Turner *et al.* 1979; Mallet and Barton 1989*b*), and mtDNA genealogies are uncorrelated with racial boundaries (Brower 1994, 1996). Even the fork in the mtDNA genealogies of both *H. erato* and *H. melpomene* in north-eastern Colombia (Brower 1996) does not coincide with a racial boundary: the mtDNA break occurs within the races *H. erato hydara* and *H. melpomene melpomene* respectively. Brower concludes that mitochondrial data provide little evidence either for or against the refuge theory. The general lack of concordance between mtDNA and colour patterns, and the discordance between mtDNA genealogies of *H. erato* and *H. melpomene*, suggests a lack of vicariance in refuges (Mallet *et al.* 1996)

Brower (1996) has put forward a 'mitochondrial parsimony' hypothesis, which suggests that disjunct red-barred races either side of the Andes are independently derived. In contrast Turner (1981, 1983) and Mallet (1993) had proposed that 'colour pattern parsimony' is more likely since genetics of the similar, although disjunct, colour patterns are nearly identical. But colour pattern parsimony is now problematic; if colour patterns evolved only once, the mtDNA divergence, estimated to be 1–2 Ma old, may be newer than the colour pattern differences. This implies a much older time of divergence for some races than the Quaternary supposed by refuge models (Turner 1976*b*; Brown 1987*a*). However, it is also possible that the evolution of mtDNA and the nuclear genome have become dissociated; this is known in many other species (Avise 1994). Brower's mitochondrial parsimony hypothesis could be tested by sequencing the colour pattern genes themselves. However, this work is unlikely to lead to a resolution of the drift versus selection argument.

Quasi-Batesian mimicry

Recent work by Speed and Turner (1997) suggests a further mechanism for ring switching. Most reasonable models of learning in vertebrate predators suggest that mildly unpalatable mimics do not behave like Müllerian mimics in the way described here, but that at certain population densities they will behave like Batesian mimics, with negative frequency-dependent selection which may cause them to become polymorphic. In this case the butterfly will have forms in two or more mimicry rings. If long-term changes in species density then cause the species to behave more like a conventional Müllerian mimic, it might revert to being monomorphic. Clearly this could result in ring switching via an intermediate stage of stable polymorphism, but it is clear neither how extensive quasi-Batesian mimicry is, nor whether many *Heliconius* species are of weak enough unpalatability: the polymorphic *H. doris* is a likely candidate (Turner 1995).

The problem of novelty

New warning patterns have to arise somehow. The biotic drift model 'under-explains' this novelty of patterns: it can permit only switching between existing patterns, and as it involves extinction, the number of patterns must diminish over time. Ultimately the system will cease to diverge even within species, as more and more patterns become extinct. Mimicry cannot explain novelty.

The shifting balance allows novel colour patterns to arise. The transparent ithomiines, the tiger ithomiines and heliconiines, and the *erato/melpomene* patterns are evolutionary novelties found nowhere else than the Neotropics. Many races of *H. erato* and *H. melpomene* indeed do not seem to mimic anything much but each other, and although similar ancestral patterns can be constructed for each species, it is not clear what selective mechanism caused the two species to co-diversify in the first place. Mallet (1993) suggests that there are no potential models known for most of the Amazonian races. But some of the extra-Amazonian races are apparently ancestral patterns which can plausibly be held to be what these two species evolved from, and the reconstructed intermediate patterns in the cladogram all resemble heliconiine patterns which still exist somewhere in South America, which might have acted as the models for divergence in the past (Turner 1983, 1984*a*). On balance, however, the huge variety and novelty of warning colour patterns is strong evidence for some sort of random initiation of divergence such as shifting balance processes. Alternative ways of explaining novelty are discussed in Turner and Mallet (1996).

16.6 Conclusions

We can summarize the differences between our two models as biotic drift, plus mimetic capture, plus allopatry in refuges on the one hand versus genetic drift, plus shifting balance and cline movement, plus parapatry without refuges on the other. The biotic drift and shifting balance models therefore initially appear diametrically opposed; but it is less surprising than at first thought that they are so difficult to distinguish. The operational distinction between natural selection and genetic drift as a cause of changing gene frequencies is likewise a very fine one (Hodge 1987; Beatty 1992; see also Chapter 7); Wright (1977, p. 455) includes random fluctuations of selection in phase I of the shifting balance.

Most of the components of each model are compatible. In both models, some of the extraordinary geographical diversity is initiated by a stochastic event which involves small population sizes and a depauperate bioton: both genetic and biotic drift are likely to be produced by the same ecological crisis, whether this is continual metapopulation turnover or the result of climatic deterioration. As genes drift so do biota. We have argued that the shifting balance itself will require not only genetic drift but some degree of biotic drift into the bargain: because natural selection on warning colour is number-dependent, a reduction in population density causes not only random genetic drift but a change in the intensity of selection, and the extinction of fellow members of the mimicry ring. We have further argued that the pure biotic drift model does not adequately explain the diversity of patterns, but have suggested that the full conditions for the shifting balance might be met rather seldom. Against this, one needs to ask how many times they need to be met in what span of time to generate the existing diversity: phase III of the shifting balance is a way of magnifying the outcome of a rare event, and in the span of evolutionary time, even an event of low probability may occur frequently enough.

In both models natural selection is extremely important, and even when a novel pattern has arisen by drift, mimetic capture must be playing a major role in further evolution: in the Amazon basin, ten or so *Heliconius* species have the same rayed pattern, which must represent mimetic capture by one or two of the species of the rest.

Likewise while the island mimicry model invokes allopatry, to which the shifting balance model was framed as a parapatric alternative, mimetic capture could perfectly well take place in parapatry (Turner 1982) and genetic drift could occur in allopatry or within quaternary refuges. While moving hybrid zones form the important phase III of the shifting balance, the junction boundaries of races originally formed in refuges by mimetic capture or other means, should move in the same way.

If it is hard to disentangle the components of biotic drift and the shifting balance, it will be hard to devise a crucial test to distinguish them. It seems very unlikely that we will find a direct historical record that shows whether genetic or biotic drift actually initiated mimetic divergence. Genetic drift in small, isolated populations might leave a signature of reduced genetic diversity in a founder population; however, expansion during phase III will quickly mop up variation from surrounding populations. We suggest that it will be more fruitful to find intermediate cases of divergence, since both hypotheses predict that local processes causing divergence occur only transitorily. It would be particularly worthwhile to record the species in many more modern-day refuges and metapopulations, and to find whether and how they have diverged: both models imply that initial, perhaps failed, divergences should be common. Studies of the dynamics of actual contact zones and invasions would produce evidence for cline movement.

Finally, perhaps it is only a lack of imagination which has led us both to propose random or chaotic causes of an apparently haphazard diversity of colour patterns. Maybe a deterministic explanation has eluded us; but we doubt it!

16.7 Summary

Species of the South American butterfly genus *Heliconius* have undergone remarkably wide racial divergence in their patterns, and most of the resulting races are Müllerian mimics. As warning coloration normally imposes stabilizing selection on the pattern, this divergence is much in need of explanation. Two models have been suggested. Brown, Sheppard, and Turner proposed that the divergence results from 'mimetic capture', the switching of patterns between adaptive peaks generated by changes in the overall composition of the local biota ('biotic drift') and hence of the mimicry rings to which each species belongs. These changes have in turn been generated by long-term patterns of species extinction in island refuges as biota became progressively isolated and contiguous during contraction and expansion of the rainforest during the Pleistocene. An alternative model, proposed by Mallet, is that truly novel colour patterns became established by mutation and random drift, becoming predominant in local areas; subsequently the novel patterns spread over wide areas by the migration of clines. Under this application of Wright's 'shifting balance', refuges are not necessary for divergence, and Müllerian mimicry evolves after divergence rather than being the driving force for race formation. Although the respective models appear diametrically opposed, there are broad areas of agreement and the hypotheses are difficult to distinguish; in both models there is an initial stochastic event, followed by natural selection for mimicry, and both will operate either in parapatry or allopatry. The novelty and diversity of warning patterns are better explained by the shifting balance.

References

Avise, J. C. (1994). *Molecular markers: natural history and evolution.* Chapman and Hall, London.

Barton, N. H. (1979). The dynamics of hybrid zones. *Heredity*, **43**, 341–59.

Barton, N. H. and Hewitt, G. M. (1989). Adaptation, speciation and hybrid zones. *Nature*, **341**, 497–503.

Barton, N. H. and Rouhani, S. (1987). The frequency of shifts between alternative equilibria. *Journal of Theoretical Biology*, **125**, 397–418.

Barton, N. H. and Rouhani, S. (1991). The probability of fixation of a new karyotype in a continuous population. *Evolution*, **45**, 499–517.

Barton, N. H. and Rouhani, S. (1992). Adaptation and the 'shifting balance'. *Genetical Research*, **61**, 57–74.

Bates, H. W. (1862). Contributions to an insect fauna of the Amazon valley. Lepidoptera: Heliconidae. *Transactions of the Linnean Society of London*, **23**, 495–566.

Bates, H. W. (1879). [Commentary on Müller's paper]. *Transactions of the Entomological Society of London*, **1879**, xxviii–xxix.

Bazykin, A. D. (1969). Hypothetical mechanism of speciation. *Evolution*, **23**, 685–7.

Beatty, J. (1992). Random drift. In *Keywords in evolutionary biology*, (ed. E. Fox Keller and E. A. Lloyd), pp. 273–81. Harvard University Press, Cambridge, MA.

Beccaloni, G. W. (1995). Studies on the ecology and evolution of Neotropical ithomiine butterflies (Nymphalidae: Ithomiinae). Unpubl. Ph.D. thesis, Imperial College, London.

Benson, W. W. (1982). Alternative models for infrageneric diversification in the humid tropics: tests with passion vine butterflies. In *Biological diversification in the tropics*, (ed. G. T. Prance), pp. 608–40. Columbia University Press, New York.

Beven, S., Connor, E. F., and Beven, K. (1984). Avian biogeography in the Amazon basin and the biological model of diversification. *Journal of Biogeography*, **11**, 383–99.

Brower, A. V. Z. (1994). Rapid morphological radiation and convergence among races of the butterfly *Heliconius erato* inferred from patterns of mitochondrial DNA evolution. *Proceedings of the National Academy of Sciences USA*, **91**, 6491–5.

Brower, A. V. Z. (1996). Parallel race formation and the evolution of mimicry in *Heliconius* butterflies: a phylogenetic hypothesis from mitochondrial DNA sequences. *Evolution*, **50**, 195–221.

Brown, J. H. (1971). Mammals on mountaintops: non-equilibrium insular biogeography. *American Naturalist*, **105**, 467–78.

Brown, K. S. (1979). *Ecologia geográfica e evolução nas florestas neotropicais*. Livre de Docencia, Universidade Estadual de Campinas, Campinas, Brasil.

Brown, K. S. (1987*a*). Biogeography and evolution of neotropical butterflies. In *Biogeography and quaternary history in tropical America*, (ed. T. C. Whitmore and G. T. Prance), pp. 66–104. Oxford University Press.

Brown, K. S. (1987*b*). Areas where humid tropical forest probably persisted. In *Biogeography and quaternary history in tropical America*, (ed. T. C. Whitmore and G. T. Prance), p. 45. Oxford University Press.

Brown, K. S. and Benson, W.W. (1977). Evolution in modern Amazonian non-forest islands: *Heliconius hermathena*. *Biotropica*, **9**, 95–117.

Brown, K. S., Sheppard, P. M., and Turner, J. R. G. (1974). Quaternary refugia in tropical America: evidence from race formation in Heliconius butterflies. *Proceedings of the Royal Society of London* B, **187**, 369–78.

Bush, M. B., Colinvaux, P. A., Wiemann, M. C., Piperno, D. R., and Liu, K.-B. (1990). Late Pleistocene temperature depression and vegetation change in Ecuadorian Amazonia. *Quaternary Research*, **34**, 330–45.

Bush, M. B., Piperno, D. R., Colinvaux, P. A., De Oliviera, P. E., Krissek, L. A., Miller, M. C., and Rowe, W. E. (1992). A 14300-yr paleoecological profile of a lowland tropical lake in Panama. *Ecological Monographs*, **62**, 251–75.

Davies, N., Aiello, A., Mallet, J., Pomiankowski, A., and Silberglied, R. E. (1997). Speciation in two neotropical butterflies: extending Haldane's rule. *Proceedings of the Royal Society of London* B. (Submitted.)

Endler, J. A. (1977). *Geographic variation, speciation, and clines*. Princeton University Press.

Fisher, R. A. (1930). *The genetical theory of natural selection*. Clarendon Press, Oxford.

Fox, R. M. (1949). The evolution and systematics of the Ithomiidae. *University of Pittsburgh Bulletin*, **45**, 36–47.

Gilpin, M. and Hanski, I. (eds.) (1991). *Metapopulation dynamics: empirical and theoretical investigations*. Academic Press, London.

Haffer, J. (1969). Speciation in Amazonian forest birds. *Science*, **165**, 131–7.

Haffer, J. (1981). Aspects of neotropical bird speciation during the Cenozoic. In *Vicariance biogeography: a critique*, (ed. G. Nelson and D. E. Rosen), pp. 371–94. Columbia University Press, New York.

Haffer, J. (1987). Quaternary history of tropical America. In *Biogeography and quaternary history in tropical America*, (ed. T. C. Whitmore and G. T. Prance), pp. 1–18. Oxford University Press.

Haldane, J. B. S. (1948). The theory of a cline. *Journal of Genetics*, **48**, 277–84.

Halkka, O., Raatikainen, M., Halkka, L., and Hovinen, R. (1975). The genetic composition of *Philaenus spumarius* populations in island habitats variably affected by voles. *Evolution*, **29**, 700–6.

Harvey, P. H. and Greenwood, P. J. (1978). Anti-predator defence strategies: some evolutionary problems. In *Behavioural ecology*, (ed. J. R. Krebs and N. B. Davies), pp. 129–51. Blackwell Scientific Publications, Oxford.

Hewitt, G. M. (1988). Hybrid zones—natural laboratories for evolutionary studies. *Trends in Ecology and Evolution*, **3**, 158–67.

Hodge, M. J. S. (1987). Natural selection as a causal, empirical, and probabilistic theory. In *The probabilistic revolution, vol. 2. Ideas in the sciences*, (ed. G. Gigerenzer, L. Krüger, and M. Morgan), pp. 313–54. MIT Press/Bradford Books, Cambridge, MA.

Iriondo, M. (1994). The Quaternary of Ecuador. *Quaternary International*, **21**, 101–12.

Iriondo, M., and Latrubesse, E. M. (1994). A probable scenario for a dry climate in central Amazonia during the late Quaternary. *Quaternary International*, **21**, 121–8.

Jiggins, C., McMillan, W. O., Neukirchen, W., and Mallet, J. (1996). What can hybrid zones tell us about speciation? The case of *Heliconius erato* and *H. himera* (Lepidoptera: Nymphalidae). *Biological Journal of the Linnean Society*, **59**, 221–42.

Johnson, M. S., Clarke, B., and Murray, J. (1990). The coil polymorphism in *Partula suturalis* does not favor sympatric speciation. *Evolution*, **44**, 459–64.

Lamas, G. (1976). Notes on Peruvian butterflies (Lepidoptera). II. New Heliconius from Cusco and Madre de Dios. *Revista Peruana de Entomología*, **19**, 1–7.

Latrubesse, E. M. and Ramonell, C. G. (1994). A climatic model for southwestern Amazonia in last glacial times. *Quaternary International*, **21**, 163–9.

Mallet, J. (1986*a*). Hybrid zones in Heliconius butterflies in Panama, and the stability and movement of warning colour clines. *Heredity*, **56**, 191–202.

Mallet, J. (1986*b*). Dispersal and gene flow in a butterfly with home range behaviour: *Heliconius erato* (Lepidoptera: Nymphalidae). *Oecologia*, **68**, 210–17.

Mallet, J. (1989). The genetics of warning colour in Peruvian hybrid zones of *Heliconius erato* and *H. melpomene*. *Proceedings of the Royal Society of London* B, **236**, 163–85.

Mallet, J. (1993). Speciation, raciation, and color pattern evolution in Heliconius butterflies: evidence from hybrid zones. In *Hybrid zones and the evolutionary process*, (ed. R. G. Harrison), pp. 226–60. Oxford University Press.

Mallet, J. and Barton, N. (1989*a*). Inference from clines stabilized by frequency-dependent selection. *Genetics*, **122**, 967–76.

Mallet, J. and Barton, N. H. (1989*b*). Strong natural selection in a warning color hybrid zone. *Evolution*, **43**, 421–31.

Mallet, J. and Gilbert, L. E. (1995). Why are there so many mimicry rings? Correlations between habitat, behaviour and mimicry in *Heliconius* butterflies. *Biological Journal of the Linnean Society*, **55**, 159–80.

Mallet, J. and Singer, M. C. (1987). Individual selection, kin selection, and the shifting balance in the evolution of warning colours: the evidence from butterflies. *Biological Journal of the Linnean Society*, **32**, 337–50.

Mallet, J., Barton, N., Lamas, G., Santisteban, J., Muedas, M., and Eeley, H. (1990). Estimates of selection and gene flow from measures of cline width and linkage disequilibrium in *Heliconius* hybrid zones. *Genetics*, **124**, 921–36.

Mallet, J., Jiggins, C. D., and McMillan, W. O. (1996). Mimicry meets the mitochondrion. *Current Biology*, **6**, 937–40.

May, R. M., Endler, J. A., and McMurtrie, R. E. (1975). Gene frequency clines in the presence of selection opposed by gene flow. *American Naturalist*, **109**, 650–76.

Nelson, B. W., Ferreira, C. A. C., da Silva, M. F., and Kawasaki, M. L. (1990). Endemism centres, refugia and botanical collection density in Brazilian Amazonia. *Nature*, **345**, 714–16.

Nichols, R. A. and Hewitt, G. M. (1986). Population structure and the shape of chromosome cline between two races of *Podisma pedestris* (Orthoptera, Acrididae). *Biological Journal of the Linnean Society*, **29**, 301–16.

Perrins, C. (1987). *Birds of Britain and Europe*. Collins, London.

Poulton, E. B. (1890). *The colours of animals: their meaning and use especially considered in the case of insects*, 2nd edn. Kegan Paul, Trench, Trübner & Co., London.

Prance, G. T. (1973). Phytogeographic support for the theory of Pleistocene forest refuges in the amazon basin, based on evidence from distribution patterns in Caryocaraceae, Chrysobalanaceae, Dichapetalaceae and Lecythidaceae. *Acta Amazonica*, **3**, 5–28.

Schubert, C., Fritz, P., and Aravena, R. (1994). Late Quaternary paleoenvironmental studies in the Gran Sabana (Venezuelan Guayana Shield). *Quaternary International*, **21**, 81–90.

Sheppard, P. M., Turner, J. R. G., Brown, K. S., Benson, W. W., and Singer, M. C. (1985). Genetics and the evolution of muellerian mimicry in Heliconius butterflies. *Philosophical Transactions of the Royal Society of London* B, **308**, 433–613.

Silva, L. M. and Araujo, A. M. (1994). The genetic structure of *Heliconius erato* populations (Lepidoptera; Nymphalidae). *Revista Brasileira de Genetica*, **17**, 19–24.

Slatkin, M. (1973). Gene flow and selection in a cline. *Genetics*, **75**, 733–56.

Speed, M. P. and Turner, J. R. G. (1997). Learning and forgetting in mimicry. Part 2. Simulations of the natural mimicry spectrum. *Philosophical Transactions of the Royal Society of London* B. (Submitted.)

Turner, J. R. G. (1965). Evolution of complex polymorphism and mimicry in distasteful South American butterflies. *Proceedings of the XII International Congress of Entomology, London, 1964*, p. 267.

Turner, J. R. G. (1968). Some new *Heliconius* pupae: their taxonomic and evolutionary significance in relation to mimicry (Lepidoptera, Nymphalidae). *Journal of Zoology (London)*, **155**, 311–25.

Turner, J. R. G. (1971*a*). Studies of Müllerian mimicry and its evolution in burnet moths and heliconid butterflies. In *Ecological genetics and evolution*, (ed. E. R. Creed), pp. 224–60. Blackwell, Oxford.

Turner, J. R. G. (1971*b*). Two thousand generations of hybridization in a *Heliconius* butterfly. *Evolution*, **25**, 471–82.

Turner, J. R. G. (1976*a*). Adaptive radiation and convergence in subdivisions of the butterfly genus *Heliconius* (Lepidoptera: Nymphalidae). *Zoological Journal of the Linnean Society*, **58**, 297–308.

Turner, J .R. G. (1976*b*). Muellerian mimicry: classical 'beanbag' evolution, and the role of ecological islands in adaptive race formation. In *Population genetics and ecology*, (ed. S. Karlin and E. Nevo), pp. 185–218. Academic Press, New York.

Turner, J. R. G. (1977*a*). Forest refuges as ecological islands: disorderly extinction and the adaptive radiation of Muellerian mimics. In *Biogéographie et évolution en Amérique tropicale*, (ed. H. Descimon), pp. 98–117. Publications du Laboratoire de Zoologie de l'École Normale Supérieure, Paris.

Turner, J. R. G. (1977*b*). Butterfly mimicry: the genetical evolution of an adaptation. *Evolutionary Biology*, **10**, 163–206.

Turner, J. R. G. (1981). Adaptation and evolution in *Heliconius*: a defence of neodarwinism. *Annual Review of Ecology and Systematics*, **12**, 99–121.

Turner, J. R. G. (1982). How do refuges produce tropical diversity? Allopatry and parapatry, extinction and gene flow in mimetic butterflies. In *Biological diversification in the tropics*, (ed. G. T. Prance), pp. 309–35. Columbia University Press, New York.

Turner, J. R. G. (1983). Mimetic butterflies and punctuated equilibria: some old light on a new paradigm. *Biological Journal of the Linnean Society*, **20**, 277–300.

Turner, J. R. G. (1984*a*). Mimicry: the palatability spectrum and its consequences. In *The biology of butterflies*, (ed. R. I. Vane-Wright and P. R. Ackery), pp. 141–61. Academic Press, London.

Turner, J. R. G. (1984*b*). Extinction as a creative force: the butterflies of the rainforest. In *Tropical rain-forest. The Leeds symposium*, (ed. A. C. Chadwick and S. L. Sutton), pp. 195–204. Leeds Philosophical and Literary Society, Leeds, UK.

Turner, J. R. G. (1984*c*). Darwin's coffin and Doctor Pangloss—do adaptationist models explain mimicry? In *Evolutionary ecology*, (ed. B. Shorrocks), pp. 313–61. Blackwell Scientific, Oxford.

Turner, J. R. G. (1987). Random genetic drift, R. A. Fisher and the Oxford school of ecological genetics. In *The probabilistic revolution, vol. 2. Ideas in the sciences*, (ed. G. Gigerenzer, L. Krüger, and M. Morgan), pp. 313–54. MIT Press/Bradford Books, Cambridge, MA.

Turner, J. R. G. (1992). Stochastic processes in populations: the horse behind the cart? In *Genes in ecology*, (ed. R. J. Berry, T. J. Crawford, and G. M. Hewitt), pp. 29–53. Blackwell, Oxford.

Turner, J. R. G. (1995). Mimicry as a model for coevolution. In *Biodiversity and evolution*, (ed. R. Arai, M. Kato, and Y. Doi), pp. 131–50. National Science Museum Foundation, Tokyo.

Turner, J. R. G. and Mallet, J. L. B. (1996). Did forest islands drive the diversity of warningly coloured butterflies? Biotic drift and the shifting balance. *Philosophical Transactions of the Royal Society of London* B, **351**, 835–45.

Turner, J. R. G., Johnson, M. S., and Eanes, W.F. (1979). Contrasted modes of evolution in the same genome: allozymes and adaptive change in *Heliconius*. *Proceedings of the National Academy of Sciences USA*, **76**, 1924–8.

Whitlock, M. C. (1995). Variance-induced peak shifts. *Evolution*, **49**, 252–9.

Wright, S. (1932). The roles of mutation, inbreeding, crossbreeding and selection in evolution. *Proceedings of the XI International Congress of Genetics*, 356–66.

Wright, S. (1977). *Evolution and the genetics of populations. Vol. 3. Experimental results and evolutionary deductions*. University of Chicago Press.

17

Adaptive plant evolution on islands: classical patterns, molecular data, new insights

Thomas J. Givnish

17.1 Introduction

The nature of plant life on oceanic islands is shaped by the remoteness of such islands, the resulting selection on plant dispersal mechanisms before arrival, and the ensuing pressures on plant ecology caused by an impoverished set of competitors, predators, pollinators, and frugivores after arrival. Island plants are thus often marked by unusually effective means of long-distance dispersal, the evolution of arborescence, extensive speciation coupled with adaptive radiation in habit and habitat, possession of unspecialized flowers, and loss of anti-herbivore defences (Darwin 1859; Wallace 1878, 1880; Guppy 1906; Rock 1919; Ridley 1930; Docters van Leeuwen 1936; Skottsberg 1956; Fosberg 1963; Carlquist 1965, 1967, 1970, 1974; van Balgooy 1971; Robichaux *et al.* 1990). These classical phenomena are widely thought to provide some of the most striking evidence for the role of ecology and natural selection in shaping life on earth.

Yet the scientific support for some of these phenomena may need re-examination. Many island plants have diverged so dramatically from putative ancestral groups that it is difficult to ascertain their relationships and, thus, to trace the pattern of their evolution (Knox *et al.* 1993; Givnish *et al.* 1994, 1995,1997; Sang *et al.* 1994; Baldwin and Robichaux 1995; Böhle *et al.* 1996; Kim *et al.* 1996; Francisco-Ortega *et al.* 1997; see also Chapter 2). Phylogenies inferred from morphology may be skewed by convergent and/or divergent evolution of selectively important traits; such skewing might be substantial in organisms undergoing 'concerted convergence', in which similar environments select for convergence in several traits simultaneously (Givnish and Sytsma 1997). Such problems may be especially severe for adaptive radiations on islands, in which phenotypic variation among species can be concentrated in the relatively few characters that underlie a specific radiation (e.g. see Grant 1986; Baldwin and Robichaux 1995; Givnish *et al.* 1995).

Molecular systematics provides an important way of circumventing these problems. Recently, several researchers have begun to use molecular techniques to infer phylogeny independently of morphological traits of interest (Givnish and Sytsma 1997), and to re-examine some of the classic cases of plant evolution on islands and island-like habitats (e.g. Baldwin *et al.* 1991; Sytsma *et al.* 1991; Baldwin 1992, 1997; Knox *et al.* 1993; Givnish *et al.* 1994, 1995, 1996, 1997; Sang *et al.* 1994, 1995; Baldwin and Robichaux 1995; Knox and Palmer 1995, 1996; Mes 1995; Böhle *et al.* 1996; Kim *et al.* 1996; Smith *et al.* 1996; Francisco-Ortega *et al.* 1996*a, b,* 1997; Sakai *et al.* 1997). This chapter provides perspective on the key results and new insights emanating from these studies, focusing on research on the Hawaiian lobelioids, perhaps the most remarkable case of explosive speciation, adaptive radiation, convergence, and the evolution of defenses against vertebrate herbivores in island plants (Fig. 17.1). I begin with a commentary on two classical patterns of insular plant evolution, involving ecological constraints on modes of seed dispersal and the evolution of arborescence. I then discuss patterns of speciation and adaptive radiation, and conclude with some comments on taxon cycles and the crucial role fleshy fruits may play in reducing gene flow and accelerating speciation in rainforest understories.

17.2 Classical patterns: dispersal syndromes

Most oceanic islands are volcanic in origin, have geologically ephemeral (< 5–20 Ma) lifetimes, and have never had a direct overland connection to mainland source areas. To colonize such islands, the seeds or spores of land plants must be capable of long-distance dispersal across vast expanses of ocean. Carlquist (1966, 1967, 1974) and van Balgooy (1971) used the seed and fruit morphology of angiosperm genera to infer their likely mode of dispersal, and then analysed the relative representation of genera with specific modes of seed dispersal on different island groups.

The most important conclusions to emerge from these comparisons are:

- Wind dispersal is effective only over relatively short distances (< *c.* 200 km), except for dust-like seeds and spores;
- Water dispersal is effective over entire ocean basins, dominant in beach and atoll floras, but mostly absent in plants of inland habitats;
- Endozoochory (dispersal of seeds in fleshy fruits inside birds or bats) is effective over long distances, but is more heavily represented in the floras of tall (> 200 m elevation) islands with relatively high rainfall;
- Ectozoochory (dispersal of barbed or sticky seeds on the feathers or skin of birds or bats) is effective over somewhat shorter distances, but more heavily represented on low islands and atolls;
- Dispersal in mud on birds' feet is effective over long distances, but relatively infrequent on most islands; and
- Dispersal of seeds via gravity is effective over the shortest distances; species with heavy, gravity- or wind-dispersed seeds (e.g. Araucariaceae, Dipterocarpaceae) are restricted to islands that were connected to mainland source areas during periods of

Fig. 17.1 Habits and habitats of representatives of the six endemic genera (subgenera in *Lobelia*) of Hawaiian lobelioids (see text). (a) *Lobelia gloria-montis* (about 1 m tall), summit bog on Pu'u Eke, Maui; (b) *Trematolobelia macrostachys* (2 m), wet forest openings, Oahu; (c) *Brighamia insignis* (about 1 m), sea-cliffs on Kauai; (d) *Delissea undulata* (about 3 m), dry forests on Hawaii; (e) *Clermontia kakeana* (2 m), mesic forest openings in 'Iao Valley, Maui; and (f) *Cyanea hamatiflora* (5 m), wet forest along Waikamoi Flume, Maui (emergent rosette tree with narrow leaves on far side of stream near middle of photograph; note human figures in stream for scale). Photos (a) and (d) from Rock (1919), reproduced courtesy of the National Tropical Botanical Garden; (c) from Carlquist (1980), reproduced courtesy of the Bishop Museum Press, Honolulu.

lower sea levels (e.g. Borneo), or that have rafted away from mainland areas via continental drift (e.g. New Caledonia).

These patterns of seed disperal in the floras of oceanic islands raise two key questions: What is the relationship between speciation rate within genera and the mechanism of seed dispersal? And, why is endozoochory favored mainly on tall islands?

Dispersal and plant speciation

The highest rates of speciation on islands and archipelagoes should occur in groups with limited dispersal capability—newly founded populations of such species are likely to be isolated genetically by modest geographic barriers, leading to allopatric speciation. It would be ideal to test this hypothesis using data for all the clades occurring on one or more archipelagoes. Unfortunately, such a test is not yet practicable—few plant lineages on any island group have been analysed to produce phylogenies and identify sister groups. In the interim I note that, of the 20 most species-rich genera of plants native to the Hawaiian Islands (Wagner *et al.* 1990; Lammers *et al.* 1994), 12 (or 60%) are endozoochorous and 5 (25%) are wind-dispersed. These figures compare with an estimated incidence of 39% endozoochory and 1% anemochory among roughly 256 founding colonists of the Hawaiian flora (Carlquist 1974).

These data should be viewed cautiously: some large genera (e.g. *Cyanea* and *Clermontia, Phyllostegia* and *Stenogyne*) are closely related and may be part of the same clade; some genera (e.g. *Cyrtandra*) are thought to reflect multiple introductions; and the minimum number of colonists required to account for the native flora has not yet been established rigorously. Nevertheless, the data are surprising in that endozoochory, a dispersal mechanism that appears very effective over long distances, also appears to be related to high rates of speciation.

It is important to recognize that 10 of the 12 large endozoochorous genera (*Cyanea* , *Cyrtandra, Pelea, Phyllostegia, Peperomia, Myrsine, Hedyotis, Labordia, Coprosma*) are primarily occupants of the shady interiors of wet rainforests and cloud forests. Givnish *et al.* (1995) argued that endozoochory might—when dependent on forest-interior birds—greatly accelerate geographic speciation, given that such birds are poor dispersers across geographical barriers caused by water or open habitats (see Diamond *et al.* 1976). This phenomenon may be an important process operating in tropical forests on both continents and islands, and deserves further investigation (see Sections 17.7 and 17.8).

Ecological constraints on endozoochory

What is the ecological basis for the association of fleshy fruits with tall islands? Carlquist (1974) observed that fleshy fruits seem more frequent in forests receiving abundant rainfall. Taller islands induce greater rainfall, at least at mid-elevations. Carlquist argued that they should thus favour plants with fleshy fruits—but left the actual advantage of endozoochory unspecified. Gentry (1982, 1988) demonstrated quantitatively that the proportion of the woody flora with fleshy disseminules increases with rainfall in the neotropics. In many neotropical forests, the incidence of endozoochory is also higher among taxa fruiting in the rainy season than among those fruiting in the dry season (Smythe 1970; Croat 1978).

Why does heavy rainfall promote endozoochory? One obvious reason is that the most common alternative in neotropical forests, wind dispersal, is relatively more effective during dry, often windy conditions (Smythe 1970; Gentry 1982). I believe that another factor may be operating, however. Most frugivorous birds are actually frugivores/gleaning insectivores (Snow 1976), requiring caterpillars and other prey as a 'protein subsidy' to complement the protein-poor fruit in their diet. Rates of insect folivory on tropical trees are highest on young, newly expanding leaves (Coley 1983). Given that rates of expansion of new leaves are much higher during wet periods in tropical forests (Frankie *et al.* 1974), it is reasonable to expect that endozoochory would be most likely to evolve in wetter forests and seasons—the protein subsidy necessary for frugivores would be greater community-wide under such conditions.

This overlooked mechanism might also lead to a paucity of endozoochory on nutrient-poor soils. Plants on such substrates should allocate heavily to anti-herbivore defenses in order to maximize expected growth (Janzen 1974; Coley 1983), so the protein subsidy available for the evolution of plant–frugivore mutualisms might be minimal. This hypothesis could be tested by comparing dispersal spectra, folivorous insect productivity, and frugivorous bird density in regions of heavy rainfall but widely differing soil fertility. High rates of speciation and local endemism in plant groups might be expected in regions with sterile substrates, based on the relative absence of an effective means of long-distance dispersal. The wet but extremely infertile summits of the tepuis (sandstone table mountains) of the Guayana Shield do indeed seem to have a low frequency of taxa with fleshy fruits (personal observation). This may contribute—together with isolation and vicariance—to the remarkable degree of narrow endemism in the tepui flora: several of the largest plant groups atop the island-like tepuis (e.g. Asteraceae, Bonnettiaceae, Bromeliaceae, Rapateaceae) have many species restricted to single mountains (Huber 1988; Berry *et al.* 1995; Givnish *et al.* 1997) and are wind dispersed or lack any obvious means of seed dispersal (see also Chapter 15).

17.3 Classical patterns: evolution of arborescence

Several groups of predominantly herbaceous plants, notably members of the Asteraceae, Apiaceae, Campanulaceae, Brassicaceae, and Boraginaceae, appear to have evolved woodiness and a tree- or shrub-like habit on different islands (Darwin 1859; Carlquist 1965, 1970, 1974). Molecular studies have provided direct evidence confirming the evolution of arborescence in the silversword alliance (Baldwin 1997), *Alsinidendron–Schidea* (Wagner *et al.* 1995; Sakai *et al.* 1997), and lobelioids (Givnish *et al.* 1995, 1996) on Hawaii; the tree lettuces of Macaronesia and Juan Fernandez (Kim *et al.* 1996); *Echium* (Böhle *et al.* 1996) and *Argyranthemum* (Francisco-Ortega *et al.* 1997) of Macronesia; and *Lobelia* and *Dendrosenecio* of the island-like East African highlands (Knox *et al.* 1993; Knox and Palmer 1996). Members of the Asteraceae appear to have evolved the tree-like habit on islands repeatedly, including the silversword alliance and three other lineages on Hawaii; the tree lettuces of Macaronesia (*Babcockia*, *Prenanthes*, *Sonchus*, *Sventenia*, *Taeckholmia*), Juan Fernandez (*Dendroseris*), Desventuradas (*Thamnoseris*), and the California Channel Islands (*Munzothamnus*); the remarkable tree cabbages, tree asters, tree sneezeweeds, and related trees of St Helena

(*Aster*, *Commidendron*, *Melanodendron*, *Petrobium*, *Psiadia*, *Senecio*); the tree thistles (*Centaurodendron*, *Yunquea*), tree sneezeweeds (*Robinsonia*, *Rhetinodendron*, *Symphyochaeta*), and tree fleabanes (*Erigeron*) of Juan Fernandez; the tree beggar's-ticks (*Bidens*, *Fitchia*) of Polynesia; the tree everlastings (*Monarrhenus*) and tree sneezeweeds (*Faujasia*) of the Mascarene Islands; and the tree sunflowers (*Scalesia*) and tree fleabanes (*Darwiniothamnus*) of the Galápagos (Carlquist 1965, 1970, 1974). Other putative origins of arborescence on islands include the lobelioids, amaranths, sandworts, violets, plaintains, and gesneriads of Hawaii; the viper bugglosses (*Echium*) of Macaronesia; the tree Fuchsia of New Zealand; and the tree celeries of Macronesia and Socotra.

Based on his observations of tree sunflowers in the Galápagos, Darwin (1859) hypothesized that trees would be unlikely to reach islands ('...whatever the cause may be'), and thus the herbaceous plants which did reach them would obtain a competitive advantage over others by growing taller and developing into shrubs and then trees. Carlquist (1965, 1974) provided a mechanistic explanation for why trees might be at a disadvantage in reaching oceanic islands, noting that many have heavy seeds, while pioneer herbs often have lighter seeds that are more easily dispersed on the wind or the feathers of birds. Wallace (1878) argued that woodiness might evolve in herbaceous colonists as a means of extending their life span, and thus, increasing their chance for sexual reproduction where pollinators may be scarce. I consider these two hypotheses in turn.

The competition hypothesis

Carlquist (1965, 1974) questioned Darwin's claim that competitive overtopping was the predominant pressure favouring the evolution of the tree-like habit, noting that the Galápagos tree sunflowers 'grow most frequently in thinly populated open situations'. He argued that many trees possess massive seeds as an adaptation to endure shade near ground level as seedlings, so that they can survive and grow into the better-lit, upper reaches of the forest. Finally, he claimed that the 'most logical explanation' for insular woodiness was that moderate climates (mean temperature between 10 and 20 °C, mean precipitation > 1000 mm) simply permit the development of plants with a tree-like growth form.

Each of Carlquist's three points is debatable. The argument against selection for greater stature in herbaceous island colonists seems to be incorrect. Givnish (1982, 1995) and Tilman (1988) have advanced detailed models and quantitative measures that, in essence, support Darwin's (1859) hypothesis, and predict that optimal plant stature should increase with the density of plant coverage. Coverage increases sharply with elevation in the Galápagos and is substantial in areas with *Scalesia*, the native tree sunflowers (Reeder and Riechert 1975; Hamann 1979; Eliasson 1984). Carlquist's (1974) claim that selection for overtopping could not account for great stature in *S. pedunculata*, because it dominates the forests in which it occurs, overlooks intra-specific competition. His impression of a very scattered coverage of shrubby *Scalesia* might reflect the destructive effects of introduced goats, horses, donkeys, and pigs, which have only recently been extirpated from several islands, or the devastating effects of El Niño (see Grant 1986; P. R. Grant personal communication).

Carlquist's argument for an advantage of large seeds in trees is erroneous. Even a fairly large seed (for example, an acorn) weighs only a few hundreds or thousands of milligrams, and cannot build a tall sapling or supply it with energy for very long. Large seeds are far more likely to provide an advantage by permitting a tree seedling's root to penetrate the leaf litter into mineral soil, or by increasing its ability to compete in a crowded below-ground environment.

With regard to his third point, moderate climates do permit tree growth (Schulze 1982; Grace 1987; Givnish 1995). However, Carlquist's (1974) argument that trees evolve in moderate climates merely because they can survive there ignores the costs of further height growth: taller plants must allocate more to support tissue, and hence have less energy available for further growth and reproduction. They can only gain an edge by increasing photosynthesis through overtopping, or in the case of paramo and Afro-alpine rosette shrubs, and plants adapted to brief fires over sterile substrates, by ameliorating leaf microclimate (Givnish *et al.* 1986; Givnish 1995). Short growing seasons or inadequate supplies of moisture/nutrients can prevent tall woody plants from achieving positive whole-plant growth under any circumstances (Givnish 1995).

It seems to have escaped notice that many of the most compelling cases of the evolution of arborescence from herbaceous ancestors involved the colonization of open or partially open habitats, mainly scrub, dry forests, and bogs. For example in the silversword alliance, basal *Argyroxiphium* and *Wilkesia* occur in alpine deserts, bogs, and dry scrub, mainland relatives occur in dry winter-rainfall habitats, while some species of *Dubautia* and *Argyroxiphium* have invaded closed forests (for phylogeny and description of ecology see Baldwin and Robichaux (1995)). In *Alsinidendron-Schiedea*, the molecular tree of Sakai *et al.* (1997) places species from mesic forest, dry coastal scrub, and dry ridges and cliffs at the base of the Hawaiian clade. The Hawaiian lobelioids appear to have evolved from an initial invasion of open, wet, upland habitats (Givnish *et al.* 1996), as have the woody Hawaiian violets (Ballard and Sytsma, unpublished). Based on their putative relatives and/or current habitats, Hawaiian *Bidens, Lipochaeta, Charpentiera, Nototrichium, Lepidium,* and *Euphorbia* seem likely to have invaded dry areas initially, as do the Macaronesian tree lettuces (Kim *et al.* 1996), tree chrysanthemums (Francisco-Ortega *et al.* 1997), and viper buglosses (Böhle *et al.* 1996), and many of the tree-like Asteraceae of St Helena, the Galápagos, Desventuradas, the Mascarene Islands, and southern Polynesia.

Open or partially open habitats are likely to be frequented by birds carrying seeds from similar environments on the mainland. These habitats thus favour colonization and persistence of sun-adapted, herbaceous colonists. I suggest that evolution of increased stature, shade adaptation, and seed mass would occur gradually along gradients of increasing coverage away from open habitats, since optimal plant height increases gradually along such gradients (see Givnish 1982, 1995; Tilman 1988). It seems unlikely that sun-adapted herbs could colonize closed communities directly, or that shade-adapted herbs (if able to disperse to an island) would be selected to increase their stature and allocation to unproductive stem tissue (see Givnish 1995).

The longevity hypothesis

Carlquist (1974) countered Wallace's (1878) argument for selection for greater length of life under conditions of pollinator scarcity, noting that many plants can self-pollinate and that suitable pollinators are likely to be either super-abundant on islands (due to lack of competition) or missing entirely (due to lack of dispersal). Nevertheless, Böhle *et al.* (1996) have recently advanced a slightly modified version of Wallace's hypothesis in a study of Macaronesian *Echium* species. They claim the species evolved the woody habit in order to promote sexual outcrossing and escape inbreeding depression, and that the woody habit made insular populations better at colonizing other islands than their herbaceous progenitors. Woody insular *Echium* are preferential outcrossers (setting only 0 to 11% fertile seeds when selfed), whereas their continental relatives (and the sole insular herbaceous species) are preferentially inbreeders (Bramwell 1972).

This argument for the evolution of woodiness is not compelling. Perennial herbs could just as easily avoid outcrossing, and many woody *Echium* are monocarpic with gigantic terminal inflorescences, ensuring small population sizes and high inbreeding rates due to small effective population sizes or geitonogamy. The greater colonizing ability of insular *Echium* may be due to shared factors other than the woody outcrossing habit, such as greater proximity to other islands, or selection on their common ancestor for greater dispersal ability. Higher rates of speciation in the insular clade might be related to some other shared trait, such as occurrence in species-poor, unsaturated island communities. Böhle *et al.*'s (1996) claim that divergent insular growth-forms are equivalent means of ensuring longevity and are not specifically adapted to different environments is belied by the remarkable convergence of *E. bourgeauanum* on the highly unusual, silvery monocarpic rosette of the Haleakala silversword growing in a similar tropical, high-elevation habitat on Maui (see Carlquist 1974).

17.4 Origin of insular plant groups

Over the past seven years, detailed analyses of the origin, adaptive radiation, and geographic pattern of speciation of several lineages of island plants have been conducted using phylogenies based on variation in nuclear and/or chloroplast DNA. These investigations provide some of the most powerful evidence to date on insular plant evolution. Almost all were conducted on groups initially thought to be monophyletic (i.e. to represent the entire set of daughter taxa derived a single common ancestor) and to have undergone extensive adaptive radiation. Monophyly has been confirmed in essentially all cases, with derivation from a specific continental group being deduced for some clades.

Comprehensive studies by Baldwin and his colleagues (Baldwin *et al.* 1990, 1991; Baldwin 1992, 1997; Baldwin and Robichaux 1995) have confirmed the initial inference by Carlquist (1959) that the Hawaiian silversword alliance was derived from a North American tarweed ancestor in the subtribe Madiinae of Asteraceae, pinpointing *Raillardiopsis* of summer-dry California as their sister group. Böhle *et al.* (1996) showed convincingly that the Macaronesian species of *Echium* were derived from a single common ancestor, with an explosive speciation in the Canary Islands followed by single invasions (and subsequent speciation) of Madeira and the Cape Verde Islands. Kim *et al.* (1996) inferred that the Macaronesian tree lettuces were derived from continental herbs in *Sonchus* subgenus *Sonchus,* which are also sister to the Juan Fernandez tree lettuces and allied taxa in New Zealand. Givnish *et al.* (1994, 1995) demonstrated that *Brighamia, Clermontia, Cyanea* (including *Rollandia*), and *Delissea* were derived from a single common ancestor; recent findings indicate that the entire group of Hawaiian lobelioids are monophyletic (Givnish *et al.* 1996).

The geological dating of individual islands (e.g. Clague and Dalrymple 1987; Coello *et al.* 1992), combined with molecular data and an assumed molecular clock, have permitted estimation of the time of origin of a few plant groups. Based on the historical pattern of inter-island dispersal inferred from a molecular phylogeny for *Cyanea* (Fig. 17.2) and observed amounts of genetic divergence, Givnish *et al.* (1995) calculated that *Cyanea* arose 8.7 to 17.4 Ma (million years) ago, on a former tall island between Nihoa and French Frigate Shoals. These areas lie 400 to 1100 km north-west of Kauai, the oldest (5.1 Ma) tall island now above water. Baldwin (1997) estimated that the Hawaiian silversword alliance originated roughly 6 Ma ago. Using a molecular-clock assumption, Kim *et al.* (1996) calculated that tree lettuces evolved in Macaronesia roughly 4.2 Ma ago, long after the formation of the Canary Islands 15 to 20 Ma ago. Böhle *et al.* (1996) estimated that *Echium* arrived on Madeira and the Cape Verde archipelago 0.75 to 3.0 Ma ago. Kim *et al.* (1996) suggested that massive extinctions in the Macaronesian Islands during the past few million years, induced by glacial cycles and the beginning of Sahara desertification, may have provided open habitats for radiation of the tree lettuces; presumably, this would be true for other groups (e.g. *Aeonium, Crambe, Echium*) as well.

17.5 Geographic pattern of speciation

Several studies, especially those involving lineages on the 'conveyor belt' of the Hawaiian Islands, reveal the sequence of dispersal and speciation events. Givnish *et al.* (1994, 1995) found that inter-island dispersal (and speciation) events in *Cyanea* were primarily down the chain, from one island to the next younger one in the sequence (Fig. 17.3). Such patterns were first reported for Hawaiian *Drosophila* (Carson 1983) based on chromosomal banding patterns; they have subsequently been reported for many other groups, including the silversword alliance (Carr *et al.* 1989; Baldwin 1992), *Alsinidendron-Schiedea* (Wagner *et al.* 1995; Sakai *et al.* 1997), based on molecular data, and the plant genera *Hesperomannia, Remya, Hibiscadelphus,* and *Kokia* (Funk and Wagner 1995), based on morphology. This pattern is thought to reflect the greater chance of successful dispersal, establishment, and subsequent speciation by colonists on nearby, newly formed, and ecologically unoccupied islands.

Exceptions to this pattern do exist. Lowrey (1995) used morphology to infer that dispersal in *Tetramelopium* (Asteraceae) occurred mainly from older to younger islands, perhaps reflecting a relatively recent colonization of the archipelago. Lammers (1995) used morphology to argue that *Clermontia* originated on Hawaii less than 0.4 Ma

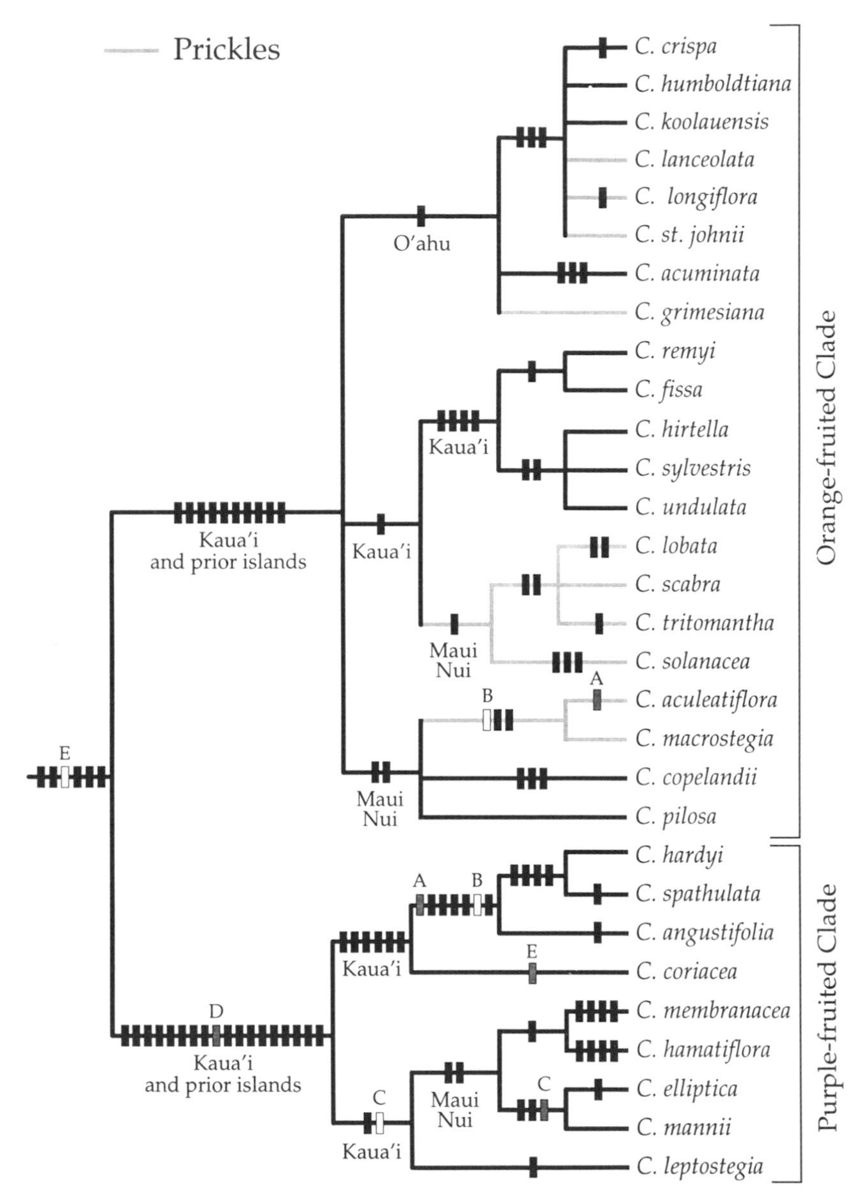

Fig. 17.2 Geographic distribution and possession of thorn-like prickles (grey lines) superimposed on the molecular phylogeny of *Cyanea* derived by Givnish *et al.* (1995), including 31 of 54 extant species. Island names represent the distribution of ancestral taxa inferred from the distribution of present-day taxa. Four terminal taxa have distributions deviating from those of closely related species, indicating a recent inter-island dispersal: *C. tritomantha* and *C. pilosa* on Hawaii, *C. membranacea* on Oahu, and *C. angustifolia* on Oahu and younger islands. Vertical bars indicate shared derived cpDNA restriction site gains or losses. Hollow bars indicate convergent site losses; grey bars, convergent site gains; and solid bars, unreversed site gains or losses. Prickles appear to have evolved on at least three occasions, on Oahu and Maui Nui.

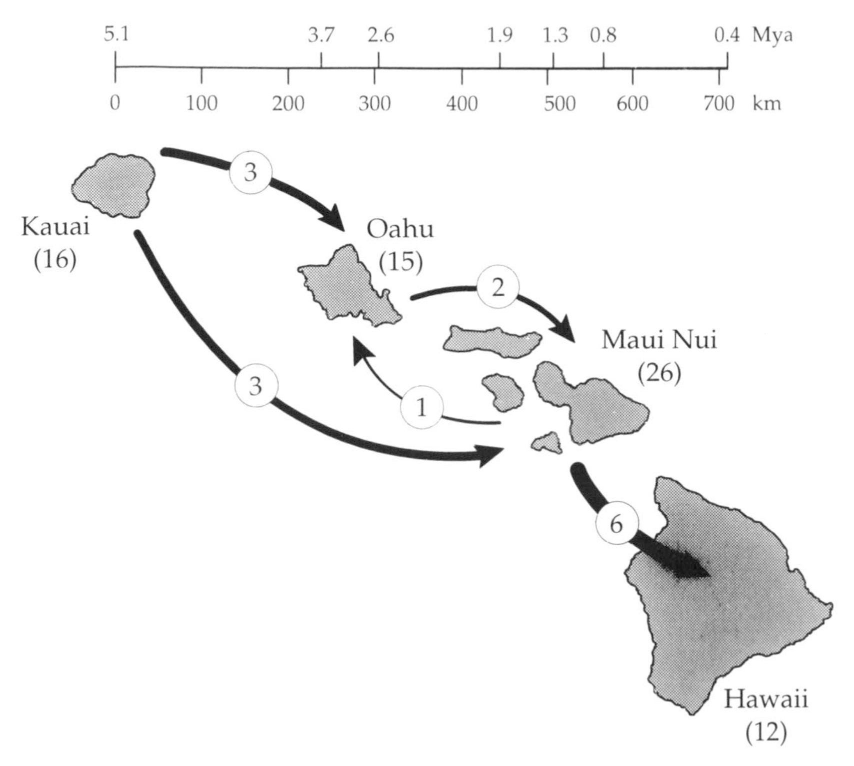

Fig. 17.3 Minimum number of inter-island dispersal events needed to account for present-day species in *Cyanea,* based on molecular and morphological data (after Givnish *et al.* 1995); note tendency for species to disperse from one island to the next younger one. The width of each arrow is proportional to the number of dispersal events between the corresponding pair of islands; the number of species found on each island or island group is indicated in parentheses.

ago, and then back-dispersed to the older islands. However, this conclusion depends on a questionable rooting of the *Clermontia* phylogeny. Givnish *et al.* (1995) concluded from molecular data that *Clermontia* diverged from *Cyanea* at least 8.7 Ma ago, and argued that many former species of *Clermontia* (an early-successional group) may have been extirpated on older islands as they became volcanically more quiescent. The patterns of genetic divergence in *Cyanea* compared with *Clermontia* (Fig. 17.4) support this view; *Clermontia* have a polled family tree from which the lower, early-diverging, branches appear to have been pruned, leaving only a crown of recently diverged twigs.

When plant groups invade successively younger terrains, within-island speciation frequently occurs. Instances include the silversword alliance, *Cyanea-Clermontia, Alsinidendron-Schiedea, Hesperomania, Hibiscadelphus, Kokia,* and *Remya* on the Hawaiian Islands; *Dendroseris* on the Juan Fernandez Islands; and the tree lettuces, *Aeonium, Argyranthemum,* and *Echium* in the Macaronesian Islands. Speciation events sometimes appear to be tied mainly to geographic isolation created by ridges or lava flows (e.g. *Cyanea giffardii*; Givnish *et al.* 1995). In other cases speciation appears to

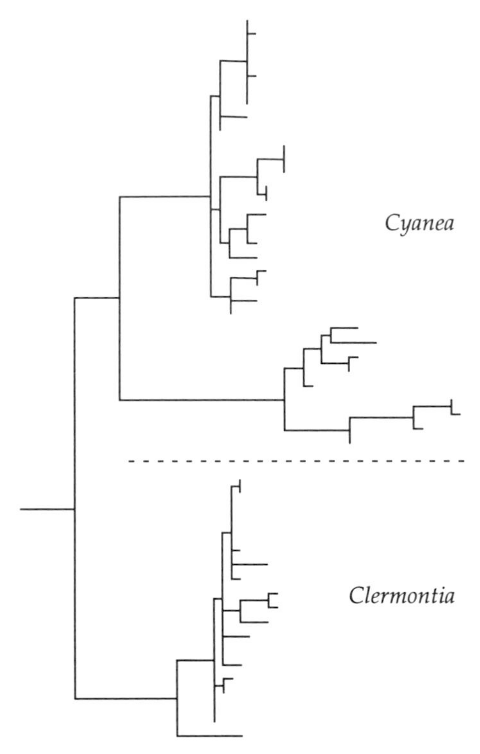

Fig. 17.4 Phylogram for *Clermontia* and *Cyanea* based on cpDNA restriction-site variation; the length of each horizontal branch is proportional to the number of site mutations inferred between nodes (Givnish *et al.* 1995 and unpublished data). Note that the phylogeny of *Clermontia* seems to be a 'polled tree', in which all but the most recently diverged lineages appear to have become extinct in the 8.7 Ma or more since the genus diverged from *Cyanea* (see text).

be coupled to ecological shifts as well; for example sister species of *Cyanea* occur at different elevations (Givnish *et al.* 1995), sister species of *Dubautia* are adapted to wet or dry environments (Baldwin and Robichaux 1995, Baldwin 1997), and on Tenerife sister species of *Argyranthemum* are adapted to different life-zones (Francisco-Ortega *et al.* 1996*a*, 1997).

17.6 Extent of genetic differentiation

Many insular plant groups show little genetic divergence between species (Crawford 1990; Baldwin *et al.* 1991; Baldwin 1992, 1997; Crawford *et al.* 1992; Givnish *et al.* 1995; Böhle *et al.* 1996; Kim *et al.* 1996; Francisco-Ortega 1996*a*, 1997; Sakai *et al.* 1997). While this pattern partly reflects low rates of molecular evolution in Asteraceae (Kim and Jansen 1995), it extends to other families as well. It has caused many investigators to rely on sequencing rapidly evolving regions, such as the ITS region of nuclear ribosomal DNA or the *trn*L–*trn*F region of cpDNA. As might be expected, there is often little sequence or restriction-site variation coupled with dramatic changes in morphology and ecology; Clarke and colleagues (Chapter 11) and Rüber and colleagues (Chapter 14) noted the same pattern of discordance in molecular and morphological studies of snails and fish respectively. Examples from island plants include major shifts of flower length and leaf shape in *Cyanea,* of growth form in *Echium,* of growth form, leaf shape, and physiology in the silversword alliance, and of habitat in the silversword alliance, *Alsinodendron-Schiedea, Argyranthemum, Cyanea, Echium,* and Macaronesian tree lettuces.

Hybridization appears to have played a major role in speciation within *Cyrtandra* on Hawaii (Smith *et al.* 1996) and *Argyranthemum* in Macaronesia (Francisco-Ortega *et al.* 1996*a*), and a less prominent role in the Hawaiian silversword alliance (Baldwin and Robichaux 1995; Carr 1995). Shifts in chromosome number and arrangement, where investigated, are generally minor or non-existent; an exception is the silversword alliance, where reciprocal translocations and aneuploidy are common and involved in partial reproductive isolation (Carr and Kyhos 1986; Baldwin 1997).

17.7 Adaptive radiation

Strong evidence for ecological divergence and the evolution of corresponding morphological and physiological adaptations in island plant groups—as opposed to random or non-adaptive change accompanying speciation—is provided to date by a few studies using molecular systematics. These include investigations of (1) ecophysiology, morphology, and biogeography in the Hawaiian silversword alliance (reviewed by Baldwin and Robichaux 1995); (2) habitat divergence in Macaronesian *Argyranthemum* (Francisco-Ortega *et al.* 1996*a*, 1997); (3) breeding systems and growth forms in Pacific *Fuchsia* (Sytsma *et al.* 1991) and Hawaiian *Alsinidendron-Schiedea* (Wagner *et al.* 1995; Sakai *et al.* 1997); (4) nutrient-capture strategies atop tepuis in *Brocchinia* (Bromeliaceae) (Givnish *et al.* 1997); and (5) stature, leaf form, floral morphology, ecology, and biogeography in the Hawaiian lobelioids (Givnish *et al.* 1994, 1995, 1996). The last provide one of the most spectacular examples of explosive plant speciation, adaptive radiation, and coevolution on islands (Carlquist 1965, 1970, 1974; Givnish *et al.* 1994, 1995, 1996), and are discussed below as a striking illustration of the insights that may be obtained from molecular systematic studies of island plants.

Hawaiian lobelioids

The Hawaiian lobelioids (six genera, about 105 species) include alpine rosette shrubs, cliff succulents, forest and bog shrubs, trees, treelets, a few epiphytes, and one vine-like species (Fig. 17.1). Within *Cyanea* (about 65 species), the largest Hawaiian genus of angiosperms, leaves range from entire to toothed, lobed, compound, and doubly compound (Fig. 17.5). Nearly one-third of *Cyanea* species bear thorn-like prickles along their stems and leaf veins; almost all of these also display a developmental shift in leaf form, in which the juvenile leaves are armed and more deeply divided. Almost all the

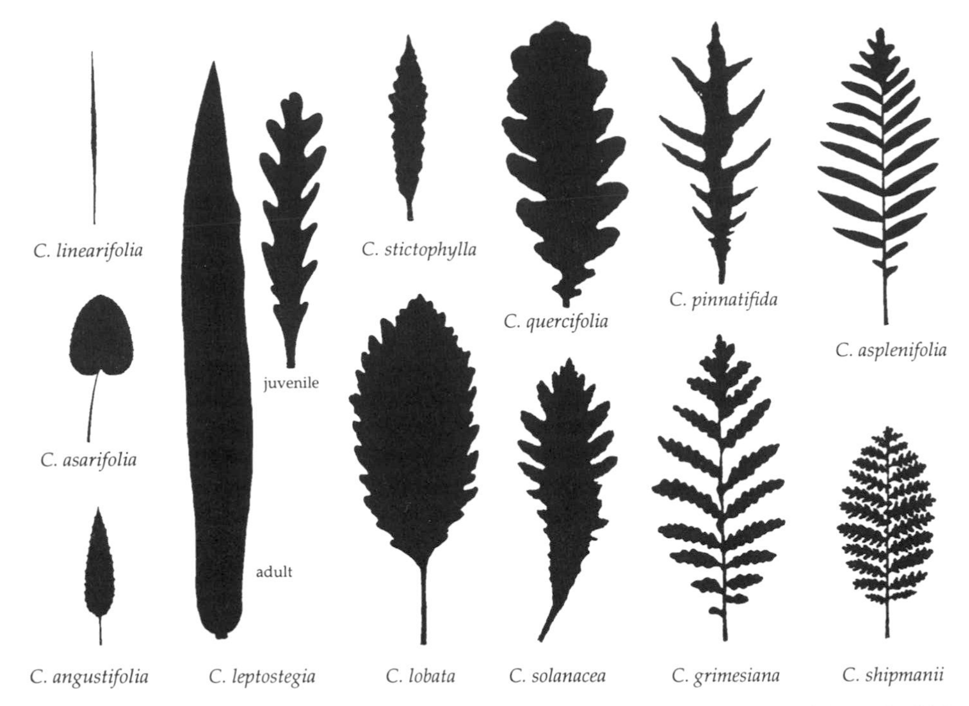

Fig. 17.5 Interspecific variation in leaf size and shape within *Cyanea* (after Givnish *et al.* 1995).

lobelioids were pollinated by honeycreepers and other native birds. They bear curved flowers 15 to 80 mm in length and coloured wine-red, purple, blue, green, or white suffused with red or purple (see Fig. 17.6). Three forest genera (*Clermontia, Cyanea, Delissea*) have fleshy fruits dispersed by birds—unusual in Lobelioideae—and were long thought to be derived from a single colonist (Rock 1919; Carlquist 1965; Lammers 1990); the remaining three genera from more open habitats have wind-dispersed seeds. The Hawaiian lobelioids comprise one-ninth of the native flora; all species and higher categories (including the two Hawaiian sections of cosmopolitan *Lobelia*) are endemic to the archipelago, and most species are restricted to single islands.

Surprisingly, an analysis based on cpDNA restriction-site variation showed that wind-dispersed *Brighamia*, a bizarre sea-cliff succulent whose systematic affinities had eluded botanists for over a century, is the closest relative of fleshy fruited *Delissea,* and that these two genera are sister to the closely related pair *Clermontia* and *Cyanea* (Givnish *et al.* 1995). *Delissea*, mainly restricted to dry forests, shares relatively large white seeds and a narrow canopy with *Brighamia,* and several species also have slightly succulent stems. *Brighamia,* in turn, shares axillary inflorescences with the fleshy fruited genera, and its fruits are fleshy early in development. Recent analyses based on chloroplast and nuclear DNA sequences show that all the Hawaiian genera form a monophyletic group, perhaps the largest plant clade on any archipelago. The open-habitat specialists of *Lobelia* sect. *Galeatella* (alpine bogs), *Lobelia* sect. *Revolutella*

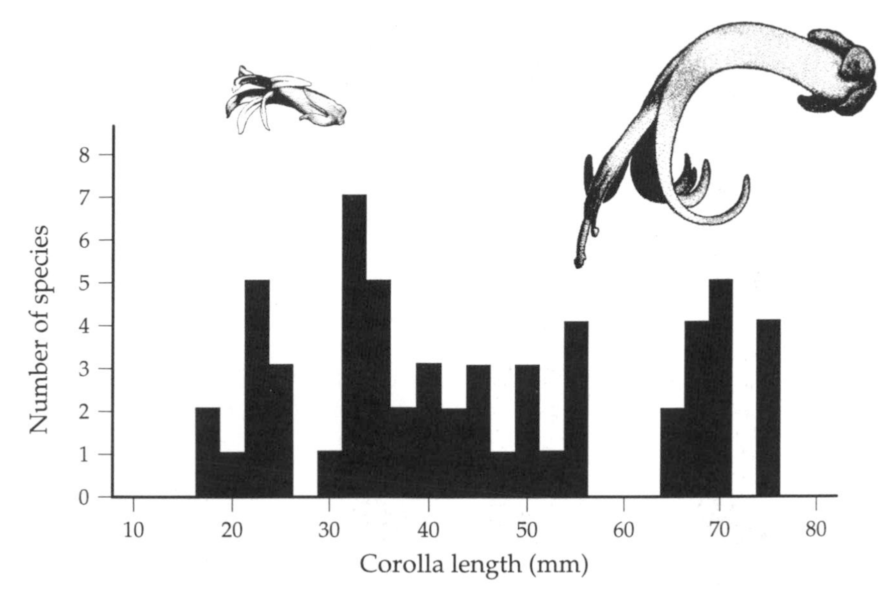

Fig. 17.6 Histogram of corolla lengths across species of *Cyanea;* values are means of minimum and maximum lengths reported by Lammers (1990*a*). Representative flowers shown are those of *C. fissa* from Kauai (left) and *C. superba* from Oahu (after Givnish *et al.* 1995).

(cliffs), and *Trematolobelia* (alpine bogs, wet forest openings) are sister to the fleshy fruited clade (Givnish *et al.* 1996). Their closest relatives are rosette shrubs from open habitats on the Society Islands, Bonin Islands, and East Africa, suggesting that the Hawaiian lobelioids have radiated into progressively shadier habitats, including cliffs (*Brighamia*), dry forests (*Delissea*), edges and openings in moist and wet forests (*Clermontia*), and the deeply shaded interior of moist and wet forests (*Cyanea*).

The genus Cyanea

Within *Cyanea,* the evolution of several traits (e.g. flower tube length, colour, leaf shape, plant stature) has been quite rapid, occurring in perhaps no more than the last 7–8 Ma and hence since the estimated origin of their important pollinators, the honeycreepers (Sibley and Ahlquist 1982; see also Tarr and Fleischer 1995). Several traditional taxonomic groups represent amalgamations of convergent species (Givnish *et al.* 1995). *Cyanea* divides into orange- and purple-fruited clades (Fig. 17.2). Members of the orange-fruited clade appear to be adapted to shady conditions in wet-forest understories, based on their unbranched habit, unusually long leaves and broad crown, and a fruit colour that is visually conspicuous under canopies. Members of the purple-fruited clade appear to be adapted to the somewhat drier and/or more open conditions which many occupy, based on their suckering habit, shorter leaves and narrow crown, and a fruit colour that may be more conspicuous to birds in well-lit habitats (Givnish *et al.* 1995). Orange-fruited species also tend to have longer flowers, paralleling a pattern seen in hummingbird-pollinated plants in the neotropics (see Feinsinger 1983), in which understory species often have long flower tubes and are pollinated by trap-lining, long-billed hermits, while gap species often have shorter flowers and are pollinated by territorial trochilines.

On each of four major islands—Kauai, Oahu, Maui Nui (including Maui, Lanai, and Molokai, all connected during lower sea levels in the Pleistocene), and Hawaii—a set of species spanning nearly the full range of flower tube lengths has evolved, although the interspecific average is significantly lower on the oldest island of Kauai (Givnish *et al.* 1995). On Kauai, the related and morphologically similar species of the Hardyi clade (*C. coriacea, C. hardyi, C. spathulata*) partition the elevational gradient between 150 and 1200 m. Another member of this clade, *C. angustifolia,* appears to have undergone ecological release on Oahu and other islands where its relatives (and perhaps closest competitors) are absent, spanning the elevations occupied by *C. coriacea* and *C. hardyi.*

Protection against browsing herbivores

The fraction of *Cyanea* species with thorn-like prickles, divided leaves, and juvenile–adult leaf dimorphism increases sharply toward the younger islands, from none on Kauai to 50% on Hawaii. The phylogenetic and geographic distribution of such species implies that they arose fairly recently, no more than 3.7 Ma ago, the age of the oldest rocks on Oahu (Fig. 17.3).

Carlquist (1962, 1965, 1970, 1974) proposed that prickles were adapted to deter grazing by native Hawaiian tree snails (Succinidae). This seems most unlikely, however, given that these snails have never been observed to consume lobelioid leaves; like the achatinellids (the largest group of native terrestrial molluscs), the succinids usually graze on epiphyllic fungi and algae (Givnish *et al.* 1994, 1995). Furthermore, prickles of *Cyanea* are so large (up to about 1 cm long) and widely spaced that it is not at all clear they would deter any small snail from climbing or feeding. This is especially true of the foliar prickles, which are restricted to the leaf veins.

A few years after Carlquist had proposed his hypothesis Olson and James (1982, 1991) discovered the sub-fossil remains of moa-nalos (giant ducks) and giant geese in lava tubes and lithified sand dunes. These flightless herbivorous birds evidently flourished in the Hawaiian Islands until they were extirpated by the invading Polynesians roughly 1600 years ago. Browsing forest species appear to have been present on each of the five largest islands except Kauai, where one moa-nalo, apparently adapted to grazing, occurred (Givnish *et al.* 1994, 1995).

Givnish *et al.* (1994, 1995) suggested that prickles and juvenile–adult leaf dimorphism evolved in *Cyanea* species as defences specifically against these large flightless birds. Prickles serve as mechanical defences against browsing of leaves and stem. Divided foliage would associate the remaining leaf tissue more closely with the vein-borne prickles, and might serve to confuse visually orienting avian herbivores (Givnish *et al.* 1994, 1995). Restriction of prickles and divided leaves to juvenile (or mechanically damaged) plants less than about 1 m tall makes adaptive sense, given the stature and reach of the moa-nalos and giant geese.

This hypothesis, developed to explain features of *Cyanea*, may be quite general, as it could account for the enigma of a relatively high incidence of heterophylly on certain other oceanic islands (Friedmann and Cadet 1976). Those islands are just the ones on which different groups of birds evolved into flightless browsers and grazers, the insular equivalents of antelopes or deer elsewhere: moas on New Zealand, *Sylviornis* on New Caledonia, elephant birds on Madagascar, and possibly the dodos, solitaires, or rails on the Mascarene Islands. Several heterophyllous species on the Mascarene Islands also bear thorns or prickles in the juvenile condition (Friedmann and Cadet 1976). Atkinson and Greenwood (1989) have persuasively argued that the densely divaricate branching of the juveniles of several New Zealand trees (55 species in 21 families), together with the visually cryptic and/or mimetic juvenile foliage in some species, served to deter browsing by moas.

Carlquist (1995) recently resuscitated the snail hypothesis, arguing that some (unknown) molluscs fed on *Cyanea* and favoured the evolution of prickles on leaves and stems in the humid layer close to the ground, where snails are more likely to be active. This does not explain the striking absence of damage by molluscs on healthy *Cyanea* foliage (unless the snails have all gone extinct), the inappropriate scale of prickles as anti-mollusc defences, the occurrence of prickles and divided leaves on damaged shoots (possibly an induced defence against avian damage), and the significance of divided leaves. In our view (Givnish *et al.* 1994) of all the elements in the Hawaiian flora, *Cyanea* species were especially likely to evolve anti-herbivore defences against vertebrate herbivores because (1) they are unbranched, and hence likely to be killed by removal of the terminal meristem; (2) they bear unusually soft, palatable, and chemically poorly protected foliage (and wood!) that today is avidly sought and consumed by pigs and other introduced mammalian herbivores; and (3) they grow in extremely shady, resource-poor environments in which heavy investment in anti-herbivore defence is favoured (Coley 1983). The snail hypothesis, in contrast, does not explain the presence of prickles on *Cyanea* but absence on other Hawaiian plants in wet-forest understories.

Carlquist's (1970) general argument that vertebrate herbivores exerted no influence on Hawaiian plant defences should be re-examined. Contrary to his claim, some Hawaiian raspberries retain thorns that, while reduced in size and density, are still effective. Furthermore, a few other sparsely branched wet-forest plants (e.g. *Cheirodendron*; Araliaceae), apparently overlooked by Carlquist, also bear foliar armaments.

Cyanea *and* Clermontia *compared*

Cyanea has more than three times as many species as *Clermontia,* its sister genus. Givnish *et al.* (1995) argued that the higher diversity ultimately reflects a crucial difference in dispersal ability and, thus, in rate of speciation via geographic isolation.

Members of both genera are native to moist and wet tropical forests, with *Cyanea* mainly being found in forest interiors, and *Clermontia* usually found in forest gaps and edges. *Clermontia* also has far larger fruits (about 25 mm diameter) than those of *Cyanea* (about 8–15 mm diameter). As a consequence of these two differences, Givnish *et al.* (1995) hypothesized that *Cyanea* would be less adept at long-distance dispersal than *Clermontia.* Forest-interior birds are poor dispersers across water barriers or habitat edges (Diamond *et al.* 1976), whereas forest-edge birds, inherently adapted to disturbance, are far more vagile. Lower dispersal in forest-interior *Cyanea* would favour lower rates of gene flow between populations, resulting in higher rates of genetic isolation and local geographic speciation. High rates of speciation and low rates of dispersal would lead to narrow geographic distributions, and ultimately favour fine patterns of resource partitioning of habitats and pollinators, involving narrow elevational ranges and highly specialized flowers. In fact, 89% of *Cyanea* species are restricted to single islands, compared with 58% of *Clermontia* species, and the average elevational range of a *Cyanea* species is 438 ± 294 m, roughly half that of a *Clermontia* species (Givnish *et al.* 1995).

The narrower distributions and more specialized flowers of forest-interior *Cyanea,* in turn, make it more vulnerable to extinction than *Clermontia.* Three years ago, 22% of all *Cyanea* species were considered extinct and 29% were considered threatened, compared with an extinction rate of 5% and an endangerment rate of 13% in *Clermontia* (Givnish *et al.* 1995). Over the ensuing period, these numbers have changed slightly due to species rediscoveries and additional losses, but the same picture remains, and *Cyanea* still has the largest number of species considered threatened, endangered, or extinct in the US flora. Apparent extinctions in *Cyanea* have been concentrated in species with unusually narrow distributions, long specialized flowers, and/or occurrence in habitats heavily impacted by human activities (Givnish *et al.* 1995).

17.8 Taxon cycles, diversity, and extinction

The invasion of shady forest interiors could have spurred diversification in *Cyanea* in two ways, through adaptive radiation and through an acceleration of local geographic speciation without adaptation. The invasion of wet forests, evolution of fleshy fruits (Section 17.2), and subsequent invasion of shady microsites may have fostered a dependence on non-vagile forest-interior frugivores and led to high speciation rates. Ultimately, this taxon cycle, leading from open habitats and low speciation rates into closed habitats with higher speciation, appears to culminate in a high probability of extinction. The very characteristics leading to great species richness in *Cyanea*, that is limited dispersal, narrow ranges, specialized flowers, are the same characteristics making its species especially prone to extinction.

This pattern may not be restricted to *Cyanea:* 10 of the 12 largest Hawaiian genera with fleshy fruits are also forest-interior specialists (Section 17.2). Indeed, many of the largest genera found in tropical rainforests elsewhere are composed mainly of understory plants with fleshy fruits, such as *Chamaedorea, Dypsis, Geonoma, Psychotria* (and Rubiaceae generally), *Peperomia, Piper,* and *Solanum.* Of course, such groups share a number of characteristics in addition to fleshy fruits: woody habit, short stature, occurrence in tropical forests, etc. Nevertheless, the proposal that geographic speciation in plants may be accelerated by endozoochory in forest understories is novel and deserves further examination. The evolution of heavy, passively dispersed, seeds in other shade-adapted lineages (e.g. *Fitchia*; Carlquist 1974) may be another, related, avenue whereby wet-forest interiors can accelerate plant speciation.

17.9 Summary

Molecular systematics provide a powerful tool for studying the origin and evolution of island plants. This chapter reviews recent studies that test long-standing hypotheses regarding (1) the derivation of insular groups from mainland herbaceous ancestors, (2) the evolution of a tree-like habit after arrival, and (3) extensive speciation and adaptive radiation in relatively short periods of time, often accompanied by little genetic change.

Arborescence is a frequently evolved trait on islands. I review alternative explanations and conclude that a modified version of Darwin's hypothesis of competition for light is most supported; arborescence frequently appears to follow colonization of open habitats by herbaceous ancestors and subsequent invasion of denser communities nearby. A related evolutionary trend towards increased seed size is explained in terms of the advantages of litter penetration and below-ground competition. Plants on tall oceanic islands often bear fleshy fruits. The prevalence of this habit in wet forests is best interpreted as reflecting the need of frugivorous birds for a protein subsidy of insects in their diet. I discuss recent findings from molecular and field studies of the Hawaiian lobelioids. Based on these studies I propose that evolution of fleshy fruits upon invasion of moist forest interiors may spur plant speciation by decreasing gene flow and increasing the (non-adaptive) process of geographic isolation, but it might also make such specialized species extinction-prone.

Acknowledgements

This research was supported by grant DEB 9509550 from the NSF Systematics and Population Biology Program. I would like to thank Ken Sytsma, my colleague on several studies of adaptive radiation and molecular evolution in Hawaii and elsewhere, for helpful discussions and calling my attention to some useful references.

References

Atkinson, I. A. E. and Greenwood, R. M. (1989). Relationships between moas and plants. *New Zealand Journal of Ecology*, **12**, 76–86.

Baldwin, B. G. (1992). Phylogenetic utility of the internal transcribed spacers of nuclear ribosomal DNA in plants: an example from the Compositae. *Molecular Phylogeny and Evolution*, **1**, 3–16.

Baldwin, B. G. (1997). Adaptive radiation of the Hawaiian silversword alliance: congruence and conflict of phylogenetic evidence from molecular and non-molecular investigations. In *Molecular evolution and adaptive radiation*, (ed. T. J. Givnish and K. J. Sytsma), pp. 103–28. Cambridge University Press.

Baldwin, B. G. and Robichaux, R. H. (1995). Historical biogeography and ecology of the Hawaiian silversword alliance (Asteraceae): new molecular phylogenetic perspectives. In

Hawaiian biogeography: evolution on a hot-spot archipelago, (ed. W. L. Wagner and V. A. Funk), pp. 259–87. Smithsonian Institution Press, Washington, D. C.

Baldwin, B. G., Kyhos, D. W., and Dvorak, J. (1990). Chloroplast DNA evolution and adaptive radiation in the Hawaiian silversword alliance (Madiinae, Asteraceae). *Annals of the Missouri Botanical Gardens*, **77**, 96–109.

Baldwin, B. G., Kyhos, D. W., Dvorak, J., and Carr, G. D. (1991). Chloroplast DNA evidence for a North American origin of the Hawaiian silversword alliance (Asteraceae). *Proceedings of the National Academy of Sciences USA*, **88**, 1840–3.

Berry, P. E., Huber, O., and Holst, B. K. (1995). Floristic analysis and phytogeography. In *Flora of the Venezuelan Guayana, Vol. 1*, (ed. P. E. Berry, B. K. Holst, and K. Yatskievych), pp. 161–92. Timber Press, Portland, OR.

Böhle, U.-R., Hilger, H. H., and Martin, W. F. (1996). Island colonization and evolution of the insular woody habit in *Echium* L. (Boraginaceae). *Proceedings of the National Academy of Sciences USA*, **93**, 11740–5.

Bramwell, D. (1972). Endemism in the flora of the Canary Islands. In *Phytogeography and evolution*, (ed. D. H. Valentine), pp. 141–59. Academic Press, London.

Carlquist, S. (1959). Studies on Madiinae: anatomy, cytology, and evolutionary relationships. *Aliso*, **4**, 171–236.

Carlquist, S. (1962). Ontogeny and comparative morphology of thorns of Hawaiian Lobeliaceae. *American Journal of Botany*, **49**, 413–19.

Carlquist, S. (1965). *Island life.* Natural History Press, New York.

Carlquist, S. (1966). The biota of long-distance dispersal. I. Principles of dispersal and evolution. *Quarterly Review of Biology*, **41**, 247–70.

Carlquist, S. (1967). The biota of long-distance dispersal. V. Plant dispersal to Pacific islands. *Bulletin of the Torrey Botanical Club*, **94**, 129–62.

Carlquist, S. (1970). *Hawaii: a natural history.* Natural History Press, New York.

Carlquist, S. (1974). *Island biology.* Natural History Press, New York.

Carlquist, S. (1995). Introduction. In *Hawaiian biogeography: evolution on a hot-spot archipelago*, (ed. W. L. Wagner and V. A. Funk), pp. 14–29. Smithsonian Institution Press, Washington, D. C.

Carr, G. D. (1995). A fully fertile intergeneric hybrid derivative from *Argyroxiphium sandwichense* ssp. *macrocephalum*x*Dubautia menziesii* (Asteraceae) and its relevance to plant evolution in the Hawaiian Islands. *American Journal of Botany*, **82**, 1574–81.

Carr, G. D. and Kyhos, D. W. (1986). Adaptive radiation in the Hawaiian silversword alliance (Compositae-Madiinae). II. Cytogenetics of artificial and natural hybrids. *Evolution*, **40**, 969–76.

Carr, G. D., Robichaux, R. H., Witter, M. S., and Kyhos, D. W. (1989). Adaptive radiation of the Hawaiian silversword alliance (Compositae-Madiinae): a comparison with Hawaiian picture-winged *Drosophila*. In *Genetics, speciation, and the founder principle*, (ed. L. V. Giddings, K. Y. Kaneshiro, and W. W. Anderson), pp. 79–97. Oxford University Press.

Carson, H. L. (1983). Chromosomal sequences and interisland colonizations in the Hawaiian *Drosophila*. *Genetics*, **103**, 465–82.

Clague, D. A. and Dalrymple, G. B. (1987). The Hawaiian-Emperor volcanic chain. In *Hawaiian biogeography: evolution on a hot-spot archipelago*, (ed. W. L. Wagner and V. A. Funk), pp. 1–54. Smithsonian Institution Press, Washington, D. C.

Coello, J., Cantagrel, J.-M., Hernán, F., Fúster, J.-M., Ibarrola, E., Ancochea, E., Casquet, C., Jamond, C., Díaz de Téran, J.-R., and Cendrero, A. (1992). Evolution of the eastern volcanic ridge of the Canary Islands based on new K-Ar data. *Journal of Vulcanology and Geothermal Research*, **53**, 251–74.

Coley, P. D. (1983). Herbivory and defensive characteristics of tree species in a lowland tropical forest. *Ecological Monographs*, **53**, 209–33.

Crawford, D. J. (1990). *Plant molecular systematics: macromolecular approaches*. Wiley, New York.

Crawford, D. J., Stuessy, T. F., Cosner, M. B., Haines, D. W., Silva, M., and Baeza, M. (1992). Evolution of the genus *Dendroseris* (Asteraceae, Lactuceae) on the Juan Fernandez Islands—evidence from chloroplast and ribosomal DNA. *Systematic Botany*, **17**, 676–82.

Croat, T. B. (1978). *Flora of Barro Colorado Island*. Stanford University Press, Palo Alto, CA.

Darwin, C. (1859). *The origin of species by means of natural selection*. John Murray, London.

Diamond, J. M., Gilpin, M. E., and Mayr, E. (1976). Species-distance relation for birds of the Solomon archipelago, and the paradox of the great speciators. *Proceedings of the National Academy of Sciences USA*, **73**, 2160–4.

Docters van Leeuwen, W. M. (1936). Krakatau, 1883–1933. *Ann. Jard. Bot. Buitenzorg*, **46–47**, 1–506.

Eliasson, U. (1984). Native climax forests. In *Galápagos*, (ed. R. Perry), pp. 101–14. Pergamon Press, Oxford.

Feinsinger, P. (1983). Coevolution and pollination. In *Coevolution*, (ed. D. J. Futuyma and M. Slatkin), pp. 282–310. Sinauer, Sunderland, MA.

Fosberg, F. R. (1963). Derivation of the flora of the Hawaiian Islands. In *Insects of Hawaii* (ed. E. C. Zimmermann), pp. 107–119. University of Hawaii Press, Honolulu.

Francisco-Ortega, Jansen, R. K., and Santos-Guerra, A. (1996*a*) Chloroplast DNA evidence of colonization, adaptive radiation, and hybridization in the evolution of the Macaronesian flora. *Proceedings of the National Academy of Sciences USA*, **93**, 4085–90.

Francisco-Ortega, Fuertes-Aguilar, J., Kim, S.-C., Crawford, D. J., Santos-Guerra, A., and Jansen, R. K. (1996*b*) Molecular evidence for the origin, evolution, and dispersal of *Crambe* (Brassicaceae) in the Macaronesian Islands. In *Abstracts of the 2nd Symposium on the Fauna and Flora of the Atlantic Islands,* p. 41. Universidad de Las Palmas de Gran Canaria, Las Palmas de Gran Canaria.

Francisco-Ortega, Crawford, D. J., Santos-Guerra, A., and Jansen, R. K. (1997). Origin and evolution of *Argyranthemum* (Asteraceae: Anthemideae) in Macaronesia. In *Molecular evolution and adaptive radiation*, (ed. T. J. Givnish and K. J. Sytsma), pp. 407–31. Cambridge University Press.

Frankie, G. W., Baker, H. G., and Opler, P. A. (1974). Comparative phenological studies of trees in tropical wet and dry forests in the lowlands of Costa Rica. *Journal of Ecology*, **62**, 881–919.

Friedmann, F. and Cadet, T. (1976). Observations sur l'hétérophyllie dans les iles Mascareignes. *Adansonia*, **15**, 423–40.

Funk, V. A. and Wagner, W. L. (1995). Biogeography of seven ancient Hawaiian plant lineages. In *Hawaiian biogeography: evolution on a hot-spot archipelago*, (ed. W. L. Wagner and V. A. Funk), pp. 160–294. Smithsonian Institution Press, Washington, D. C.

Gentry, A. H. (1982). Patterns of neotropical plant species diversity. *Evolutionary Biology*, **15**, 1–84.

Gentry, A. H. (1988). Changes in plant community diversity and floristic composition on environmental and geographical gradients. *Annals of the Missouri Botanical Gardens*, **75**, 1–34.

Givnish, T. J. (1982). On the adaptive significance of leaf height in forest herbs. *Amererican Naturalist*, **120**, 353–81.

Givnish, T. J. (1995). Plant stems: biomechanical adaptation for energy capture and influence on species distributions. In *Plant stems: physiology and functional morphology*, (ed. B. L. Gartner), pp. 3–49. Chapman and Hall, New York.

Givnish, T. J. and Sytsma, K. J. (1997). Homoplasy in molecular vs. morphological data: the likelihood of correct phylogenetic inference. In *Molecular evolution and adaptive radiation*, (ed. T. J. Givnish and K. J. Sytsma), pp. 55–100. Cambridge University Press.

Givnish, T. J., McDiarmid, R. W., and Buck, W. R. (1986). Fire adaptation in *Neblinaria celiae* (Theaceae), a high-elevation rosette shrub endemic to a wet equatorial tepui. *Oecologia*, **70**, 481–5.

Givnish, T. J., Sytsma, K. J., Smith, J. F., and Hahn, W. J. (1994). Thorn-like prickles and heterophylly in *Cyanea:* adaptations to extinct avian browsers on Hawaii? *Proceedings of the National Academy of Sciences USA*, **91**, 2810–14.

Givnish, T. J., Sytsma, K. J., Hahn, W .J., and Smith, J. F. (1995). Molecular evolution, adaptive radiation, and geographic speciation in *Cyanea* (Campanulaceae, Lobelioideae). In *Hawaiian biogeography: evolution on a hot-spot archipelago*, (ed. W. L. Wagner and V. A. Funk), pp. 299–337. Smithsonian Institution Press, Washington, D. C.

Givnish, T. J., Knox, E., Patterson, T. B., Hapeman, J. R., Palmer, J. D., and Sytsma, K. J. (1996). The Hawaiian lobelioids are monophyletic and underwent a rapid initial radiation roughly 15 million years ago. *American Journal of Botany*, **83**, 159.

Givnish, T. J., Sytsma, K. J., Smith, J. F., Hahn, W. J., Benzing, D. H., and Burkhardt, E. M. (1997). Molecular evolution and adaptive radiation in *Brocchinia* (Bromeliaceae: Pitcairnioideae) atop tepuis of the Guayana Shield. In *Molecular evolution and adaptive radiation*, (ed. T. J. Givnish and K. J. Sytsma), pp. 259–301. Cambridge University Press.

Grace, J. (1987). Climatic tolerance and the distribution of plants. *New Phytol.*, **106(Suppl.)**, 113–30.

Grant, P. R. (1986). *Ecology and evolution of Darwin's finches.* Princeton University Press.

Guppy, H. B. (1906). *Observations of a naturalist in the Pacific between 1896 and 1899. Vol. II: Plant dispersal.* Macmillan, London.

Hamann, O. (1979). Dynamics of a stand of *Scalesia pedunculata* Hooker fil., Santa Cruz Island, Galápagos. *Botanical Journal of the Linnean Society*, **11**, 101–22.

Huber, O. (1988). Guayana Highlands vs. Guyana Lowlands: a reappraisal. *Taxon*, **37**, 595–614.

Janzen, D. H. (1974). Tropical blackwater rivers, animals, and mast fruiting in the Dipterocarpaceae. *Biotropica*, **6**, 69–105.

Kim, K. J. and Jansen, R. K. (1995). *ndh*F sequence evolution and the major clades in the sunflower family. *Proceedings of the National Academy of Sciences USA*, **92**, 10379–83.

Kim, S.-C., Crawford, D. J., Francisco-Ortega, J., and Santos-Guerra, A. (1996). A common origin for woody *Sonchus* and five related genera in the Macaronesian Islands: molecular evidence for extensive radiation. *Proceedings of the National Academy of Sciences USA*, **93**, 7743–8.

Knox, E. B. and Palmer, J. D. (1995). Chloroplast DNA variation and the recent radiation of the giant senecios (Asteraceae) in the tall mountains of Eastern Africa. *Proceedings of the National Academy of Sciences USA*, **92**, 10349–53.

Knox, E. B. and Palmer, J. D. (1996). The origin of *Dendrosenecio* within the Senecioneae (Asteraceae) based on chloroplast evidence. *American Journal of Botany*, **82**, 1567–73.

Knox, E., Downie, S. R., and Palmer, J. D. (1993). Chloroplast genome rearrangements and the evolution of giant lobelias from herbaceous ancestors. *Molecular Biology and Evolution*, **10**, 414–30.

Lammers, T. G. (1990). Campanulaceae. In *Manual of the flowering plants of Hawai'i*, (ed. W. L. Wagner, D. R. Herbst, and S. H. Sohmer), pp. 420–89. Bishop Museum Publications, Honolulu, HI.

Lammers, T. G. (1995). Patterns of speciation and biogeography in *Clermontia* (Campanulaceae, Lobelioideae). In *Hawaiian biogeography: evolution on a hot-spot archipelago*, (ed. W. L. Wagner and V. A. Funk), pp. 338–62. Smithsonian Institution Press, Washington, D. C.

Lammers, T. G., Givnish, T. J., and Sytsma, K. J. (1994). Merger of the endemic Hawaiian genera *Cyanea* and *Rollandia* (Campanulaceae: Lobelioideae). *Novon*, **3**, 437–41.

Lowrey, T. K. (1995). Phylogeny, adaptive radiation, and biogeography of Hawaiian *Tetramelopium* (Asteraceae, Astereae). In *Hawaiian biogeography: evolution on a hot-spot archipelago*, (ed. W. L. Wagner and V. A. Funk), pp. 195–220. Smithsonian Institution Press, Washington, D. C.

Mes, T. H. M. (1995). Origin and evolution of the Macaronesian Sempervivoideae (Crassulaceae). Unpubl. Ph.D thesis, University of Utrecht, Netherlands.

Olson, S. L. and James, H. F. (1982). Fossil birds from the Hawaiian Islands: evidence for wholesale extinction by man before Western contact. *Science*, **217**, 633–5.

Olson, S. L. and James, H. F. (1991). Descriptions of thirty-two new species of birds from the Hawaiian Islands: Part I. Non-Passeriformes. *Ornithological Monographs*, **45**, 1–88.

Reeder, W. G. and Riechert, S. E. (1975). Vegetation change along an altitudinal gradient, Santa Cruz Island, Galápagos. *Biotropica*, **7**, 162–75.

Ridley, H. N. (1930). *The dispersal of plants throughout the world*. L. Reeve, Ashford.

Robichaux, R. H., Carr, G. D., Liebman, M., and Pearcy, R. W. (1990). Adaptive radiation of the Hawaiian silversword alliance (Compositae-Madiinae): ecological, morphological, and physiological diversity. *Annals of the Missouri Botanical Garden*, **77**, 64–72.

Rock, J. F. (1919). A monographic study of the Hawaiian species of the tribe Lobelioideae, family Campanulaceae. *Memoirs of the Bishop Museum*, **7**, 1–394.

Sakai, A. K., Weller, S. G., Wagner, W. L., Soltis, P. S., and Soltis, D. E. (1997). Phylogenetic perspectives on the evolution of dioecy: adaptive radiation in the endemic Hawaiian genera *Schiedea* and *Alsinidendron* (Caryophyllaceae: Alsinoideae). In *Molecular evolution and adaptive radiation*, (ed. T. J. Givnish and K. J. Sytsma), pp. 455–73. Cambridge University Press.

Sang, T., Crawford, D. J., Kim, S.-C., and Stuessy, T. F. (1994). Radiation of the endemic genus *Dendroseris* (Asteraceae) on the Juan Fernandez Islands: evidence from sequences of the ITS regions of nuclear ribosomal DNA. *American Journal of Botany*, **81**, 1494–1501.

Sang, T., Crawford, D. J., Stuessy, T. F., and Silva O. M. (1995). ITS sequences and the phylogeny of the genus *Robinsonia* (Asteraceae). *Systematic Botany*, **20**, 55–64.

Schulze, E.-D. (1982). Plant life forms and their carbon, water, and nutrient relations. In *Encyclopedia of plant physiology, new series, vol. 12A,* (ed. O. L. Lange, P. S. Nobel, C. B. Osmond and H. Ziegler), pp. 615–76. Springer Verlag, Berlin.

Sibley, C. G. and Ahlquist, J. E. (1982). The relationships of the Hawaiian honeycreepers (Drepanidini) as indicated by DNA-DNA hybridization. *Auk*, **99**, 130–40.

Skottsberg, C. (1956). Derivation of the flora and fauna of Juan Fernandez and Easter Island. *Nat. Hist. Juan Fern. Easter Isl.*, **1**, 193–438.

Smith, J. F., Burke, C. C., and Wagner, W. L. (1996). Interspecific hybridization in natural populations of *Cyrtandra* (Gesneriaceae) on the Hawaiian Islands – evidence from RAPD markers. *Plant Systematics and Evolution*, **200**, 61–77.

Smythe, N. (1970). Relationships between fruiting season and seed dispersal methods in a neotropical forest. *American Naturalist*, **104**, 25–36.

Snow, D. (1976). *The web of adaptation: bird studies in the American tropics*. Quadrangle, New York.

Sytsma, K. J., Smith, J. F., and Berry, P. E. (1991). The use of chloroplast DNA to assess biogeography and evolution of morphology, breeding systems, flavonoids, and chloroplast DNA in *Fuchsia* sect. *Skinnera* (Onagraceae). *Systematic Botany*, **16**, 257–69.

Tarr, C. L. and Fleischer, R. C. (1995). Evolutionary relationships of the Hawaiian honeycreepers (Aves, Drepanidinae). In *Hawaiian biogeography: evolution on a hot-spot archipelago*, (ed. W. L. Wagner and V. A. Funk), pp. 147–59. Smithsonian Institution Press, Washington, D. C.

Tilman, D. (1988). *Resource competition and community structure*. Princeton University Press.

van Balgooy, M. M. J. (1971). Plant-geography of the Pacific based on a census of phanerogam genera. *Blumea*, **6**, Supplement, 1–222.

Wagner, W. L., Herbst, D. R., and Sohmer, S. H. (eds.) (1990). *Manual of the flowering plants of Hawai'i*. Bishop Museum Publications, Honolulu, HI.

Wagner, W. L., Weller, S. G., and Sakai, A. K. (1995). Phylogeny and biogeography in *Schiedea* and *Alsinidendron* (Caryophyllaceae). In *Hawaiian biogeography: evolution on a hot-spot archipelago*, (ed. W. L. Wagner and V. A. Funk), pp. 221–58. Smithsonian Institution Press, Washington, D. C.
Wallace, A. R. (1878). *Tropical nature and other essays*. Macmillan, London.
Wallace, A. R. (1980). *Island life*. Macmillan, London.

18

Epilogue and questions

Peter R. Grant

18.1 The need for further research

The study of patterns and processes of evolution on islands has played an important role in the development of general theories of how and why evolution occurs. Such studies are continuing to be exciting new fields for investigation, as theoreticians attack questions of genetic change in island models and empirical biologists seek explanations of adaptive radiations in archipelagos. This book has attempted to survey our current knowledge, conveying, at the same time, the excitement of this important and developing subject.

Despite the impressive body of knowledge and understanding built up since the days of Wallace and Darwin, it is difficult not to be struck by our ignorance on matters both of detail and general principle. Regarding detail, the point can be made with a single example. In this century alone Mendelian genetics have been (re)discovered, DNA has been identified, the genetic code cracked and whole genomes sequenced, not to mention that humans have climbed Mount Everest and walked on the moon, yet scarcely half the species of *Drosophila* in the Hawaiian archipelago have been named. Collecting activity on islands and taxonomic description have been going on for three centuries. Why have the patterns of variation in this very important group not been thoroughly and exhaustively catalogued and described? After all, to understand processes of evolution on islands we ought to first clearly establish the patterns they have produced. The answer is that the task of finding, collecting, documenting, describing, and naming is very large, and demands time, money, and effort. The demands have simply not been met.

Beyond this example of inadequate description there are several issues outstanding and questions unresolved. Some of them have already been raised in various chapters. Others I now briefy discuss as a stimulus and encouragement for further research. The issues have a general significance, since they apply to evolution in all environments. There will be more questions asked than answers offered.

18.2 Establishing the patterns

Our knowledge of evolutionary patterns on islands is far from complete. Even though incompleteness has not stopped us from identifying trends there is an obvious need to finish the task of describing the biological diversity on islands. Rigorous quantitative analysis of the improved data that are being ordered in improved phylogenies will do a better job of first confirming or denying the existence of those trends, and second of characterizing them quantitatively. Of particular interest is the need to establish similar trends among different organisms occupying the same or different islands (Chapters 1 and 5), and to ascertain whether they are general and widespread, or taxonomically or geographically restricted.

Geographical variation in evolution on islands?

The global geography of evolution on islands has received scant attention, in contrast to the situation on continents. Do evolutionary changes on islands vary geographically in a regular and predictable manner, and if so how?

A confident answer can be given for variation along one geographical axis; distance from source. Evolutionary divergence is not so pronounced on islands close to large land masses as on more remotely situated ones, because the biota on near islands are constituted from a larger fraction of the mainland biota, and are subject to more frequent disturbances (immigrations, extinctions, changes in isolation due to changes in sea level, etc.) than on far islands. In contrast relatively few species reach far islands, they experience markedly different conditions from those on the mainland, and change in response to the different selection pressures, relatively unaffected by the disturbing factors of the near islands. There may be an autocatalytic element in the evolution of biota on far islands, with evolutionary change of species late in a colonizing sequence facilitated and enhanced by evolutionary change in species arriving early in the sequence.

Along another axis, latitude, there is no clear answer to the question of regular variation, only hints that morphological trends within continents may be magnified on islands. Flightlessness in beetles is especially prevalent on high-latitude islands (Roff 1990). Birds show contrasting tendencies to become even larger than their large mainland relatives on high-latitude islands, and even smaller than their small mainland relatives on low-latitude islands (Grant 1968).

The total continental land mass is larger in the northern than in the southern hemisphere, mainly through inequalities in the temperate regions. Without counting them I suspect the largest number of marine islands is in the equatorial one-third of the globe, bounded by 30° north and south of the equator. Vast numbers occur in the south-east Pacific, remote from large continental areas. They contain a wealth of biological diversity. Witness the 17 000 islands that make up Indonesia; they are occupied by 10% of the world's plant species, 12% of the world's bird species, and 35% of its fish species (Sawhill 1997). There is much less biodiversity generally at higher latitudes than at lower ones. Thus evolution on islands has produced its most marked effects in or close to the tropics. Of course climate change has affected the high latitudes more than the low ones, resulting in shifts in the distribution of continental organisms, or extinction; a latitudinal shift in the distribution of island organisms is the same as extinction. The record of evolution is therefore richer at low latitudes partly because of its origin and partly because of its preservation. *If* evolution in the tropics is different from evolution in the temperate zone the disproportionate number of islands in the tropics could introduce a bias in the assessment of (global) island trends (temperate zone islands may be better known, however). But *is* evolution on tropical islands different from evolution on temperate islands in any way that could not be predicted from known latitudinal contrasts on continents?

How can events in the past be inferred?

Much of evolutionary biology involves making inferences about the past, in ignorance of the exact time when an important evolutionary event occurred and the environmental circumstances prevailing at that time. Identifying the correct temporal sequence of events on an island can be of crucial importance in determining whether evolution has occurred there as opposed to elsewhere, the circumstances under which it occurred and hence the interpretation of it. In the absence of datable fossils, relative time has been estimated on the basis of such indirect measures as frequency of occurrence of a taxon in different habitats, per cent endemism, distribution, degree of morphological diversification within a group, and distinctness from presumed relatives. All of these are unsatisfactory.

Modern molecular methods of characterizing genetic differences between taxa, coupled with the application of some estimate of the rate of divergence and cladistic methods for ordering the data, are now in favour (Chapters 5, 8, 11, 14, and 17). They are not without uncertainties and problems associated with estimating rates of change in different parts of the genome, reticulation as a result of introgressive hybridization (Chapter 6), and so forth. Confidence intervals on the estimates of temporal events are often large. Much better methods are needed to estimate time in phylogenies, especially for fairly recent or slowly evolving ones, by callibrating molecular divergence against datable geological events.

Callibration of a molecular clock can be accomplished when (1) populations become isolated on islands as sea level rises or land subsides, and the animals or plants are incapable of naturally crossing water barriers, (2) such geological events can be dated, and (3) for a group of organisms there is variation in dates of isolation and measures of genetic divergence. A good example is provided by frogs (*Rana esculenta* group) in the Aegean sea, isolated on islands for periods of time ranging from 0.012 to 5.2 Ma, and unable to cross sea water because their skin is permeable to salt and water (Beerli *et al.* 1996). Studies of island organisms can contribute to improving methods of reconstructing the past, and in turn they benefit from the improvements.

Of equal importance to the interpretation of evolution on islands is the need to reconstruct the environments in which it occurred (Chapters 9, 14, and 16). Tectonic plates move, sea level fluctuates, islands are born, grow, and die, and local and global climates change. Modern conditions are not necessarily a good guide to those experienced by an evolving taxonomic lineage, which means that important selective agents may now be missing and the correct ones misidentified. Reconstruction of past climates lags behind the reconstruction of physical features of islands. For a good example of the latter see Carson and Clague (1995).

Where did evolution occur?

Lemurs are now restricted to Madagascar. Did they evolve there or were they once more widely ranging and have since become restricted to the island? Did the tortoises of Galápagos and the Seychelles evolve their exceptional size there, on the islands, or were they giants when they arrived? Answering such questions of evolutionary origin requires fossils and a phylogeny. Both are in short supply. Examples of using a molecularly based phylogeny to answer questions of taxonomic or trait origin were given in Chapters 11, 13, 14, and 17. Figure 18.1 shows how a phylogenetic reconstruction can provide answers to questions of origins (see also Fig. 18.2). Such phylogenies are few now but will surely proliferate in the future. I wish the discovery of enlightening fossils of island organisms could keep pace with the construction of molecular phylogenies.

How many colonizations?

Often a question arises as to whether one or a few colonizations gave rise to a group of island species; i.e. was there a common ancestor from a single colonization event that gave rise to all of them, or did two or more species arrive on the islands (Chapters 9 and 17)? In principle this is an easy question to answer by comparing island species with mainland relatives. It has been found, for example, that *Echium* colonized the islands of Macaronesia once (Fig. 18.2), and the Galápagos were colonized by two different stocks of *Tropidurus* lizards and three stocks of *Phyllodactylus* geckos (Wright 1983). It is possible that Hawaii was colonized as many as 17 times by the fern *Asplenium adiantum-nigrum* (Ranker *et al.* 1995). But if two (or more) mainland ancestral stocks were involved and one (or more) has since become extinct the wrong conclusion will be drawn; the conclusion will be that the islands were colonized only once, or perhaps twice but by a double invasion of the same ancestral stock. The more ancient the colonization the more difficult it is to identify the origin or origins of an island taxon, partly as a result of evolved distinctiveness and partly for reasons of extinction of mainland (and island) relatives. Thus questions of ancestral origins (paragraph above) and numbers of colonizations are interrelated.

Molecular studies are producing evidence that colonization of an island may be followed by additional immigrations of the same species, in both plants (Section 2.3) and animals (Section 9.4). Observational field studies are finding the same (e.g. Grant and Grant 1995*a*). Multiple intraspecific immigrations are likely to elevate the level of genetic variation in an island population, with important consequences for its persistence and speciation potential (Chapter 6). Important, yet unknown, variables are the interval between successive immigrations and the degree of genetic difference between residents and new immigrants.

Evolution observed

A footnote to these questions about evolution in the past: direct studies of observable evolution on islands are surprisingly rare, despite their importance as a complement to the essentially static analysis of evolution in the past, and the dynamic analysis of

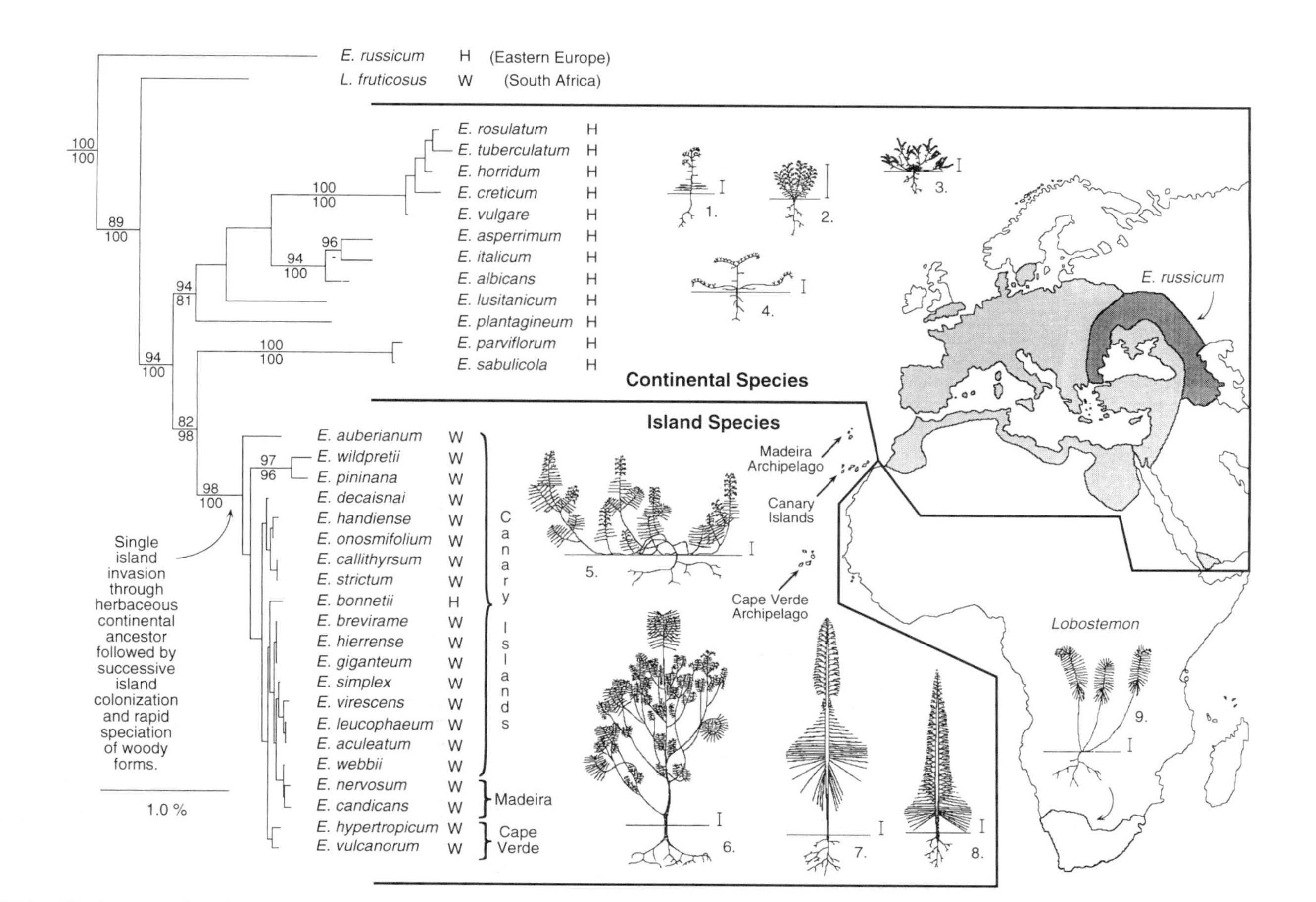

Fig. 18.1 Phylogeny of *Echium* (see Fig. 1.2) and its association with biogeography on islands of Macaronesia, as well as Europe and Africa. (Original photos: *H. H. Hilger* and *H. Kuerschner*, Berlin; graphics *S. B. Tautz*, Braunschweig). From Böhle *et al.* (1996) with permission.

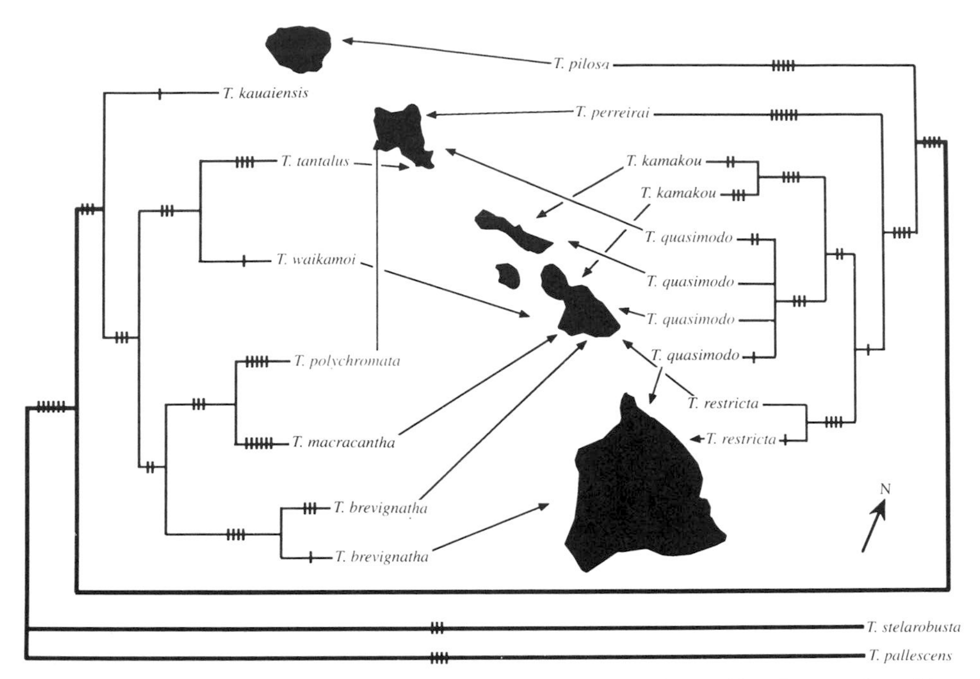

Fig. 18.2 Phylogeny and biogeography of the spiny leg clade of Hawaiian *Tetragnatha* spiders, reproduced with permission from Gillespie and Croom (1995, their Fig. 8.5). Habitat or behavioural changes in the phylogeny are shown by wide connecting lines. Morphological character changes are indicated by small vertical bars. The diagram shows that some of the changes can be geographically identified, and that most but not all speciation events occurred on neighbouring islands.

evolution in laboratory or greenhouse. Clarke and colleagues interpret their findings on hybridization of *Partula* snails on Moorea as evidence of speciation in progress (Section 11.5). Evolutionary responses to oscillating directional selection on morphological traits in a population of Darwin's finches have been measured (Grant and Grant 1995*b*), and the directional selection forces have been used by extrapolation to reconstruct speciation (Price *et al.* 1984). Oscillating selection at gene loci has been studied in island populations of mammals (Chapter 4; see also Berry and Murphy 1970) and butterflies (Ford 1975). There is a largely unexploited potential for experimental investigation of sexual and natural selection in island populations (e.g. see Chapter 5, and Halkka and Raitikainen 1975; Halkka *et al.* 1975). More attention to present-day evolution on islands is warranted: description, elucidation of causes, and extrapolation to the larger patterns of evolution.

18.3 Speciation

Ernst Mayr received the Japan Prize in 1992, and the reason was explained to readers of *Science* with the words: 'Mayr cleared up an area that had confused Darwin—how new species arise.' (Gibbons 1994). Had he done so, literally, a third of this book would not have been written. He has, however, clarified the area, and amongst his many contributions has been the stimulating theory of speciation by founder effects (Mayr 1942, 1954, 1963, 1992). Problems with it have not yet been resolved (Chapters 6–9), and they will probably accompany us into the next century. Two other unresolved issues, alluded to briefly or indirectly in the chapters, are the interrelated questions of whether speciation can occur in the face of gene flow, and whether it can occur entirely (but gradually) sympatrically. These are matters of concern to students of island biology. Island studies can contribute to their resolution.

Speciation without complete genetic isolation

A species colonizes an island, and diverges through genetic drift and directional selection. Unless the source population becomes extinct, or dispersal and colonization is extremely rare because the island is so strongly isolated, additional immigration is likely. If immigration recurs, can speciation proceed fast enough to prevent interbreeding with the next set of immigrants? According to one view repeated gene flow is expected to retard or prevent speciation from occurring (Fosberg 1963; Mayr 1963; Diamond 1977). An example of how the effects of gene flow on divergence are inferred from studies of island populations is given by Barrett in Chapter 2. Mainland–island differences in plant population breeding characteristics are a positive function of the geographical distance between them and hence possibly, though not necessarily, a negative function of the gene flow between them.

As a means of counteracting the stochastic drifting of gene frequencies in small populations the role of gene flow is uncontroversial, but the power of gene flow to counteract the effects of directional selection is not so clear. Experimental evidence shows that it is weak. In fact under some circumstances gene flow could enhance the responsiveness of a population to directional selection (Chapter 9), rather than retard it. The outcome depends on the particular genes (alleles) that are flowing, how often and in what numbers they flow, and the selective coefficients associated with them in the newly entered environment (see also Section 3.3).

Rice and Hostert (1993) have written that divergence-with-gene-flow speciation has been neglected because it can be 'completely deduced from experimental results, it is simple, it is obvious, and hence is not theoretically interesting, despite its potential importance in nature'. If so, then surely the divergence-without-gene-flow speciation is even simpler, more obvious (more likely) and hence is theoretically even less interesting! Conclusion: lack of theoretical attention is no reflection of lack of importance.

Sympatric speciation

Theories are not dead until they are buried, and even then they have a habitat of coming to life again. Apparently dead 35 years ago, the theory of sympatric speciation is enjoying a new life. It has been resurrected and resuscitated by a combination of field studies, showing the implausibility of divergence in allopatry in some cases (Bush 1975; Tauber and Tauber 1989; Feder *et al.* 1994; Schliemann *et al.* 1994), theoretical studies, showing that difficulties arising from considerations of the disruptive effects of gene flow are not as severe as was previously thought (Maynard Smith 1966; Felsenstein 1981; Diehl and Bush 1989), and experimental studies, actually demonstrating the inability of gene flow to inhibit sympatric divergence (Rice and Salt 1990; Rice and Hostert 1993). Sympatric speciation has been suggested from time to time to explain evolutionary diversification on islands (Grant and Grant 1979, 1989; Asquith 1995), as has the similar phenomenon of parapatric speciation (Cook 1996). Similarly, the large number of related species of fish in the island-like environment of lakes has been accounted for by a sympatric process of speciation (Schliemann *et al.* 1994; McCune and Lovejoy 1997; also Chapters 10, 12, and 14). Sympatric speciation is no longer the *enfant terrible* of the field; founder effect speciation has taken over (for some).

18.4 Adaptive radiation

The world's biota can be thought of as one enormous adaptive radiation, that is diversification guided (non-exclusively) by selection. Within this radiation, nested at different hierarchical levels, are radiations of smaller taxonomic units and reduced biological scope. Island radiations are at the lower end of the hierarchy. To the extent that their environments remain relatively undisturbed, studying them can throw light on the evolutionary origin and causes of diversification at higher levels of the hierarchy.

Schluter stressed in Chapter 10 that a knowledge of selection pressures arising from the environment is needed to understand adaptive radiations. This is the province of ecology. There are books on the ecology of populations and communities on islands (Gorman 1979; Vitousek *et al.* 1995), largely independent of evolutionary concerns (Williamson (1981) is an exception). Evolutionary ecology is needed to answer the following questions.

Why do some taxa undergo extensive radiations and others do not?

Why have the Hawaiian honeycreeper finches and Darwin's finches speciated so spectacularly, and other passerine species in the same archipelagos have not? Is it just a matter of who got there first and had the greatest range of ecological opportunities within their reach, or do some taxonomic groups have more genetic (evolutionary) potential than others? A deeper knowledge of genetic architectures, ecology, and molecular phylogenies would help to resolve this issue. Of the three, molecular phylogenies are the most closely within reach. If rates of molecular divergence can be assumed to be the same in the different taxa being compared one does not have to know how fast the molecular clock ticked to be able to support or reject the hypothesis of first-come, first-radiate.

Radiations are not inevitable, and long residence in an archipelago is no guarantee that they will occur. Iguanas have occupied the Galápagos for 10–20 Ma, yet only twice have they divided into separate lineages; the first split gave rise to marine (*Amblyrhynchus*) and land (*Conolophus*) forms, and the second gave rise to two species of land iguanas from one (Rassmann 1997). Given that radiations have occurred, there are two ways to view them: as responses to pre-existing ecological opportunities, or as responses to opportunities created by the evolutionary invention of key innovations (see Section 13.6 for a similar idea expressed about the *Anolis* lizard radiations). Key innovations make further evolutionary change more likely. They have been invoked to explain the diversification, apparently rapidly, of cichlid fish (pharyngeal jaws; Liem 1973), columbines (nectar spurs; Hodges and Arnold 1995), and swallowtail butterflies (cytochrome P450 monooxygenases that metabolise fouranocoumarins in food plants; Berenbaum *et al.* 1996).

On islands the evolution of arborescence may have similarly been a key innovation facilitating diversification of plants in the genus *Echium* (Böhle *et al.* 1996; but see Chapter 17). Twenty seven species are endemic to Canary, Madeira, and Cape Verde islands. All have a common ancestor, the closest relative of which is not in the nearby continental regions but in South Africa (Fig. 18.1). The colonists were herbaceous, woodiness evolved early in the colonization history, and all but two of the derived species are woody. This is seen as an important feature allowing the rapid species multiplication in the group. In fact Böhle *et al.* (1996, p. 11 743) write 'the critical and rate-limiting step in the evolutionary success (measured in terms of both new species and forms per unit time) of the contemporary island species was the origin of the perennial woody habit.' Givnish (Chapter 17) is doubtful of this claim.

Swarms of flightless beetle (Paulay 1985; Roff 1994; Finston and Peck 1995) and cricket (Shaw 1996) species on islands lend themselves to an interpretation of flightlessness being a key innovation. If flightlessness is equated with sedentariness, and that in turn is equated with geographical restriction, small demes, isolation and enhanced speciation potential, then flightlessness may be justly described as a key factor in diversification. Whether it should be considered a key innovation or not becomes a matter of definition. If flightlessness does not aid the entry of a new ecological habitat or niche it is not strictly a key innovation.

Is there a consistent temporal pattern of species accumulation within a radiation?

This question arises because individual islands as well as archipelagos have a finite life. For example volcanic islands formed over a hotspot grow in height, coalesce with neighbours, subside, separate, and finally disappear (Carson and Clague 1995). Islands become joined to each other or to the mainland, or separate, according to fluctuations in sea level (Chapter 9). Rainfall and temperature regimes change. Thus the scope for diversification changes, but in general that scope should increase and then decrease.

Is this scope reflected in the temporal pattern of speciation and species accumulation within adaptive radiations? As more and more improved molecular phylogenies become available it will be feasible to examine the possibility of a repeated pattern of species accumulation. This in turn will allow inferences to be made about the temporal pattern of speciation, to a degree dependent (through fossil evidence) on knowledge of the extinction of species. I see this as one approach towards the development of an evolutionary theory of radiations that will complement the essentially non-evolutionary theory of island biogeography (MacArthur and Wilson 1967).

Two obvious candidate patterns are a sigmoid progression, with an initial exponential increase giving way to a dampened rate, and bursts or spurts in species accumulation separated by pauses. The first is expected from the hypothesis that radiations are governed simply by available ecological opportunities. Diversification is not limited by genetic potential, and when all the opportunities have been exploited, i.e. the niches filled, the radiation ceases. This hypothesis has a formal similarity to population growth models based on biotic potential and carrying capacity. It is recognized by its effects; gradual radiation.

The second pattern is expected from the hypothesis that the extent of radiations is determined by opportunities created by the initial evolutionary steps in the radiation. This assumes that some niches are more difficult to enter than others. In terms of a Simpsonian adaptive landscape some peaks are further away than others. Niches that are essentially variations on a theme, close peaks in an adaptive landscape, may be entered or reached easily, whereas qualitatively different niches can only be exploited with a more profound evolutionary change, such as key innovations. These require substantial time to realize, hence the pauses (see also Section 1.5). The hypothesis is therefore one of radiation in spurts. Mini-radiations nested within the macro-radiation are to be expected.

All models are caricatures of nature. My proposals may be too simple for the complexities of the real world. For example, they are strictly ecological and make no reference to possible influences of founder effect, and epigamic or cultural speciation. There may be as many temporal patterns as radiations. Empirical studies are needed to determine whether this is true or whether there are regularities, however weak the signal may be above the noise.

Is evolution on islands especially rapid?

On the basis of allozyme and molecular data and assumed clock-like average constancy in base-pair substitutions, it has been estimated that 13 species of Darwin's finches evolved on the Galápagos islands in a period of about 3 Ma, and approximately four times as many species of honeycreeper finches evolved in the Hawaiian archipelago over a somewhat longer period (Section 9.6). Over roughly the same period of time (3 Ma) 19 species of horses evolved in a rapid continental radiation (MacFadden and Hulbert 1988), yet only one or two, possibly three, species were formed in our own genus. What governs the rate of diversification, and are those rates especially high on islands?

There has been no thorough and extensive comparison of speciation rates in insular and continental environments, but the large numbers of species produced on islands and in lakes in an apparently short time suggest that speciation rates are especially high there (Chapter 14). Relatively short genetic distances between congeneric species on islands implies the same. For example, the mean age (estimated from allozyme data) since separation from a common ancestor is 0.4 Ma for congeneric pairs of Darwin's finch species, and 2.0 Ma for congeneric pairs of other emberizids (*Zonotrichia* and *Melospiza* species) in North and South America; the comparison is based on data in Yang and Patton (1981) and Zink (1982), and Marten and Johnson's (1986) callibration of an assumed molecular clock.

Employing a different approach, McCune (1997) has used the metric 'time for speciation', equivalent to the average time to double the number of species (Sections 9.9 and 12.3), to compare terrestrial island endemic birds and arthropods with lake fish. The times for lacustrine fish vary from 0.0015 to 0.3 Ma, and are much shorter than the times for the island endemics (0.6 to 1.0 Ma) by her calculations. Archipelagos would seem to be particularly favourable for allopatric speciation, and hence high rates are to be expected, but even so the higher rates of speciation within single lakes call for a special explanation. Although some spatial and temporal heterogeneity factors within lakes promote allopatric speciation (Section 14.5), they are not considered sufficient to explain all speciation, and sympatric speciation has been invoked instead (e.g. Schliemann *et al.* 1994; McCune and Lovejoy 1998). There is a shortage of ecological information to test these ideas (Section 10.4), and a need to carefully define the criteria by which alternatives to sympatric speciation can be rejected. Hanging over all debates about speciation is uncertainty over whether (1) introgressive hybridization and (2) extinction have clouded and confused the modern patterns that we attempt to explain.

How are adaptations to be interpreted?

Alexander Weinstein, a physicist, wrote a letter to Sewall Wright in 1931, following publication of Wright's paper on a genetical (mathematical) theory of evolution, in which he said 'Thanks to the mathematical work on selection and the x-ray work on mutation we have at last emerged from what I call the KatydidKatydidn't period when evolutionary discussion consisted largely of statements that selection could or couldn't accomplish results.' (p. 282 in Provine 1986). Evolutionary discussion is certainly more sophisticated now, but Katydidism has not disappeared. Indeed, as Berry (Chapter 3), Mallet and Turner (Chapter 16), and several other authors in this book show, asking whether selection could or could not accomplish results is an integral part of making attempts to interpret evolutionary history. It is part of the process of seeking answers to 'why' questions (e.g. see Reeve and Sherman 1993), and must be done in a manner cognizant of the pitfalls (Chapter 3), using a knowledge of mechanisms revealed in answer to 'how' questions as fully as possible.

A problem in the study of evolution on islands (and elsewhere) is the problem of disentangling correlated factors (traits) and exposing those of adaptive significance. For example wind pollination predominates in some island floras, and is associated with woodiness and dioecy (Section 2.2). Is this one adaptive syndrome, or is there a single adaptive component (and if so which is it?) that has evolved under selection, passively carrying with it a set of correlated associates? Reference has been made repeatedly to evolutionary trends on islands that may be independent or interdependent (e.g. Section 1.5). Their proper description and interpretation requires a more integrated study than has been customary.

The problem arises in quite another guise when two alternative hypotheses are offered to explain the same phenomenon. Darwin's crowding hypothesis and Wallace's longevity hypothesis to explain the evolution of woodiness on islands is a case in point. Both may apply in some circumstances, because both of the relevant sets of responsible factors co-occur, but for other plant taxa in the same or different environments one or the other hypothesis may be the correct one. Identifying the applicability of simple and compound (combined) hypotheses is a challenge.

18.5 The past and the future

While not formally dedicated to anyone, this book is a tribute to two exceptionally influential (and long-lived) thinkers of this century; Sewall Wright and Ernst Mayr. Sewall Wright probably walked on few islands, but contributed enormously to the genetical theory of evolution in island-like settings. Ernst Mayr, like Alfred Russel Wallace of the last century, had first-hand experience of many islands in the Pacific south-west, and drew upon it to formulate major concepts about evolution on islands. Several chapters refer to them. Their writings constitute the intellectual backbone of this book, and provide much of the blue-print for the future.

The future that Wright and Mayr inherited from Wallace is different from the one they bequeath to their successors of the twenty-first century. Consider the words of Wallace (1865) as he began an article on the swallowtail butterflies of Malaysia: 'When the naturalist studies the habits, the structure, or the affinities of animals, it matters little to which group he especially devotes himself; all alike offer him endless materials for observation and research.' There is a comfortable Victorian certitude in this sentence, implying that the world will continue to provide 'endless materials for observation and research'. Contrast this with the dire modern predicament, with Wilson (1992) estimating that more than 25 000 species of organisms become extinct each year through various forms of human activity.

I have raised the spectre of extinction several times in this book. Extinctions, natural and human-caused, create problems in the interpretation of evolution, from the natural course of taxon cycles to the identification of ancestors and ancestral traits, from the reconstruction of the past to the interpretation of current adaptations. For example the story of evolution as revealed by the Hawaiian honeycreeper finches can be likened to a book which is missing most of its pages, indeed some whole chapters. Similarly, for Cook (1996) extinctions of Madeira snails are like lost pages of a diary. There are many arguments to be made for conserving our biological heritage by preventing further unnatural extinctions (Reaka-Kudla *et al.* 1996). In the present context the main argument is that unless we halt extinctions we will not be able to answer the questions raised in this chapter. Losses of biodiversity associated with human activities are propelling us to a state of permanent ignorance and impoverishment. This would be a great pity as there is still much to be learned about evolution on islands, as a subject of interest in its own right, as a model for evolution in general, and possibly for the insights we may gain about our own origins and evolution.

The last words are Ernst Mayr's:

> Islands are an enormously important source of information and an unparalleled testing ground for various scientific theories. But this very importance imposes an obligation on us. Their biota is vulnerable and precious. We must protect it. We have an obligation to hand over these unique faunas and floras with a minimum of loss from generation to generation. What is once lost is lost forever because so much of the island biota is unique. Island faunas offer us a great deal scientifically and aesthetically. Let us do our share to live up to our obligations for their permanent preservation.
>
> (Mayr 1967, p. 374)

References

Asquith, A. (1995). Evolution of *Sarona* (Heteroptera, Miridae): speciation on geographic and ecological islands. In *Hawaiian biogeography: evolution on a hot-spot archipelago*, (ed. W. L. Wagner and V. A. Funk), pp. 90–120. Smithsonian Institution Press, Washington, D. C.

Beerli, P., Hotz, H., and Uzzell, T. (1996). Geologically dated sea barriers calibrate a protein clock for Aegean water frogs. *Evolution*, **50**, 1676–87.

Berenbaum, M. R., Favret, C., and Schluer, M. A. (1996). On defining 'key innovations' in an adaptive radiation: cytochrome P450S and Papilionidae. *American Naturalist*, **148** Supplement, S139–S155.

Berry, R. J. and Murphy, H. M. (1970). The biochemical genetics of an island population of the house mouse. *Proceedings of the Royal Society of London* B, **176**, 87–103.

Böhle, U.-R., Hilger, H. H., and Martin, W. F. (1996). Island colonization and evolution of the insular woody habit in *Echium* L. (Boraginaceae). *Proceedings of the National Academy of Sciences USA*, **93**, 11740–5.

Bush, G. L. (1975). Modes of animal speciation. *Annual Review of Ecology and Systematics*, **6**, 339–64.

Carson, H. L. and Clague, D. A. (1995). Geology and biogeography of the Hawaiian islands. In *Hawaiian biogeography: evolution on a hot-spot archipelago*, (ed. W. L. Wagner and V. A. Funk), pp. 14–29. Smithsonian Institution Press, Washington, D. C.

Cook, L. M. (1996). Habitat, isolation and the evolution of Madeiran landsnails. *Biological Journal of the Linnean Society*, **59**, 457–70.

Diamond, J. M. (1977). Continental and insular speciation in Pacific land birds. *Systematic Zoology*, **26**, 263–8.

Diehl, S. R. and Bush, G. L. (1989). The role of habitat preference in adaptation and speciation. In *Speciation and its consequences*, (ed. D. Otte and J. A. Endler), pp. 345–65. Sinauer, Sunderland, MA.

Feder, J. L. *et al.* (1994). Host fidelity is an effective premating barrier between sympatric races of the apple maggot fly. *Proceedings of the National Academy of Sciences USA*, **91**, 7990–4.

Felsenstein, J. (1981). Skepticism towards Santa Rosalia, or Why are there so few kinds of animals? *Evolution*, **35**, 124–38.

Finston, T. L. and Peck, S. B. (1995). Population structure and gene flow in *Stomium*: a species swarm of flightless beetles of the Galápagos Islands. *Heredity*, **75**, 390–7.

Ford, E. B. (1975). *Ecological genetics*, (4th edn). Chapman and Hall, London.

Fosberg, F. R. (1963). Plant dispersal in the Pacific. In *Pacific basin biogeography*, (ed. J. L. Gressitt), pp. 273–81. Bishop Museum Press, Honolulu, HI.

Gibbons, A. (1994). Ernst Mayr wins the Japan Prize. *Science*, **266**, 365.

Gillespie, R. G. and Croom. H. B. (1995). Comparison of speciation mechanisms in web-building and non-web-building groups within a lineage of spiders. In *Hawaiian biogeography: evolution on a hot-spot archipelago*, (ed. W. L. Wagner and V. A. Funk), pp. 121–46. Smithsonian Institution Press, Washington, D. C.

Gorman, M. (1979). *Island ecology*. Chapman and Hall, London.

Grant, B. R. and Grant, P. R. (1979). Darwin's finches: population variation and sympatric speciation. *Proceedings of the National Academy of Sciences USA*, **76**, 2359–63.

Grant, P. R. (1968). Beak size, body size, and the ecological adaptations of bird species to competitive situations on islands. *Systematic Zoology*, **17**, 319–33.

Grant, P. R. and Grant, B. R. (1989). Sympatric speciation and Darwin's finches. In *Speciation and its consequences*, (ed. D. Otte and J. A. Endler), pp. 433–57. Sinauer, Sunderland, MA.

Grant, P. R. and Grant, B. R. (1995*a*). The founding of a new population of Darwin's finches. *Evolution*, **49**, 229–40.

Grant, P. R. and Grant, B. R. (1995*b*). Predicting microevolutionary responses to directional selection on heritable variation. *Evolution*, **49**, 241–51.

Halkka, O. and Raatikainen, M. (1975). Transfer of individuals as a means of investigating natural selection in action. *Hereditas*, **80**, 27–34.

Halkka, O., Raatikainen, M., Halkka, L., and Hovinen, R. (1975). The genetic composition of *Philaenus spumarius* populations in island habitats variably affected by voles. *Evolution*, **29**, 700–6.

Hodges, S. A. and M. L. Arnold. (1995). Spurring plant diversification: are floral nectar spurs a key innovation? *Proceedings of the Royal Society of London* B, **262**, 343–8.

Liem, K. F. (1973). Evolutionary strategies and morphological innovations: cichlid pharyngeal jaws. *Systematic Zoology*, **22**, 425–41.

MacArthur, R. H. and Wilson, E. O. (1967). *The theory of island biogeography*. Princeton University Press.

MacFadden, B. J. and Hulbert, R. C. (1988). Explosive speciation at the base of the adaptive radiation of Miocene grazing horses. *Nature*, **336**, 466–8.

Marten, J. A. and Johnson, N. K. (1986). Genetic relationships of North American cardueline finches. *Condor*, **88**, 409–20.

Maynard Smith, J. (1966). Sympatric speciation. *American Naturalist*, **100**, 637–50.

Mayr, E. (1942). *Systematics and the origin of species*. Columbia University Press, New York.

Mayr, E. (1954). Change of genetic environment and evolution. In *Evolution as a process*, (ed J. Huxley, A. C. Hardy, and E. B. Ford), pp. 157–80. Allen and Unwin, London.

Mayr, E. (1963). *Animal species and evolution*. Belknap Press, Harvard University, Cambridge, MA.

Mayr, E. (1967). The challenge of island faunas. *Australian Natural History*, **15**, 359–74.

Mayr, E. (1992). Controversies in retrospect. *Oxford Surveys in Evolutionary Biology*, **8**, 1–34.

McCune, A. R. (1998). How fast is speciation: molecular, geological, and phylogenetic evidence from adaptive radiations of fishes. In *Molecular evolution and adaptive radiation*, (ed. T. J. Givnish and K. J. Sytsma), pp. 585–610. Cambridge University Press.

McCune, A. R. and Lovejoy, N. R. (1997). The relative rate of sympatric and allopatric speciation in fishes: tests using DNA sequence divergence between sister species and among clades. In *Endless forms: species and speciation*, (ed. D. J. Howard and S. H. Berlocher). Oxford University Press. (In press.)

Paulay, G. (1985). Adaptive radiation on an isolated oceanic island: The Cryptorhynchinae (Curculionidae) of Rapa revisited. *Biological Journal of the Linnean Society*, **26**, 95–187.

Price, T. D., Grant, P. R., and Boag, P. T. (1984). Genetic changes in the morphological differentiation of Darwin's ground finches. In *Population biology and evolution*, (ed. K. Wöhrmann and V. Loeschcke), pp. 49–66. Springer, Berlin.

Provine, W. B. (1986). *Sewall Wright and evolutionary biology*. University of Chicago Press.

Ranker, T. A., Floyd, S. K., and Trapp, P. G. (1995). Multiple colonizations of *Asplenium adiantum-nigrum* onto the Hawaiian archipelago. *Evolution*, **49**, 1364–70.

Rassmann, K. (1997). Evolutionary age of the Galápagos iguanas predates the age of the present Galápagos islands. *Molecular Phylogenetics and Evolution*, **7**, 158–72.

Reaka-Kudla, M. L., Wilson, D. E., and Wilson, W. O. (eds.) (1996). *Biodiversity II. Understanding and protecting our biological resources.* Joseph Henry (National Academy) Press, Washington, D. C.

Reeve, H. K. and Sherman, P. W. (1993). Adaptation and the goals of evolutionary research. *Quarterly Review of Biology*, **68**, 1–32.

Rice, W. R. and Hostert, E. E. (1993). Perspective: laboratory experiments on speciation: what have we learned in forty years? *Evolution*, **47**, 1637–53.

Rice, W. R. and Salt, G. W. (1990). The evolution of reproductive isolation as a correlated character under sympatric conditions: experimental evidence. *Evolution*, **44**, 1140–52.

Roff, D. A. (1994). The evolution of flightlessness: is history important? *Evolutionary Ecology*, **8**, 639–57.

Sawhill, J. C. (1997). Pushing the boundaries. *Nature Conservancy*, **47**, 5–6.

Schliemann, U. K., Tautz, D., and Pääbo, S. (1994). Sympatric speciation suggested by monophyly of crater lake cichlids. *Nature*, **368**, 629–32.

Shaw, K. L. (1996). Sequential radiations and patterns of speciation in the Hawaiian cricket genus *Lapaula* inferred from DNA sequences. *Evolution*, **50**, 237–55.

Tauber, C. A. and Tauber, M. L. (1989). Sympatric speciation in insects: perception and perspective. In *Speciation and its consequences*, (ed. D. Otte and J. A. Endler), pp. 307–44. Sinauer, Sunderland, MA.

Vitousek, P. M., L. Loope, L., and Anderson, N. (1995). *Islands: biological diversity and ecosystem function.* Springer, New York.

Wallace, A. R. (1865). I. On the phenomena of variation and geographical distribution as illustrated by the Papilionidæ of the Malayan region. *Transactions of the Linnean Society*, **25**, 1–71 and eight plates.

Wilson, E. O. (1992). *The diversity of life.* Harvard University Press, Cambridge, MA.

Wright, J. W. (1983). The evolution and biogeography of the lizards of the Galapagos archipelago: evolutionary genetics of *Phyllodactylus* and *Tropidurus* populations. In *Patterns of evolution in Galápagos organisms*, (ed. R. I. Bowman, M. Berson, and A. E. Leviton), pp. 123–55. American Association for the Advancement of Science, Pacific Division, San Francisco, CA.

Yang, S.-Y. and Patton, J. L. (1981). Genic variability and differentiation in Galápagos finches. *Auk*, **98**, 230–42.

Zink, R. M. (1982). Patterns of genic and morphologic variation among sparrows in the genera *Zonotrichia*, *Melospiza*, *Junco*, and *Passerella*. *Auk*, **99**, 632–49.

Index

Note: page numbers in *italics* refer to figures and tables

abnormal abdomen system of *Drosophila* 128
adaptation 11–12
  small mammals 43
  to local conditions 13–14
adaptationist programme, critique 35
adaptive landscape
  evolution 104
  gene combinations 88
  inter-island variation in topography 89–90
  mean fitness of population 107–8
  peaks 88, 89, 103, 107, 115, 198
    mobility 89
  radiations 314
  ridges 88–9, 115, *116*, 117–18, 119
  specialization 198
  strong isolation 118–19
  valley 108, 115, 117
adaptive radiation 28, 30, 163, 312–16
  Anolis lizard 197–8
  cichlid fish 198–200, 225
    eretmodine 230
  concerted convergence 281
  ecological opportunities 200, 312
  ecology 197–8
  history of Lake Tanganyika 234
  insular plant groups 293–8
  lakes 198–200
  plants 204–6
  sigmoid progression 314
  speed of evolution 314–15
  spurts 314
  temporal pattern of species accumulation 313–14
adenosine deaminase, Soay sheep 57–8
allele frequency 12
  distribution after bottleneck 108, *109*, 110
  drift 108
  introduced house mice 43–4
  mean fitness *116*
alleles
  loss of ancestral 118
  neutral 85
  recessive 24, 113
allochthony 20
allopatry 88, 89, 96
  Anolis lizards 219–20, 221
  Darwin's finches 145, 146
  divergence of resource exploiting traits 90
  fish in Lake Tanganyika 226, 234–5
  fish in postglacial lakes 166, 170
  hybridization 149
  intralacustrine 226
    speciation 234
  island mimicry model 275
  partial 146
  race formation in island refuges 266
  reproductive isolation 90, 142
  speciation
    of Darwin's finches 155, 157
    with founder effects 151
  within-island 197
allozyme locus
  fluctuations 44
  genetic variation reduction 29
allozyme variation 30
allozymic diversity 13
altitude 2
Amazonia *see* rainforest of Amazonia
anagenesis 142
analysis of variance, multivariate 78
aneuploidy, silversword alliance 293
anole, Dominican 14, 70–1, *72*, 73, *74*, 75
  colouration 71, *72*
  cytochrome b sequence data 73, *74*
  geographic variation 73
  mtDNA 73
  phylogenetic distance matrix 73
  phylogenetic tree 73, *74*
  scalation 71, *72*
Anolis lizard 14, 196, 197, 210–11
  adaptive radiation 197–8
  allopatry 219–20, 221
  Caribbean 210–11, 213, *214*, 215
  colonization 211, 216, 218
  dispersal 217
    capabilities 215
  distance effect 216
  evolutionary diversification 216–18
  extinction 216
  habitat types 220–1
  island class heterogeneity 213, 215
  island size and speciation 217–22
  Lesser Antillean 76–7
  lineage distribution 210–11
  over-water dispersal 215
  range shifts 217
  rescue effect 216
  species 210
    diversity in Greater Antilles 218
    numbers 215
    occurrence 216

Anolis lizard (*cont.*):
species-area relationship 213, *214*, 215, 220, 221, 222
sympatry 219–20
thermal microclimates 220
within-island diversification 216–17
within-island speciation 216, 217–18
Anolis marmoratus 76–7
Anolis oculatus 70–1, *72*, 73, *74*, 75, 76–7
generation time 77
morphological characters 77–8
natural selection 77–9
selection intensity 78
survival relationship to ecotype 78–9
Apodemus spp. 36, 42
arborescence 5, 299
evolution 285–8, 313
colonization of open habitat 287
Hawaiian archipelago 205
artificial selection, divergence 120
autochthony 20, 21
autogamy 24

background selection 30
bacteria, colicin production 103
Bahamas 212
Baker's rule 8, 12
dioecy 27
self-compatibility 21
base-pair substitutions 314
beak
Darwin's finches 146, *147*
size 9–10, 148
bees, pollination 20
beetles
flightlessness 3, *4*, 5
swarms of flightless 313
wingless 2, 3, *4*, 5
behavioural traits, birds 87
benthic habitat, divergence of fish in postglacial lakes 167, 168, 170
Bidens spp. 27
biotic drift 262, 264, *265*, 275, 276
Heliconius butterflies 269
biotic drift model 264–6, 275
natural selection 275
warning patterns 274
birds
beak size 9–10
body size 10
clutch size 9, 10
ecological segregation 91
evolution timescale 154–5
flightlessness 3
founder effects 151–2, *153*
founder effect speciation 87–8
fruit dispersal 206
ground-living 205
Hawaiian flightless herbivores 296
Hawaiian lobelioid fruit dispersal 294
hybridization 142–3, 149–51
invasions by ancestral species 91
peripatric speciation 151–2, *153*
pre-mating barriers to gene exchange 93
reduced wings 3
satellite islands 146, 158
seed carrying 287
seed dispersal in mud 282
sexual selection 154
song characteristics 63, 92
speciation 142–5
tameness 10–11
body size
birds 10, 87
Lesser Antillean anoles 76
lizards 68
mammals v
Bombina spp. 104, *105*
breeding habitat, fish 170, 171
Brighamia spp. 294
Brown–Sheppard–Turner model 264, 266
buntings 151
butterflies
Celebesian 1–2, 9, 14
see also Heliconius butterflies

caatinga 201
Amazonia 247, *248*, 249
woody species *248*
campina 201, 248–9
woody species *248*
Canary Islands, western 67
Caribbean islands
community ecology 136–7
database 212
geology 130
isolated 215
landbridge 215
map *212*
Carocar spp. 253–4, 255
cerrados 245–6
Certhidea sp. 155
Cervus elaphus 51
Chalcides spp. 75–6
Chorthippus spp. 118, *119*
chromosome rearrangement 118
heterozygotes 103
cichlid fish v
adaptive radiation 198–200, 225
allopatric colour morphs 199
coloration 199
dietary diversification 199, 200
ecological diversification 199, 200
genetic variation 199
hybridization 200
mate attraction 199
mtDNA classification 199–200

patterns 199
rocky habitat 226
sexual selection 199
species flocks 225
sympatric speciation 169, 200
cichlid fish, rock-dwelling 225, 226
allopatry 234
intralacustrine 226
diversification periods 235
habitat availability 234
intralacustrine speciation 234
lake level 234
mating systems 236
parental care patterns 236
phylogeographic patterns 235
speciation rate 226
swimbladder 227
Tanganyikan flock 225–6
see also eretmodine cichlids
cladogenesis 142
Clermontia spp. 291, *292*
dispersal 297–8
species numbers 297
climate 2
tree growth 287
climatic factors, natural selection 14
cline movement 267, *268*
Heliconius butterflies 270–1
cline width 104, *105*
clutch size 9, 10
coadaptation 115
coadapted gene complexes 84–5
cold hypothesis 5
colicins 103
colonization 30, 39–40
Anolis lizards 211, 216, 218
archipelagos 5, 144
Caribbean *Drosophila* 87, 132, 133, 136, 137
Darwin's Finches 144
evolutionary patterns 308
inter-island
by Hawaiian *Drosophila* 86
sequence 79
mtDNA data 148
open habitats 287
repeated by Hawaiian *Drosophila* 126–7
small mammals 36–7
speciation 44
successful 19
colonizers
alleles 43
genetic composition 42–3
coloration
cichlid fish 199
Dominican anole 71, *72*
field mouse 43
floral 6, 8
Heliconius butterfly 203
pelage of Soay sheep 55
see also warning patterns
colour pattern
Canary Island skinks 75–6
crypsis 69–70
density trough trapping 267
display 70
Gallotia galloti 68, 69–70, *71*
geographical diversity 262
Heliconius butterflies 262, 269
inheritance 268–9
Lesser Antillean anoles 76, 77
moving clines 267
novel 276
parsimony 273
shifting balance process 274
co-mimic distribution 272
competition
hypothesis 146, 148, 286–7
intraspecific 56
small mammals 39–40
competitive interaction for space 39–40
competitive pressure, intraspecific 2
competitors, scarcity 2
congeneric pairs 315
conservation
genetic variation 63
planning 63
convergence
concerted 281
Partula snails 191
Cope's rule 6
Coregonus spp. 166
courtship response to song characteristics 92
covariation, ecological change with evolutionary change 152
crowding hypothesis of Darwin 316
crypsis 69–70
Cyanea spp. 289, *290*, 291, *292*
browsing herbivore protection 296–7
clades 295–6
corolla length *295*
dispersal 297–8
diversification 298
extinction vulnerability 298
flowers 293, 298
flower tube length 296
leaves 293, *294*, 296, 297
prickles 296
snail theory 296, 297
species numbers 297
trait evolution 295
cytochrome b sequence
Dominican anole 73, *74*
Gallotia galloti 70
Lesser Antillean anoles 76–7

deer, red 53–4
culling 53
fecundity 54, 57
female fertility 57

deer, red (*cont.*):
  intraspecific competition 56
  isocitrate dehydrogenase 56
  juvenile mortality 57
  juvenile survival 54
  mannose phosphate isomerase 57
  molecular investigations 55–6
  population fluctuations 56
  population size *54*
density trough trapping 267
developmental instability 28
diabetes insipidus, nephrogenic 43
dichogamy 26
dietary diversification, cichlid fish 199, 200
differentiation
  ecological of fish in postglacial lakes 170–1
  founder 45
  genetic 292–3
  island populations 12
  morphological 170–1
  mtDNA data 148
dioecy 8, 25
  evolution 26–7
  wind pollination 21
discordant variation, *Partula* snails 190–1
dispersal
  Anolis lizards 217
  long-distance 3, 18, 21
  reduced 2–3, *4*, 5
  syndromes in island plant evolution 282–5
display 70
distance effect 216
distyly 22
divergence 41
  allopatric 90
  artificial selection 120
  ecological and sympatry 154
  evolutionary 198, 306
  gene exchange effects 104
  introgressive hybridization 94
  mate recognition systems 92
  mean fitness of hybrids 106
  reproductive isolation 85
  speciation 106
  sympatry 90
diversification 10
  ecological in cichlid fish 199, 200
  evolutionary in *Anolis* lizards 216–18
  habitat islands 200–2
dodo v, *vi*, 206
double invasion phenomenon 91, 148, 169, 308
Drosophila
  abnormal abdomen system 128
  founder effect speciation 86–7
  giant polytene chromosomes 86
  head shape 128
  oviposition site 93
  population bottleneck 128
  pre-mating isolation 92–3
Drosophila, Caribbean 86–7, 124, 129–33, *134*, 135–7
  abdominal pigmentation 131, 132–3
  cardini group 131, *133*, *134*, *135*, 136
  colonization 132, 133, 136, 137
  common ancestor 132
  differences from Hawaiian *Drosophila* 136
  dunni subgroup 131–2, 135–6
  fertile hybrids 133
  fossil representatives 131
  founder effect speciation 135
  general purpose genotype 135
  genetic relationships 131
  invasions 87
  process of speciation 135–7
  related continental groups 129–30
  relationship assignment 131–2, *134*, *135*
  reproductive isolation 133
  speciation patterns 130–3, *134*, *135*
Drosophila, Hawaiian v, 86, 124–9
  differences from Caribbean *Drosophila* 136
  picture-winged species group 125–8
  relationship network 126
  repeated colonization 126–7
  single-island endemism of species 86

Echium spp. *7*, 288–9
  colonization 308
  growth form 293
  phylogeny *309*
  speciation 313
ecological change, speciation 152
ecological generalization 10
ecological niche
  availability 2
  broad 9–10
ecological segregation, birds 91
ecotype, survival relationship 78–9
ectozoochory 282
Eichhornia paniculata 22, 28–9
  autogamy 24
  floral morphs 22, *23*
  genetic load 24
  long-styled morph 24
  mating system evolution 22, *23*, 24
  mid-styled morph 24
  selfing rate 24
Elephant bird 205
endemism
  Amazonian rainforest islands 201
  Heliconius butterflies 271–2
endozoochory 206, 282, 299
  ecological constraints 284–5
  nutrient-poor soils 285
  rainfall association 284–5
  speciation 284
  tall islands 283
  understorey rainforest plants 298
epigamic diversification, cichlid fish 199

epigamic theory of speciation 92
epistasis 85, 113
  additive genetic variance 113
  relative fitness of less fit homozygotes 115
epistatic genes, new combinations 84
equilibrium model of island species 40
  see also island biogeography
eretmodine cichlids 227–30, *231*, *232*, 233–4
  age estimates 228–9
  dental features 235, 236
  diversification periods 235
  intralacustrine distribution of mtDNA clades 230, 233–4
  morphology based taxonomy 235–7
  mtDNA phylogeny 235–6, 237
  neighbour-joining tree 230, *231*
  parsimony sequence analysis 230, *232*
  phylogeny 229–30, *231*, *232*, 233–4
  phylogeographic patterns 235
  radiation 230
  sequence divergence 230
  sympatry 236
Erysimum spp. 28
establishment 39–40
  successful 19
Eupithicia vi
evolution
  directional 148
  replicated v
  shifting balance theory 103
  timescale 154–5
evolutionary patterns 305–8, *309*, 310
  colonization 308
  divergence 198, 306
  environmental reconstruction 307
  geographical variation 306–7
  inferences about the past 307
  location 308
  observable 308, 310
  phylogeny 308, *309*, *310*
  temporal sequences 307
evolutionary trends 1
  microevolutionary beginnings 11–14
extinction vii
  Anolis lizards 216
  disorderly 264, *265*, 269
  island races 44
  latitudinal shift in organism distribution 306
  predictability 216
  prevention 316
  rate 316
    small mammals 40
  seed distribution mechanism 206
  specialists 10
  vulnerability 30

fast evolving loci 39
faunal relaxation 216
fecundity
  red deer 54, 57
  Soay sheep 58
feeding efficiency of postglacial fish 172, 173
fertility
  hybrid 118–19
  red deer 57
Ficedula spp. 151
finch
  tree 148
  warbler 155
  see also honeycreeper finch
finches, Darwin's 10–11, 35–6
  allopatric speciation 155, 157
  beak differences 146, *147*
  colonization 155
  divergence of resource exploiting traits 90
  diversification 155
  genetic variation 63
  geological history of Galápagos 155, *156*, *157*
  hybridization 145–6, *147*, 149
  introgression 146, 150
  phenogram *143*
  radiation time 151, 155
  reinforcement 145–6
  sexual selection 154
  speciation 90, 143–5, 312
    allopatric model 142–3
    cycle 144–5
fish
  genetic similarity between coexisting species of fish 94–5
  niche differentiation 93, 176
  parallel speciation 91
  pre-mating isolation 93
fish, postglacial 163–4, *165*
  allozyme frequency 167, 168
  assortative mating 172–3, 174, 176
  breeding habitat 170, 171
  ecological differentiation 170–1
  ecological speciation 172–4
  fertility of hybrids 176
  foraging habitat 170, 171
  gene flow 166–70, 175
  high viability hybrids 171
  hybrid fertility 171
  hybrid viability 175, 176
  morphological differentiation 170–1
  morphological traits 173
  niche differentiation 176
  parallel speciation 173–4
  persistence in sympatry 176
  postmating isolation 172
  reproductive isolation 164, 166
  speciation in lakes 91, 93
  species pairs 164, *165*

fitness
  backcross 106
  fixed 106
  hybrid 106
  hybridization 151
  loss in interbreeding 146
  mean 103, 104, 106, 107–8, *116*
fixation probability *109*, 110
flightlessness 3, *4*, 5, 197, 313
floral biology 20
floral colour 6, 8
floral syndromes 19
flowers
  conspicuousness 6, 8
  size in pollinator-mediated selection 20
flycatchers 151
food availability 2
foraging habitat of fish in postglacial lakes 170, 171
forest
  galley 255
  refuge model 266
  see also rainforest
founder
  differentiation 45
  heterogeneity 43
  population 41
  selection 41, 43
founder effect 36–9
  allozyme loss 120
  birds 151–2, *153*
  DNA variation loss 120
  reproductive isolation 85
  small mammals 40–4
  speciation 41–2, 84–5, 115, 128, 135, 311
    birds 87–8
    Drosophila 86–7
    genetic drift 85
    in Hawaiian *Drosophila* 126–7
    models 107
    theories 83
founder event 18, 28–9, 41, 44
  coadapted gene complexes 84–5
  epistatic genetic variance triggering 128
  fixation probability *109*, 110
  genetic reorganization 85, 136
  mitochondrial genes 127–8
  mutation 117
  neutral trait sensitivity 128
  new fitness equilibria 129
  peak shifts with genetic changes 129
  random genetic drift 108, *109*, 110–11, *112*, 113, *114*
  seed transport 206
  shift 117
  small mammals 45
  speciation 86
founder-flush model 84, 127
founding genome persistence 46
founding population, importance 37
founding propagule 42
frequency-dependent selection 103, 267
fruit dispersal, birds 206

Galápagos islands
  geological history *153*, 155, *156*, *157*
  self-compatible plants 21–2
Gallotia galloti 67
  altitude 69, 70
  cloud layer vicariance 70
  colour pattern 68, 69–70, *71*
  cytochrome b sequence data 70
  geographic proximity 69, 70
  geographic variation 69–70
  haplotypes 70
  historical separation 69, 70
  molecular affinities 70
  mtDNA 70, *71*
Gambusia holbrooki 38
gene flow 27–8
  barrier 104, *105*, 106
  distance-dependent 89
  fish in postglacial lakes 166–70, 175
  Partula snails 192
  repeated 311
  restricted 30
  speciation 311
gene frequency
  random variability 41
  stochastic drifting 311
genes
  combinations in adaptive landscape 88
  exchange in pre-mating barriers 93
  mitochondrial 127–8
  nuclear 127–8
genetic bottlenecks 25
genetic change, random 44
genetic differences in house mouse populations 38
genetic differentiation 292–3
genetic distances, congeneric species 315
genetic drift 13, 18, 28, 30, 44
  biotic drift relationship 275
  directional effects 154
  founder speciation 85
  post-colonization 39
  random 83
  shift to adaptive peaks 128
  see also random genetic drift
genetic hitch-hiking 171
genetic load, *Eichhornia paniculata* 24
genetic polymorphism 51–3
genetic processes, speciation 83
genetic reorganization, founding event 85
genetic restriction 142
genetic revolution 40, 41, 44
  inbreeding level 127
genetics 27–30

genetic structure of population 12
  island 45
genetic transilience 107
  model 84, 127
genetic variation 29–30
  additive 113, 128–9, 149
  cichlid fish 199
  conservation 63
  fluctuations 111
  founding event 38–9
  hybrid fitness 119
  hybrid populations 119
  loss prevention 56
  loss retarding 58–60
  maintenance 13
  population
    bottleneck 111, 113
    reduction 63
    size reductions 52–3
  random genetic drift 111
  reduction 29
  selfing organisms 30
genotype 103
  connection by fit intermediates 115
geographic variation
  Canary Island skinks 75–6
  lizard species 75
  within-island patterns 79
Geospiza spp. 145–6
glaciation 30
grasshopper
  chromosome races 106
  male fertility 118, *119*
gravity, dispersal 282–3
Great Bahamas Bank 215, 216
Greater Antilles 130, 212
  Anolis species diversity 218, 220, 221
  species number 197
  within-island speciation 217
Great Lakes of Africa 198–200
  see also Tanganyika, Lake
gynodioecy 25, 27

habitat
  anthropogenic fragmentation 30
  breeding for fish in postglacial lakes 170, 171
  exploitation 3
  foraging of fish in postglacial lakes 170, 171
  islands
    diversification 200–2
    refugia 201, 202
  long-distance dispersal 3
  niche hierarchy 221–2
  pioneer 22
  silversword alliance 293
  see also niche; niche differentiation
Hawaiian flora 204–6
  biogeographical analysis 26–7
  browsing mammal absence 205
  prickles 205
  speciation rate 206
  thorns 205
  wind pollination 206
Hawaiian islands *126*
  community ecology 136–7
  fossil flightless birds 296
  geology 125
  inter-island dispersal 289, 291
hawk, Galápagos 10, *11*
Heliamphora spp. 252, *253*
Heliconius butterflies 202–4, 262
  biotic drift 269
  cline movement 270–1
  coloration 203
  coloration pattern 269
    evolution 264–5
    inheritance 268–9
  disorderly extinction 269
  endemism centres 271–2
  extinctions in island refuges 266
  fossil evidence of refuges 272–3
  genetic architecture 268–9
    wing patterns 204
  geographical variation 262, *263*, 264
  hybrid zone mobility 271
  local change 270
  mimetic capture 269
  mimetic pattern conversion 264
  mimicry 269
    rings 269, 272
  mitochondrial parsimony hypothesis 273
  mtDNA genealogies 273
  Müllerian mimics 262–3, 264, 274, 276
  postman pattern 269
  races 202, 203
  racial differentiation 266
  random genetic drift 270
  range edges 272
  ring capture 269
  ring switching 203
  shifting balance model 266–7, *268*
  warning patterns 264–5
herkogamy 22, 25
hermaphroditism 19
heterophylly 205, 293
heterostyly 8
heterozygosity, reduced 13
Hirta (Outer Hebrides) *52*, 53, 55
Hirtella ciliata 245–6
hitch-hiking 30
  effect 104
  genetic 171
homostyly 22
honeycreeper finch *vii*
  coexisting species 148
  Cyanea pollination 295
  plumage 154
  radiation 151
  speciation 312

horn, Soay sheep
  morphs 57
  phenotype 55
Huntington's disease 43
hybrid contact zones 272
hybrid fertility 118–19
  fish in postglacial lakes 171, 175, 176
hybrid fitness
  fish in postglacial lakes 175
  mean 106
  recovery 118
  reduction for isolation 110
hybridization 8–9, 93–6
  birds 142, 149–51
  cichlid fish 200
  Darwin's finches 145–6, *147*, 149
  double invaders 148
  fitness 151
  genetic similarity 95
  insular plant groups 293
  introgressive 8–9, 93, 94, 96, 148, 150
  island species pairs 148
  Partula snails 95, 191, 193
  phylogenetic history 95
  potential 150
  speciation 106
  sticklebacks 93
  without fitness loss 151
hybrid populations, genetic variance 119
hybrids
  fertile of Caribbean *Drosophila* 133
  fitness 108
  genetic divergence effects on mean fitness 106
  inviability 137, 138
  salmon 172
  sterility 137, 138
    measures 175
  stickleback 173
  viability 175, 176
    postglacial fish 171
hybrid zones, mobility 271

iguanas, Galápagos 313
immigration recurrence 311
inbreeding 30
  depression 24–5
    Bidens spp. 27
  effective size 108
  genetic revolution 127
  natural selection against 145
  selective value of alleles 127
  small mammals 44
insects
  flightlessness 3, *4*, 5
  Hawaiian Islands 19
  scant fauna 21
  size 6, 8
inselbergs 241, 249
interbreeding of Darwin's finches 146
introductions, Galápagos rats 36
introgression
  Darwin's finches 146, 150
  hybridization 148
invasion hypothesis
  multiple 148
  Partula snails 188, *189*, 190
island biogeography theory 37
  equilibrium 210
  non-evolutionary 197
island effect, maritime 38
island mimicry model 275
island plant evolution, dispersal syndromes 282–5
island population loss vii
islands
  marine 306
  oceanic 146, 148
  satellite 146, 148
  tall 283, 284
  terrestrial habitat 30
island size 42
  Anolis lizard speciation 217–22
  small 217
  species variation 197–8
island species pairs 148
isocitrate dehydrogenase, red deer 56
isolation, post-zygotic 150–1

juvenile mortality, red deer 57
juvenile survival, red deer 54

key innovation 313
Kilmory Glen (Rum) *52*
Komodo dragon v

Lacerta sicula 9
lakes, postglacial 164, *165*, 166–71
latitude 306
Lesser Antilles 131, *132*, 212
  within-island speciation 217
Licania spp. 254
lifetime reproductive success of Soay sheep 60
limnetic habitat, divergence of fish in postglacial lakes 167, 168, 170
lineage transformation 142
linkage disequilibrium 104
lizards
  adaptation to local conditions 13–14
  body size 68
  clutch size 9
  geographic variation pattern similarity between species 75
  matrix correspondence tests 68, 69–71, 73, 75
  microgeographical variation 68–9
  molecular studies 68, 69–71, *72*, 73, *74*, 75
  morphology variation 13
  size 6

tail autotomy 75–6
wall 9
western Canary Island 67–9
see also anole, Dominican; *Anolis* lizard
llanos 242, 245
lobelioids, Hawaiian 205, 282
adaptive radiation 293–8
cpDNA restriction-site variation 294
flowers 294
fruit dispersal 294
habitat *283*
invasion 287
leaves 293, *294*
monophyly 289, 294
local change, *Heliconius* butterflies 270
local trends 1–2
locus-specific protein, paternity analysis 59
longevity hypothesis of Wallace 288, 316
Loxops spp. 148

mammals
change in size of island populations 42
see also deer, red; rodents; sheep, Soay; small mammals
mannose phosphate isomerase, red deer 57
maritime island effect 38
marker alleles, hybrid zones 104
mate attraction in cichlid fish 199
mate choice
fish in postglacial lakes 175
parallel speciation 174
phenotype-based 173
sexual imprinting 92
mate recognition systems 92
mating
assortative 172–3, 174, 176
probability 173
mating success
fluctuations 63
genetic variation loss retarding 58–60
Soay sheep 59, 60, *61*, *62*
mating systems
cichlids 236
evolution 21–2, *23*, 24–7
matrix correspondence
partial regression 69–70, 71
tests 79
meiosis, heterozygotes for chromosome rearrangements 103
meiotic drive system, feral mouse 92
Metrosiderus–Freycinetia forest *viii*
microsatellite markers, paternity analysis 59
Microtus arvalis orcadensis 36
mimetic capture 265, 276
Heliconius butterflies 269
mimetic pattern conversion 203
mimicry
Bates' theory 262
Heliconius butterflies 202–4, 269
island model 275
parallel 262, *263*
population bottleneck 270
quasi-Batesian 273–4
mimicry ring 264–5
catching 265–6
Heliconius butterflies 269, 272
Müllerian 266
species composition 266
mimics
Batesian 274
conversion by natural selection 264
Müllerian 103, 262–3, 264, 274, 276
assemblages 203
mitochondrial parsimony hypothesis 273
molecular clock calibration 307, 315
molecular leakage 181–2
Partula snails 191, 192
Moorea (French Polynesia) 95, 181
morph fixing 103
morphological characters, *Anolis oculatus* 77–8
morphological differentiation 170–1
morphological features of island populations 9
mosquito fish 38–9
mountain tops 2, 241, 249
mouse
feral 92
field 43
house 36
allozyme locus fluctuations 44
competitive interaction for space 39–40
genetic composition of colonizers 42–3
inherited traits in introduced population 43–4
island races 38
size 42
skeletal defect in Skokholm population 43
t haplotype 92
ong-tailed field 36
size 42
mtDNA
directional evolution 148
Dominican anole 73
Gallotia galloti 70, *71*
multiple invasion hypothesis 148
Mus musculus faeroensis 36
mutation
founder event 117
rate of deleterious genes 25

natural selection 102–3
abdominal pigmentation of Caribbean *Drosophila* 132
against inbreeding 145
Anolis oculatus 77–9
biotic drift model 275
divergent 174
existing genetic variation 87
flush–crash–founder variant 84

natural selection (*cont.*):
  island biotas 37
  role in evolution 79
  shifting balance process 275
  speciation 128
Nesospiza spp. 151
nest building efficiency 43
neutral allele loss 85
niche
  availability 2
  broad 9–10
  exploitation 314
  hierarchy 221–2
niche differentiation by fish 93, 176
  postglacial lakes 176
  stickleback 91
Nigella spp. 22

oceanic islands
  satellite islands 146, 148
  speciation cycle 148
Opuntia 7
outcrossing
  adaptive significance 26
  evolution 25–7
  woodiness evolution 288
outcrossing-advantage hypothesis 27
overtopping 286, 287
oviposition site, *Drosophila* 93
Ovis aries 51
owl, short-eared 11

parallel variation 75–7
parapatry 106
  fish in postglacial lakes 166
  mimetic capture 275
  shifting balance model 266
parasites, plumage traits 154
parental care patterns, cichlids 236
Partula snails 95, 181–2
  allele frequency 190–1
  convergence 191
  discordant variation 190–1
  fauna vii
  gene flow 192
  gene frequencies 187
  hybridization 191, 193
  invasion hypothesis 188, *189*, 190
  molecular leakage 191, 192
  Moorean species 182–3
  neighbour-joining tree *189*
  northern populations 190
  sampling 183–4, *185–6*, 187
  shell
    coil 95, 103, 183
    size/shape 182–3
  southern populations 188, *189*, 190
  speciation 181–2
  sympatry 193
  variation patterns 187–8, *189*, 190
paternity analysis, Soay sheep 59–60
patterns
  cichlid fish 199
  mimetic 202, 203
  pelage of Soay sheep 55
  see also colour pattern; warning patterns
peripatry 197
persistence, shrews 40
phenotypes, adaptive landscape 88
phylogenetic distance matrix, Dominican anole 73
phylogeny
  evolutionary patterns 308, *309*, *310*
  molecular based 79, 308, *309*
  reconstruction 86, 95–6, 310
pioneer habitat availability 22
plant evolution on islands
  adaptive 281–2
  arborescence 285–8
  competition hypothesis 286–7
  coverage 286
  dispersal syndromes 282–5
  longevity hypothesis 288
plant groups, insular 288–9
  adaptive radiation 293–8
  genetic differentiation 292–3
  hybridization 293
  origins 288–9, *290*
  speciation 289, 291–2
plants
  genetics 18–19
  pollination biology 19–21
  reproductive biology 18–19
  size 5
  stature hypotheses 286
plumage 87
  traits 154
Podisma pedestris 106
pollination 8, 19–21
  altered conditions 20
  bees 20
  inadequacy 12
  insufficient 22
  mainland and island populations 19–20
  wind 2, 21
pollinators 8, 19
  diversity 20
  inferior service 27
polymorphisms 51–3, 103
  persistence of ancient 120
  symmetric overdominance 118
population bottleneck 29, 41
  additive genetic variance 128–9
  additive trait prediction *114*
  allele frequency distribution 108, *109*, 110
  coadapted gene complexes 85
  Drosophila 128
  founder events 108
  genetic variance 111, 113

heterozygosity loss 119
mimicry 270
reproductive isolation 110, *112*
speciation 42
population crash 84
Soay sheep 55, 58, 60, *61*, *62*
survivors 84
population density 2
rise 84
population dynamics 12
population fluctuations 12, 52
population size
island populations 51–2
reductions 52–3
population variation, enhanced 9
porphyria variegata 43
post-colonization selection 12
post-mating isolation 137
fish in postglacial lakes 172, 176
post-zygotic isolation 89
predation, natural selection 14
predator scarcity 2
pre-mating isolation
Drosophila 92–3
fish 93
stickleback 91
prickles 296, 297
Puerto Rico *132*

racial integrity 44
radiation 196–7
forest groups on Amazonian islands 253–4
mimics 202–4
see also adaptive radiation
rails v, 3
rainforest, understorey plants 298
rainforest of Amazonia 200
conservation planning 255
evolutionary change 204
expansion/contraction 201
species distribution 255, *257*
vegetational cover 241
rainforest of Amazonia, islands 241
caatinga 247, *248*, 249
campina 248–9
inselbergs 249
radiation of forest groups 253–4
refugia 254–5, *256*
sandstone mountains 249, *250*, 251
savanna 242, *243–4*, 245–7
tepuis 249, *250*, 251–3
white-sand areas 247, 249
Rana esculenta 307
random genetic drift 12–14, 83–5, 88, 102–3
adaptive peaks 107
divergence 115
founder events 108, *109*, 110–11, *112*, 113, *114*
genetic variance 111
Heliconius butterflies 270
partially isolated hybrid population 118
populations of steady size 107–8
quantitative trait 110
reproductive isolation 102, 107–8, *109*, 110–11, *112*, 113, *114*
sterility 174
strong isolation 118
range edges, *Heliconius* butterflies 272
rats, Galápagos 36
recombinants 85
recombination, epistatic genes 84
refuge hypothesis of tropical diversity 201–2, 204
refuges, fossil evidence 272–3
refugia 254–5, *256*
centres of endemism 254–5, *256*
diversification 204
galley forest 255
habitat islands 201, 202
post-glacial 164
reinforcement hypothesis, Darwin's finches 145–6
reproductive barriers, speciation 44
reproductive isolation
allopatry 90
Caribbean *Drosophila* 133
defining 104, *105*, 106
divergence 85
evolution 92, 142
fish in postglacial lakes 164, 166
founder effects 85
genetic model 103
genetic variance 111
high fitness path 115
hybrid fitness reduction 110
interactions between recessive alleles 120
population bottleneck 110, *112*
post-mating 137, 172, 176
post-zygotic 85
random genetic drift 102, 107–8, *109*, 110–11, *112*, 113, *114*
speciation mechanisms 175
strong 118–19
sympatry 90
reproductive success, lifetime 60
rescue effect 216
resource
exploitation 172
reallocation 11–12
retinitis pigmentosa 43
ring capture 269
Heliconius butterflies 269
ring switching 268
quasi-Batesian mimicry 273–4
Robertsonian translocations 43, 44
rodents
differentiation of island populations 12
see also mouse; rats, Galápagos; shrew; vole, Orkney

Rum, Isle of (Inner Hebrides) 53
ruts, Soay sheep 60, *62*

salmon 166
  assortative mating 173
  breeding habitat 170, 171
  hybrids 172
  parallel speciation 174
  phylogeny 166, *167*
salmonids 166
satellite islands 146, 148
  speciation cycle 148
savanna, Amazonian 242, *243–4*, 245–7
  cerrados 245
  distribution *243*
  llanos-type 242, 245
  neotropical species *245*
  non-flooded grassland 245
  plant dispersal 246–7
  seasonally flooded areas 242
  white sand soil 242, *244*
scalation
  Dominican anole 71, *72*
  Lesser Antillean anoles 76, 77
Schiedea spp. 27
sea-level fluctuations 152, *153*, 197
seeds
  dispersal 282, 283
  distribution 206
  size v, 3, 286, 287, 299
    Hawaiian flora 205
  structure 3
  transport in founding event 206
selection
  disruptive *113*, *114*
  epistatic 108
  frequency-dependent 103, 267
  intensity in *Anolis oculatus* 78
  mean fitness 103
  oscillating directional 13, 310
  pressure factors 2
  sexual 14, 91–2, 154
self-compatibility 2, 8, 12, 19
  Baker's rule 21
  floral traits promoting 22
self-fertilization evolution 21–2, *23*, 24–5
self-incompatibility 8
selfing 12
selfing rate
  Bidens spp. 27
  Eichhornia paniculata 24
self-pollination 12
sequence divergence 230
sexual dimorphism 12
  origin 26
sexual imprinting, mate choice 92
sexual selection
  birds 154
  cichlid fish 199
  existing genetic variation 87
  honeycreeper finch 154
sheep, Soay 51, *52*, 53, 55
  adenosine deaminase 57–8
  fecundity 55
  horn morphs 57
  horn phenotype 55
  lifetime reproductive success 60
  mating success 59, 60, *61*, *62*
  molecular investigations 55–6
  nematode parasitism 55
  paternity analysis 59–60
  pelage colour/pattern 55
  population crash 55, 58, 60, *61*, *62*
  ruts 60, *62*
  sibship size 60
  young rams 59, 60, *61*
shifting balance model 266–7, *268*
  frequency-dependent selection 267
shifting balance process 275
  colour patterns 274
shifting balance theory 83, 103, 107, 115
shrew
  metacentric karyotypes 118
  persistence 40
signalling space, competition 272
silversword alliance 8–9, 287
  aneuploidy 293
  habitat 293
  Hawaiian 288
  hybridization 293
  reciprocal translocation 293
  within-island speciation 291
size
  change 5–6
  island 42, 197–8, 217–22
skinks, Canary Island 75–6
slow evolving loci 39
small islands, speciation 217
small mammals
  adaptive traits 43
  colonization 39–40
  founder effect 36–9, 40–4
  founder event 45
  inbreeding 44
  island forms 36
  relict populations 37
snails
  size 6
  see also Partula snail
song characteristics of birds
  genetic variation 63
  transmission 92
Sorex araneus 118
Sorex spp. 40
sparrow, genetic variation of song 63
specialization 10, 198, 206
speciation 83–4, 311–12
  adaptive landscape 88–9

allopatric 87, 89, 90, 155, 157
model 142
barriers 45
biological character influence 227
birds 142–5
Caribbean *Drosophila* 130–3, *134*, 135–7
cladogenetic 216
coadapted gene complexes 84
colonization 44
cultural theory 92
cycle 144–5, 148
Darwin's finches 90, 143–5, 312
dispersal capability 284
divergence 106
divergence-within-gene-flow 311
ecological 163–4
ecological change 152
ecological shifts 291–2
ecology 90–1
endozoochory in understorey rainforest plants 298
epigamic theory 92
explosive in cichlid fish 225
fish in post-glacial lakes 91, 93, 172–4
fit recombinant 85
founder 84–5
founder effect 41–2, 115, 128, 135, 151–2, *153*, 311
theories 83
founder-flush 107
gene flow 311
genetic architecture of traits 137
genetic processes 83
genomic theory 92
geographic pattern 289, 291–2
honeycreeper finch 312
hybrid 118–19
hybridization 93, 106
natural selection 128
non-ecological 163–4
obstacles 103, 114–15, *116*, 117–19
parallel of fish in postglacial lakes 91, 173–4
parapatric 106
patterns for Hawaiian *Drosophila* 125–7
peripatric 107, 151–2, *153*, 197
population divergence 90
process for Caribbean *Drosophila* 135–7
process for Hawaiian *Drosophila* 127–9
random genetic drift 117
rate 314–15
rate and seed dispersal 283
reproductive barriers 44
sympatric 87, 89, 200, 311, 312
within-island 216, 217–18
without complete genetic isolation 311
see also allopatry; founder effect, speciation; parapatry; sympatry
species
accumulation 313, 314
genotype 103
Sphenodon v, *vi*
spina bifida occulta, Skokholm house mouse 43
sterility, random genetic drift 174
stickleback
benthic 170
body size 173
breeding habitat 170
hybridization 93
hybrids 173
lacustrine species 91
lake colonization by marine species 91
limnetic 170
niche differentiation 91
pre-mating isolation 91
threespined 168, *169*
stochastic processes 12, 27–9
sunflower, tree 286
survival relationship to ecotype, *Anolis oculatus* 78–9
swimbladder, reduced 227
sympatry 88, 89, 96
Anolis lizards 219–20
divergence 198
ecological 154
resource exploiting traits 90
eretmodine cichlids 236
fish in postglacial lakes 166, 167, 169
persistence 176
Partula snails 193
reproductive isolation 90, 142
speciation 311, 312

Tabebuia spp. 246
tail autotomy, lizards 75–6
tall islands
endozoochory 283
rainfall 284
tameness 10–11
Tanganyika, Lake *229*, *233*
cichlid fish 225–6
geology 226–7
history influence on cichlid adaptive radiation 234
lake level 234
water level 226, 227
taxon cycle hypothesis 148, 206, 298–9
Tepuianthus spp. 252, *253*
tepuis 201, 241, 249, *250*, 251–3
dispersal 251
endemism 251–2, 285
forest 253
habitat range 253
species diversity 253
terrestrial habitat islands 30
see also refugia
Tetragnatha spp., phylogeny *310*
thermal microclimates, *Anolis* lizards 220

thorns 296, 297
trait
  additive *113*, *114*, 115
  genetic architecture 128, 137
translocation
  reciprocal 293
  Robertsonian 43, 44
tree growth, climate 287
tree lettuce 289
tristyly 22
  inheritance 29
Tropheus spp. 235
tropical diversity, refuge hypothesis 201–2, 204
trout 166
tuatara *vi*
Turnera ulmifolia 22

variance
  effective size 108
  hybridizational 149
visitation rates 20
vole, Orkney 36

wallaba forest *248*
warning patterns 202, 264
  biotic drift model 274
  ring switching 264
  shifting balance model 266
water dispersal 282
white-eye 151
whitefish 166, 167, *168*
wind
  dispersal 282
  hypothesis 3
  pollination 21, 206
wings, reduction 3
Wolbachia 103
woodiness 8
  evolution 288
  hypotheses 316
  insular 5–6, *7*
  wind pollination 21

Zosterops spp. 151